"Government debt and global climate change are two of the great problems of our time. Carbon taxation uniquely has the potential to address both. But as always the devil is in the details. This important volume is lucid, comprehensive and acute on all the issues bearing on the decision to implement carbon taxes. It will be an essential resource in the debates to come."

Larry Summers, *President Emeritus and Charles W. Eliot Professor, Harvard University, USA*

"There is little doubt in my mind that for dealing for global climate change, the best policy includes a tax on carbon emissions. This new volume edited by Parry, Morris, and Williams dives into the details to help make this simple and sensible policy a reality."

N. Gregory Mankiw, *Professor of Economics, Harvard University, USA*

"This important and timely book of papers by a distinguished bipartisan group of economists lays out the case for a carbon tax as the most effective and efficient way to reduce carbon emissions while generating a large fiscal dividend. The papers tackle the most challenging questions about a carbon tax including its size, and how to offset its potential negative effects on low-income households, economic growth and competitiveness. The book's overarching conclusion is that a well-designed carbon tax would have significant benefits for the environment, for the long-run fiscal outlook, and for economic growth."

Laura Tyson, *Professor of Business Administration and Economics, and Director, Institute for Business & Social Impact, Haas Business and Public Policy Group, University of California at Berkeley, USA*

IMPLEMENTING A US CARBON TAX

Although the future extent and effects of global climate change remain uncertain, the expected damages are not zero, and risks of serious environmental and macroeconomic consequences rise with increasing atmospheric greenhouse gas concentrations. Despite the uncertainties, reducing emissions in the United States now makes sense, and a carbon tax is the simplest, most effective, and least costly way to do this. At the same time, a carbon tax would provide substantial new revenues which may be badly needed, given historically high debt-to-GDP levels, pressures on social security and medical budgets, and calls to reform taxes on personal and corporate income.

This book is about the practicalities of introducing a carbon tax in the United States, set against the broader fiscal context. It consists of 13 chapters, written by leading experts, covering the full range of issues policymakers would need to understand. These include, for example, the revenue potential of a carbon tax, how the tax can be administered, the advantages of carbon taxes over other mitigation instruments, the environmental and macroeconomic impacts of the tax, and opportunities for reforming broader fiscal, regulatory, and technology policies.

A carbon tax can work in the United States. This volume shows how, by laying out sound design principles and feasible solutions to specific implementation challenges.

Ian Parry is Principal Environmental Fiscal Policy Expert in the Fiscal Affairs Department of the IMF.

Adele Morris is a Fellow and Policy Director for Climate and Energy Economics at the Brookings Institution.

Roberton C. Williams III is a Professor in the Department of Agricultural and Resource Economics at the University of Maryland, Senior Fellow and Director of Academic Programs at Resources for the Future, and a Research Associate of the National Bureau of Economic Research.

Routledge Explorations in Environmental Economics

Edited by Nick Hanley
University of Stirling, UK

13 Modeling Environment-Improving Technological Innovations under Uncertainty
Alexander A. Golub and Anil Markandya

14 Economic Analysis of Land Use in Global Climate Change Policy
Thomas Hertel, Steven Rose and Richard Tol

15 Waste and Environmental Policy
Massimiliano Mazzanti and Anna Montini

16 Avoided Deforestation
Prospects for mitigating climate change
Edited by Stefanie Engel and Charles Palmer

17 The Use of Economic Valuation in Environmental Policy
Phoebe Koundouri

18 Benefits of Environmental Policy
Klaus Dieter John and Dirk T.G. Rübbelke

19 Biotechnology and Agricultural Development
Robert Tripp

20 Economic Growth and Environmental Regulation
Tim Swanson and Tun Lin

21 Environmental Amenities and Regional Economic Development
Todd Cherry and Dan Rickman

22 New Perspectives on Agri-Environmental Policies
Stephen J. Goetz and Floor Brouwer

23 The Cooperation Challenge of Economics and the Protection of Water Supplies
A case study of the New York City Watershed Collaboration
Joan Hoffman

24 The Taxation of Petroleum and Minerals
Principles, problems and practice
Philip Daniel, Michael Keen and Charles McPherson

25 Environmental Efficiency, Innovation and Economic Performance
Massimiliano Mazzanti and Anna Montini

26 Participation in Environmental Organizations
Benno Torgler, Maria A. García-Valiñas and Alison Macintyre

27 Valuation of Regulating Services of Ecosystems
Pushpam Kumar and Michael D. Wood

28 Environmental Policies for Air Pollution and Climate Change in New Europe
Caterina De Lucia

29 Optimal Control of Age-Structured Populations in Economy, Demography and the Environment
Raouf Boucekkine, Natali Hritonenko and Yuri Yatsenko

30 Sustainable Energy
Edited by Klaus D. John and Dirk Rubbelke

31 **Preference Data for Environmental Valuation**
Combining revealed and stated approaches
John Whitehead, Tim Haab and Ju-Chin Huang

32 **Ecosystem Services and Global Trade of Natural Resources**
Ecology, economics and policies
Edited by Thomas Koellner

33 **Permit Trading in Different Applications**
Edited by Bernd Hansjürgens, Ralf Antes and Marianne Strunz

34 **The Role of Science for Conservation**
Edited by Matthias Wolff and Mark Gardener

35 **The Future of Helium as a Natural Resource**
Edited by W. J. Nuttall, R.H. Clarke and B.A. Glowacki

36 **The Ethics and Politics of Environmental Cost-Benefit Analysis**
Karine Nyborg

37 **Forests and Development**
Local, national and global issues
Philippe Delacote

38 **The Economics of Biodiversity and Ecosystem Services**
Edited by Shunsuke Managi

39 **Analyzing Global Environmental Issues**
Theoretical and experimental applications and their policy implications
Edited by Ariel Dinar and Amnon Rapoport

40 **Climate Change and the Private Sector**
Scaling up private sector response to climate change
Craig Hart

41 **Game Theory and Fisheries**
Essays on the tragedy of free for all fishing
Ussif Rashid Sumaila

42 **Government and the Environment**
The role of the modern state in the face of global challenges
Edited by Laura Castellucci

43 **Is the Environment a Luxury?**
An inquiry into the relationship between environment and income
Edited by Silvia Tiezzi and Chiara Martini

44 **Implementing a US Carbon Tax**
Challenges and debates
Edited by Ian Parry, Adele Morris and Roberton C. Williams III

IMPLEMENTING A US CARBON TAX

Challenges and debates

Edited by Ian Parry, Adele Morris and Roberton C. Williams III

First published 2015
by Routledge
2 Park Square, Milton Park, Abingdon, Oxon OX14 4RN

and by Routledge
711 Third Avenue, New York, NY 10017

Routledge is an imprint of the Taylor & Francis Group, an informa business

British Library Cataloguing in Publication Data
A catalogue record for this book is available from the British Library

Library of Congress Cataloging-in-Publication Data
Implementing a US carbon tax : challenges and debates / edited by Ian Parry, Adele Morris and Roberton Williams.
pages cm
1. Carbon taxes—United States. 2. Emissions trading—United States. 3. Climatic changes—Law and legislation—United States. I. Parry, Ian, editor. II. Morris, Adele Cecile, 1963– editor. III. Williams, Roberton C., 1972– editor.
HJ5321.I47 2015
363.738'7460973—dc23
2014022330

ISBN: 978-1-138-81415-8 (hbk)
ISBN: 978-1-138-82536-9 (pbk)
ISBN: 978-1-315-74768-2 (ebk)

Typeset in Bembo
by Apex CoVantage, LLC

CONTENTS

FIGURES

TABLES

FOREWORD

Accumulation of carbon emissions and other greenhouse gases in the atmosphere is expected, on present trends, to warm the planet by around 3.5°C by the end of the century, posing considerable risks (not least from instabilities in the global climate system) to the United States and all countries alike. While we are seeing fledgling efforts to enact comprehensive carbon mitigation policies in China, the European Union, and elsewhere, the world inevitably looks to the United States for leadership in this (as in other) areas.

As widely recognized, carbon pricing policies are potentially, by far, the most effective instruments for reducing emissions, while providing the longer-term price signals needed to advance clean technology investments. At the same time, pricing policies could contribute substantially to easing fiscal problems in the United States. In fact, the US tax system is ripe for fundamental overhaul. Not only is the long-term fiscal outlook unsustainable on current policies (with projections of ever rising debt-to-GDP ratios) but numerous loopholes erode the base of individual and company taxation, keeping tax rates unnecessarily high and causing all sorts of distortion to spending and investment behavior.

Following the failure to enact federal emissions trading legislation, and growing appreciation of the need for comprehensive tax reform, debate, at least in informal policy circles, has shifted to the possibility of a US carbon tax. At a technical level, there is considerable unanimity across the spectrum on the role of carbon taxation. In fact, a carbon tax would be a straightforward application of basic tax principles – building a carbon charge into existing (easily administered) excise duties for motor fuels and applying similar charges to other refinery products, coal, and natural gas.

Given strong likelihood that carbon pricing will be needed in the United States at some point, this volume collects policy notes (presented at a November 2012 conference) discussing likely implementation issues that will arise, should that time come.

The contributions, written by leading experts in the field, discuss many issues such as the impact of a carbon tax on the budget, emissions, the overall economy, low-income households, trade-exposed firms, and so on, and whether accompanying measures are needed to address the sensitivities. Opportunities for simultaneous reform of broader fiscal, energy, environmental, and technology policies, and how domestic carbon taxes compare with other countries' mitigation efforts, are also covered.

While the views expressed here should not be attributed to our institutions, we nonetheless hope policymakers take away the message that a carbon tax (with judicious use of revenues) is not only a sound, and pressing, policy from a fiscal and environmental perspective, but also that the practical challenges (while not to understate them) are manageable.

Sanjeev Gupta
Acting Director
Fiscal Affairs Department
International Monetary Fund

Ted Gayer
Vice President and Director
Economic Studies Program
Brookings Institution

Molly Macauley
Vice President
Resources for the Future

CONTRIBUTORS

Joseph E. Aldy is an Assistant Professor of Public Policy at the Harvard Kennedy School, a Visiting Fellow at Resources for the Future, and a Faculty Research Fellow at the National Bureau of Economic Research. His research focuses on climate change policy, energy policy, and mortality risk valuation. He also serves as the Faculty Chair of the Mossavar-Rahmani Center for Business and Government Regulatory Policy Program. In 2009–2010, he served as the Special Assistant to the President for Energy and Environment at the White House. Aldy previously served as a Fellow at Resources for the Future (2005–2008), worked on the staff to the President's Council of Economic Advisers (1997–2000), and served as the Co-Director of the Harvard Project on International Climate Agreements and Co-Director of the International Energy Workshop.

Samuel Brown is a former research associate at the Brookings Institution. He has a Master's in Public Administration from the Maxwell School of Syracuse University and is an alumnus of Willamette University in Salem, Oregon.

Dallas Burtraw is one of the nation's foremost experts on environmental regulation in the electricity sector. For two decades, he has worked on creating a more efficient and politically rational method for controlling air pollution. He also studies electricity restructuring, competition, and economic deregulation. He is particularly interested in incentive-based approaches for environmental regulation, the most notable of which is a tradable permit system, and recently has studied ways to introduce greater cost-effectiveness into regulation under the Clean Air Act. Burtraw's current areas of research include analysis of the distributional and regional consequences of various approaches to national climate policy. He also has conducted analysis and provided technical support in the design of carbon dioxide emissions trading programs in the Northeast states, California, and the European Union.

Burtraw and his colleagues recently completed a major project on estimating benefits of the value of natural resources in the Adirondack Park through surveying area residents on their willingness to pay for improvements. Also with colleagues, he studied the cost-effectiveness of various policies for promoting renewable energy.

Jack Calder. As a consultant working for the IMF and other organizations Jack Calder has advised governments in a wide range of developing countries on the administration of their natural resource revenues. He previously had a long career in the UK Inland Revenue, in the course of which he occupied various senior positions, including latterly that of Deputy Director of the Oil Taxation Office.

Leon C. Clarke is a Senior Research Economist at the Pacific Northwest National Laboratory (PNNL), and he is a staff member of the Joint Global Change Research Institute (JGCRI), a collaboration between PNNL and the University of Maryland at College Park. Dr. Clarke's current research focuses on the role of technology in addressing climate change, scenario analysis, and integrated assessment model development. Dr. Clarke coordinated the U.S. Climate Change Science Program's emissions scenario development process, and he was a contributing author on the Working Group III contribution to the IPCC's Fourth Assessment Report. Prior to joining PNNL, Dr. Clarke worked for RCG/Hagler, Bailly, Inc. (1990–1992), Pacific Gas & Electric Company (1992–1996), and Lawrence Livermore National Laboratory (2002–2003). He was also a research assistant at Stanford's Energy Modeling Forum (1999–2002), where he worked on issues related to technological change and integrated assessment modeling. Dr. Clarke received B.S. and M.S. degrees in Mechanical Engineering from University of California, Berkeley, and M.S. and Ph.D. degrees in Engineering Economic Systems and Operations Research at Stanford University.

Terry Dinan is a senior advisor at the Congressional Budget Office. She has written about a variety of environmental and energy issues, including the design of climate-change policies and their implications for households and businesses in the United States, the costs and consequences of higher fuel-economy standards, and the costs and effects of policies aimed at subsidizing energy sources and technologies. She has testified before Congress on those topics, published in a variety of professional journals, served as an associate editor for the *Journal of Environmental Economics and Management,* and served on the board of the Association of Environmental and Resource Economists. She has a Ph.D. in economics from Iowa State University. Before joining CBO, she worked at the Environmental Protection Agency and at Oak Ridge National Laboratory.

Allen A. Fawcett is the Chief of Climate Economics Branch of the U.S. Environmental Protection Agency. He led the Agency's economic analyses of the leading climate change legislative proposals in the U.S. Congress. He has been extensively involved with the Stanford Energy Modeling Forum (EMF), and recently co-edited the *Energy Journal* Special Issue on the EMF-24 exercise, which explored the market-based and regulatory approaches to U.S. GHG reductions under different technology futures.

Carolyn Fischer works primarily on policy mechanisms and modeling tools that cut across environmental issues, from allowance allocation in emissions trading schemes to wildlife management in Zimbabwe. In the areas of climate change and energy policy, she has published articles on designing cap-and-trade programs, fuel economy standards, renewable portfolio standards, energy efficiency programs, technology policies, the Clean Development Mechanism, and the evaluation of international climate policy commitments. A current focus of her research is the interplay between international trade and climate policy, options for avoiding carbon leakage, and the implications for energy-intensive, trade-exposed sectors. In areas of natural resources management, her research addresses issues of wildlife conservation, invasive species, and biotechnology, with particular emphasis on the opportunities and challenges posed by international trade.

William G. Gale is the Arjay and Frances Miller Chair in Federal Economic Policy in the Economic Studies Program at the Brookings Institution. His research focuses on tax policy, fiscal policy, pensions, and saving behavior. He is Co-Director of the Tax Policy Center, a joint venture of the Brookings Institution and the Urban Institute. He is also Director of the Retirement Security Project. From 2006 to 2009, he served as Vice President of Brookings and Director of the Economic Studies Program. Prior to joining Brookings in 1992, he was an Assistant Professor in the Department of Economics at the University of California, Los Angeles, and a Senior Economist for the Council of Economic Advisers under President George H. W. Bush.

Donald B. Marron is Director of Economic Policy Initiatives and Institute Fellow at the Urban Institute, Washington, DC. His work focuses on American economic and fiscal policy. Marron previously directed the Urban-Brookings Tax Policy Center, served as a Member of the President's Council of Economic Advisers, and served as Acting Director of the Congressional Budget Office. He has also taught at the University of Chicago Graduate School of Business and the Georgetown Public Policy Institute. Marron is the editor of *30-Second Economics,* a short introduction to 50 of the most important theories in economics.

Aparna Mathur is a Resident Scholar in Economic Policy Studies at the American Enterprise Institute in Washington, DC. She received her Ph.D. in Economics from the University of Maryland, College Park, in 2005. At AEI, her research has focused on income inequality and mobility, tax policy, labor markets, and small businesses. She has published in highly esteemed scholarly journals, testified several times before Congress, and published numerous articles in the popular press on issues of policy relevance. Her work has been cited in academic journals as well as in leading news magazines such as the *Economist,* the *Wall Street Journal, Financial Times,* and *Business Week*. Government organizations such as the Congressional Research Service and the Congressional Budget Office have also cited her work in their reports to Congress. She has been an Adjunct Professor at Georgetown University's School of Public Policy, and has taught economics at the University of Maryland.

Richard Morgenstern's research focuses on the economic analysis of environmental issues with an emphasis on the costs, benefits, evaluation, and design of environmental policies, especially economic incentive measures. His analysis also focuses on climate change, including the design of cost-effective policies to reduce emissions in the United States and abroad. Immediately prior to joining Resources for the Future, Morgenstern was Senior Economic Counselor to the Undersecretary for Global Affairs at the U.S. Department of State, where he participated in negotiations for the Kyoto Protocol. Previously he served at the U.S. Environmental Protection Agency, where he acted as Deputy Administrator (1993); Assistant Administrator for Policy, Planning, and Evaluation (1991–93); and Director of the Office of Policy Analysis (1983–95). Formerly a tenured professor at the City University of New York, Morgenstern has taught recently at Oberlin College, the Wharton School of the University of Pennsylvania, Yeshiva University, and American University. He has served on expert committees of the National Academy of Sciences and as a consultant to various organizations.

Adele Morris is a Fellow and Policy Director for Climate and Energy Economics at the Brookings Institution. Her expertise and interests include the economics of policies related to climate change, energy, natural resources, and public finance. She joined Brookings in July 2008 from the Joint Economic Committee (JEC) of the U.S. Congress, where she spent a year as a Senior Economist covering energy and climate issues. Before the JEC, Adele served nine years with the U.S. Treasury Department as its chief natural resource economist, working on climate, energy, agriculture, and radio spectrum issues. On assignment to the U.S. Department of State in 2000, she was the lead U.S. negotiator on land use and forestry issues in the international climate change treaty process. Prior to joining the Treasury, she served as the Senior Economist for Environmental Affairs at the President's Council of Economic Advisers during the development of the Kyoto Protocol. She began her career at the Office of Management and Budget, where she conducted regulatory oversight of agriculture and natural resource agencies. She holds a Ph.D. in Economics from Princeton University, an M.S. in Mathematics from the University of Utah, and a B.A. from Rice University.

Richard G. Newell is the Gendell Professor of Energy and Environmental Economics at the Nicholas School of the Environment, Duke University, and Director of the Duke University Energy Initiative. In 2009 he was confirmed by the Senate as the head of the U.S. Energy Information Administration, the agency responsible for official U.S. government energy statistics and analysis, where he served until 2011. Dr. Newell has also served as the Senior Economist for Energy and Environment on the President's Council of Economic Advisers. He is on the Board of Directors and is a University Fellow of Resources for the Future, where he was previously a Senior Fellow. He is a Research Associate of the National Bureau of Economic Research and has provided expert advice and consulted with many private, governmental, non-governmental, and international institutions.

Karen L. Palmer has been a researcher at Resources for the Future for more than 20 years and is the first recipient of the Darius Gaskins Chair. She specializes in the economics of environmental and public utility regulation, particularly on issues at the intersection of air quality regulation and the electricity sector. Her work seeks to improve the design of incentive-based environmental regulations that influence the electric utility sector, including controls of multi-pollutants and carbon emissions from electrical generating plants. To this end, she identifies cost-effective approaches to allocating emissions allowances, explores policies targeting carbon emissions and other air pollutants, and efficient ways to promote use of renewable sources of electricity.

Ian Parry is Principal Environmental Fiscal Policy Expert in the IMF's Fiscal Affairs Department. He received a Ph.D. in economics from the University of Chicago and prior to joining the IMF in 2010 he worked for 15 years at Resources for the Future. Parry has written numerous articles on environmental, energy, and transportation policies in different countries, emphasizing the critical role of fiscal instruments for mitigating externalities. Recent (co-edited or co-authored) books include *Fiscal Policy to Mitigate Climate Change: A Guide for Policymakers* (IMF 2012), *Getting Energy Prices Right: From Principle to Practice* (IMF 2014), *Issues of the Day: 100 Commentaries on Environmental, Energy, Transportation, and Public Health Policy* (Resources for the Future 2010), and *Toward a New National Energy Policy: Assessing the Options* (Resources for the Future 2010).

William A. Pizer is Professor at the Sanford School and Faculty Fellow at the Nicholas Institute, both at Duke University. His current research examines how public policies to promote clean energy can effectively leverage private sector investments, how environmental regulation and climate policy can affect production costs and competitiveness, and how the design of market-based environmental policies can address the needs of different stakeholders. From 2008 until 2011, he was Deputy Assistant Secretary for Environment and Energy at the U.S. Department of the Treasury, overseeing Treasury's role in the domestic and international environment and energy agenda of the United States. Prior to that, he was a researcher at Resources for the Future for more than a decade. He has written more than 30 peer-reviewed publications, books, and articles, and holds a Ph.D. and M.A. in Economics from Harvard University and B.S. in Physics from the University of North Carolina at Chapel Hill.

Nathan Richardson is an attorney and has been a researcher at Resources for the Future since 2009, specializing in environmental law and economics. His research has examined environmental liability, environmental federalism, and the relationship between law, regulatory institutions, and policy design. He has published research on law and policy related to climate change, including EPA regulation of greenhouse gas emissions under the Clean Air Act. Other research areas include regulation and liability rules related to oil and gas development. Richardson is also managing editor of RFF's environmental policy and economics blog, Common Resources.

Fernando Saltiel is a Ph.D. student in Economics at the University of Maryland, College Park, and a former research assistant at the Brookings Institution. He has a Master's in Public Policy and an undergraduate degree in Economics from the University of Maryland. His research focuses on public finance and environmental policy in developing countries.

Kenneth A. Small, Professor Emeritus of Economics at University of California, Irvine, specializes in transportation, urban, and environmental economics. He is especially known for research on urban highway congestion, value of time and reliability, and effects of energy policies for automobiles. He is on the editorial boards of four professional journals, and previously served as Associate Editor of *Transportation Research Part B–Methodological* and as co-editor of *Urban Studies.* He is a Fellow of Regional Science Association International. At Irvine, he previously served as Chair of Economics and Associate Dean of Social Sciences. Recent research writings include articles on the gasoline tax, transit pricing, fuel efficiency standards, and a book, *The Economics of Urban Transportation,* with Erik Verhoef (Routledge 2007).

Eric Toder is an Institute Fellow at the Urban Institute and Co-Director of the Urban-Brookings Tax Policy Center. Dr. Toder has authored numerous articles on tax policy, tax administration, and retirement policy. Prior to joining the Urban Institute, he held a number of positions in tax policy offices in the U.S. government and overseas, including service as Deputy Assistant Secretary for Tax Analysis at the U.S. Treasury Department, Director of Research at the Internal Revenue Service, Deputy Assistant Director for Tax Analysis at the Congressional Budget Office, and consultant to the New Zealand Treasury. He received his Ph.D. in Economics from the University of Rochester in 1971.

John P. Weyant is Professor of Management Science and Engineering, Director of the Energy Modeling Forum (EMF), and Deputy Director of the Precourt Institute for Energy Efficiency at Stanford University. He is also a Senior Fellow of the Precourt Institute for Energy and the Freeman-Spolgi Institute for International Studies at Stanford. Prof. Weyant earned a B.S./M.S. in Aeronautical Engineering and Astronautics, M.S. degrees in Engineering Management and in Operations Research and Statistics, all from Rensselaer Polytechnic Institute, and a Ph.D. in Management Science with minors in Economics, Operations Research, and Organization Theory from University of California, Berkeley. He also was also a National Science Foundation Post-Doctoral Fellow at Harvard's Kennedy School of Government. His current research focuses on analysis of global climate change policy options, energy efficiency analysis, energy technology assessment, and models for strategic planning. He currently serves as co-editor of the journal *Energy Economics.*

Casey J. Wichman is a doctoral candidate in Agricultural and Resource Economics at the University of Maryland. His research interests lie at the intersection of environmental and public economics with an emphasis on examining the ways in which individuals make decisions in response to environmental policies. In particular, his work analyzes the effects of price as an instrument of conservation when price information is unclear, the welfare impacts of environmental regulation, and the role of information in the design of regulatory mechanisms. Some of his other research interests include the economic and environmental effects of public transportation, the role of water scarcity in the energy sector, and evaluating the validity of program evaluation techniques.

Roberton C. Williams III studies both environmental policy and tax policy, with a particular focus on interactions between the two. In addition to his role at Resources for the Future, he is a Professor at the University of Maryland, College Park, and a Research Associate of the National Bureau of Economic Research. He also serves as a co-editor of the *Journal of Public Economics,* editorial council member (and former co-editor) of the *Journal of Environmental Economics and Management* and member of the editorial board of the *B.E. Journal of Analysis & Policy.* He was previously an Associate Professor at the University of Texas, Austin, a Visiting Research Scholar at the Stanford Institute for Economic Policy Research, and an Andrew W. Mellon Fellow at the Brookings Institution.

SUMMARY FOR POLICYMAKERS

Ian Parry

In the absence of mitigating measures, rising atmospheric concentrations of carbon dioxide (CO_2) and other greenhouse gases (GHGs) are projected to warm the planet by around 3.0 to 4.0°C by 2100 relative to pre-industrial times (IPCC 2013). Temperature increases of this magnitude (and substantially higher temperature increases cannot be ruled out) are very large by historical standards and pose considerable, and poorly understood, risks.

At the same time, the United States faces substantial fiscal challenges. In the absence of fiscal consolidation (beyond the sequester) the federal debt-to-GDP ratio – already (at 73 percent) well above historical levels – is projected (after a slight decline) to rise over the medium to longer term because of higher interest costs and growing spending for Social Security and the government's major health care programs.[1] As indicated in Figure I.1, general government net debt (which includes the debt of sub-national governments) as a percent of GDP was higher in the United States than in the Euro area, advanced economies as a whole, and other country groupings (though some individual countries, like Japan, have much higher net debt levels).

A carbon tax – that is, a tax on the carbon content of fossil fuels (or on their carbon emissions) – could help to address both of these problems. Carbon taxes are potentially the most effective and cost-effective policies for reducing CO_2 emissions, which are otherwise set to gradually rise (see Figure I.2) despite high oil prices, shifting from coal to natural gas, and regulatory initiatives (e.g., on vehicle fuel economy), though emissions are not projected to reach their previous peak level in 2006. These taxes could also raise substantial revenues for easing fiscal pressures and/or funding reductions in other taxes that discourage and distort economic activity.

The possibility of a US carbon tax has received scrutiny recently, given the enticing possibility of including it in a more comprehensive tax reform effort, and the

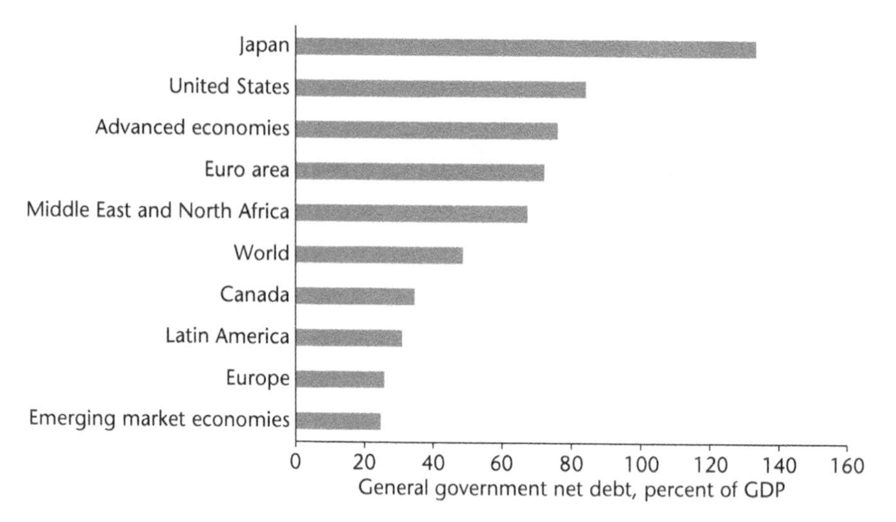

FIGURE I.1 General government net debt in selected countries and regions, 2012
Source: IMF (2013), Table 2.

Note: General government net debt includes government debt at both the national and sub-national levels, less financial assets.

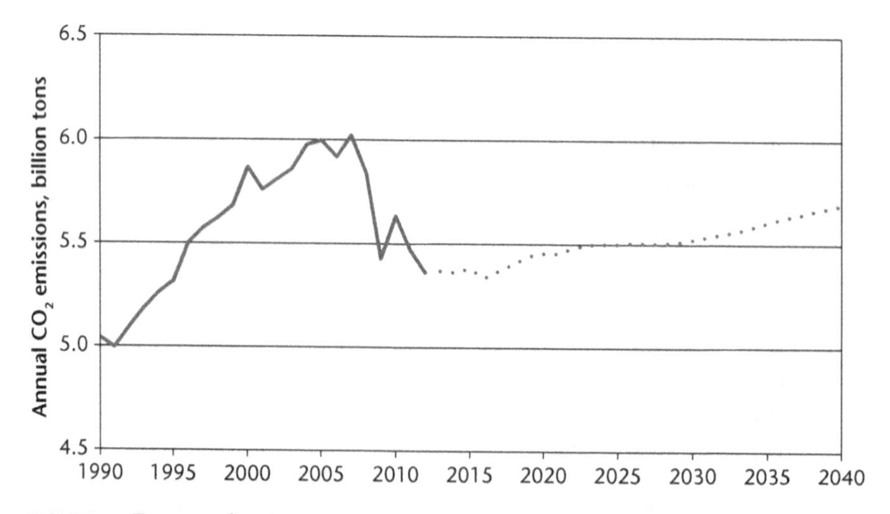

FIGURE I.2 Energy-related CO_2 emissions in the United States, 1990–2040
Source: EIA (2013a, b).

failure of attempts, culminating in 2009, to introduce a federal emissions trading system (ETS). In principle, the choice between an ETS and a carbon tax is less important than implementing one of them, but getting the design basics right (see below). Prior ETS proposals departed from some of these principles, however, most important, only a tiny portion of the emissions allowances were to be auctioned, which precluded most of the fiscal dividend, implying a much greater overall burden on the economy.

At the time this book was produced, the prospects for imminent adoption in the United States of a carbon tax – or anything like it – are remote. But politics can change. The economics of the case for carbon pricing, however, will not: there is near-universal agreement among economists that it will be essential if US emissions are ultimately to be rolled back at reasonable cost. So building consensus over appropriate policy design will be critical. In fact, a number of other countries are grappling with similar design issues in preparation for launching carbon pricing schemes in 2015 or thereabouts.[2] Having a sense of a well-designed carbon tax also provides a critical benchmark against which other (interim) mitigation measures, and other deficit control options, can be compared in terms of environmental, fiscal, economic, distributional, and other impacts, to illuminate tradeoffs and promote more informed dialogue about policy options.

The purpose of this volume is therefore to clarify the rationale for a US carbon tax and provide practical guidance on sound carbon tax design, accounting for the broader fiscal context and political sensitivities (e.g., impacts on vulnerable firms and households).[3]

At a basic level, carbon tax design is largely common sense. Insofar as possible:

- *The tax should be comprehensive* to promote all options for reducing (energy-related) CO_2 emissions. This can be achieved, for example, by levying charges on all fuel supplies in proportion to carbon content.
- *The fiscal dividend should be exploited* by using revenues for deficit reduction, cutting other highly distorting taxes, or other economically productive purposes.
- *The carbon price should be credible, stable, and rising over time* to provide clear signals for clean technology investment.
- *The price should strike a reasonable balance between economic and environmental concerns* by setting it to reflect environmental damages (see Box I.1).

Scratching deeper, however, there are numerous, more subtle questions. These include the role of carbon taxes in broader deficit reduction packages; how much more effective and cost effective are carbon taxes relative to other mitigation options; how the tax might be administered; its impacts on emissions and the overall economy; how burdens on households and firms might be handled; the role of accompanying technology policy; opportunities for reforming policies currently affecting energy and transport systems; how mitigation efforts could be compared across countries; and many others.

This volume brings together 13 policy notes, written by leading fiscal and environmental experts, to shed light on these and other issues.[4] The rest of this summary chapter briefly distills some lessons for policymakers from individual chapters (which are up to date as of 2012).

BOX I.1 CHOOSING A PRICE FOR CO_2 EMISSIONS

There are two economic approaches to gauging the appropriate price on CO_2 emissions (e.g., Bosetti et al. 2012, Griffiths et al. 2012, Nordhaus 2013).

One is to assess least-cost emissions pricing trajectories (at an international level) that are broadly consistent with ultimate goals for global climate stabilization. These price projections are highly sensitive to future uncertainties (e.g., emissions growth without mitigation, fuel prices, costs of emissions-saving technologies). Nonetheless, near-term prices in the ballpark of $30 (in year 2012) per ton of CO_2 (rising annually at about the rate of interest) appear to be broadly consistent with keeping mean projected warming (over pre-industrial levels) to 2.5°C. A 2.0°C target may still be technically feasible, though it would likely require development and global deployment of technologies to remove CO_2 from the atmosphere (to help stabilize atmospheric GHG concentrations at current levels).

The second approach is to assess the 'social cost of carbon' (SCC), or the environmental damage per additional ton of CO_2 emissions. This is estimated using models capturing links between emissions and future atmospheric CO_2 concentrations, between those concentrations and climate variables, and between these variables and various damages (e.g., agricultural, sea level, health, and ecological impacts) with all damages discounted to the present. Again uncertainties abound, but a recent US government report (IAWG 2013) puts the SCC at about $40 per ton for 2015 (in year 2012), significantly higher than in an earlier report (IAWG 2010), with the price rising at about 2 to 3 percent a year in real terms. These estimates remain highly contentious, however, given their sensitivity to different assumptions for discount rates (especially important, given the intergenerational nature of climate impacts) and alternative ways of modeling extreme risks.

Tax theory suggests that carbon taxes should be largely set on environmental (rather than fiscal) grounds with remaining government revenue needs met through broader taxes (e.g., on consumption, labor income, or products with immobile bases).

Carbon taxes in a fiscal context (Chapter 1)

As revenue is one of the two key motivations for a carbon tax, an important question discussed by William Gale, Samuel Brown, and Fernando Saltiel is how a carbon tax compares with other fiscal options simply in terms of its revenue potential.

For perspective, the permanent tax increases and/or spending reductions needed to stabilize the federal debt-to-GDP ratio at its 2012 level (72.5 percent of GDP) through 2089 – let alone reduce this ratio to historical levels – amount to 3–5 percent of GDP. There are numerous possibilities, for example, two proposals discussed in Chapter 1 to limit itemized deductions and tax expenditures could each raise roughly 0.3 percent of GDP, and even a modest carbon tax of $15 per ton could raise 0.5 percent of GDP (net of its indirect impacts on lowering other revenue sources).

But unlike most other tax options, which – leaving aside the use to which the proceeds are put – only impose costs on the economy (such as the detrimental effects of taxes on income and saving for work effort and investment), a carbon tax also generates important environmental benefits. The tax corrects for a 'market failure' – emissions that would be too high if there were no charge for their environmental damages or that would otherwise need to be addressed through alternative (more costly and less efficient) measures.

Carbon taxes in an environmental context (Chapter 2)

The other key motivation for the carbon tax is environmental – as the tax is reflected in the prices for fossil fuels, electricity, and so on, it encourages the entire range of possibilities for reducing CO_2 emissions.

Other climate-related policies are (by themselves) far less effective because, as discussed by Ian Parry, they focus on a much narrower range of these opportunities. For example, incentives for renewable generation fuels (which do not reduce electricity demand, do not reduce emissions outside of the power sector, and do not promote shifting from coal to natural gas or from these fuels to nuclear) miss around 70 percent of the emissions reduction opportunities that would be exploited by carbon pricing. A package of complementary regulations can be more effective but would still miss some opportunities (if anything, fuel efficiency standards, for instance, encourage people to drive more rather than less), would require a lot of credit trading provisions to provide firms with flexibility on how to reduce emissions, and would not raise revenue.

A carbon tax can also achieve a given emission reduction at the lowest overall cost to the economy, and conceivably a negative cost: but this hinges critically on productive use of revenues – if revenues are used for low-valued spending or excessive compensation schemes, overall policy costs are considerably higher.

Tax administration (Chapter 3)

Imposing carbon taxes on fuel supply (based on carbon content) has significant administrative advantages over downstream charging systems (based on emissions). As Jack Calder discusses, the former involves monitoring only around 1,500 taxpayers, comprehensively covers emissions, and can build on well-established procedures for administering motor fuel and coal excise taxes. Pricing embodied carbon in imported products, or providing other protection for trade-exposed/energy-intensive firms, significantly complicates administration, however (see below).

Taxing fossil fuel CO_2 emissions should take first priority, as they are by far the largest source of GHGs. If carbon capture and storage (CCS) at power plants and measurement of stored carbon in forests become practical, a simple system of government payments for sequestered carbon (to complement the carbon tax) would be administratively straightforward.

Achieving emissions targets (Chapter 4)

The emissions implications of alternative carbon tax scenarios are important, not least given the Administration's pledge to reduce emissions by 17 percent below 2005 levels by 2020. According to Allen Fawcett, Leon Clarke and John Weyant recent modeling exercises suggest a CO_2 price of around $25 per ton (in year 2012) is broadly consistent with meeting this objective, accounting for recent regulatory initiatives (e.g., for vehicles and power plants). Most reductions come from reducing coal use, a smaller amount from reducing petroleum, with net impacts related to uses of natural gas unclear.

Various combinations of technologies (e.g., CCS, renewables, and energy-saving technologies) can greatly contain carbon mitigation costs though no individual technology is likely to be enough on its own, underscoring the need for advancing a diverse portfolio of technological possibilities. 'Carbon budgets' (specifying allowable cumulative emissions over an extended period and requiring carbon tax increases if emissions are not on track to stay within budget) can help to reconcile carbon taxes with internationally negotiated targets for emissions reductions.

Macroeconomic effects (Chapter 5)

Roberton Williams and Casey Wichman suggest that carbon taxes of the size considered here should not have large impacts on annual GDP growth rates – perhaps in the order of 0.03 percent (prior to revenue use). But even these effects can be largely offset if revenues are used to lower labor income taxes, and perhaps even more than offset (leaving an, albeit modest, net economic gain, even without counting environmental benefits) if revenues are used to lower corporate income taxes (which are often found to be the most costly in terms of distorting economic activity).

Evaluating alternative uses of carbon tax revenues for public spending is difficult (without more information on the extra spending and its value to society) though well-targeted investments (e.g., in research, education, and infrastructure) could have high payoffs. As for fiscal consolidation, by lessening the need for future tax increases this could yield even larger benefits over the longer term than cutting current taxes, though minimizing macroeconomic risks suggests fiscal tightening should be gradual.

Distributional effects (Chapter 6)

Understanding who bears the burden of a carbon tax is critical for evaluating its fairness and designing possible compensation schemes. As discussed by Adele Morris and Aparna Mathur, most of the burden (prior to revenue use) is borne by households in the form of higher prices for energy in particular and products in general. And low-income households are especially vulnerable due to their relatively high budget shares for energy. However, disparities among income groups are less pronounced when burdens are expressed relative to consumption rather than income (it is not clear whether income or consumption is the better measure of household well-being) and induced changes in income (e.g., from lower rates of return to capital) are considered.

Using carbon tax revenues to fund simple proportionate reductions in other federal taxes will (for the most part) disproportionately help higher income groups. However, numerous combinations of adjustments to rate schedules, tax brackets, and rebates might be used to make the policy more distribution neutral overall, though this might strengthen the case for including the tax as part of a more comprehensive (and badly needed) overhaul of the tax and social safety net system. The burden of a federal carbon tax would also vary regionally, but the policy implications from this are unclear (e.g., states that currently use a lot of coal tend to have relatively low prices for electricity, so the tax would tend to even out electricity prices across the country).

Targeted measures to help low-income households (Chapter 7)

Even if a carbon tax is part of a fiscal package that is broadly distribution neutral, policymakers may still wish to provide some relief to low-income households for the burden of higher energy prices. The tension is that targeted measures (e.g., the – previously termed – food stamp program) inherently reduce the benefits from other possible uses of the additional revenue raised (e.g., the improved work and investment incentives from lowering income taxes).

However, as Terry Dinan shows, with well-designed measures the erosion of revenue gains is limited – households in the lowest income quintile, for example, could be fully compensated with about 12 percent of the carbon tax revenues, leaving 88 percent of the revenue for more economically efficient purposes. And where possible, compensation should involve schemes (such as payroll tax rebates) that both help low-income households and improve economic incentives. Additionally, compensation schemes would ideally build off existing administrative capabilities (e.g., for electronic benefit transfer systems) rather than adding a new layer of institutional structure.

Reforming corporate income taxes (Chapter 8)

Donald Marron and Eric Toder suggest that support for a carbon tax might be broadened by bundling it with corporate income tax reforms of the kind that many have urged for the United States. Lowering the statutory corporate tax rate, now among the highest in the world, and making permanent some (expiring) tax preferences with widespread political support (e.g., the R&D tax credit or small business reliefs), while eliminating others (e.g., support for traditional and alternative energy) potentially made redundant by a carbon tax, would help economic performance by attracting more investment to the United States and reduce incentives for corporations to shift reported income to overseas jurisdictions.

The concern here is that corporate income tax reductions disproportionately benefit the better off, underscoring the need for other measures (e.g., reduced tax preferences for the wealthy, targeted support for the poor) in a broader, distributionally balanced, package of fiscal reforms.

Addressing competitiveness impacts (Chapter 9)

Carolyn Fischer, Richard Morgenstern, and Nathan Richardson discuss the impacts of a carbon tax on the competitiveness of energy-intensive/trade-exposed industries (e.g., steel and aluminum producers), which has been an ongoing concern in efforts to price carbon. Loss of competitiveness can also cause 'emissions leakage' if firms relocate overseas or lose world markets to others not subject to the same carbon pricing. Fortunately these impacts are not especially large – industries with energy expenditures in excess of 5 percent of the value of their output account for less than 2 percent of US GDP – because dealing with them is not so easy (though efficient allocation of productive capacity implies that firms unable to compete when energy is properly priced should be allowed to go out of business over the longer term).

One possibility is to exempt these firms from carbon pricing, but this forgoes some low-cost mitigation opportunities, and reduces revenue (by around 10–15 percent). Broad reductions in corporate income taxes are not well suited either, as most of the benefits leak away to firms that are not in the energy-intensive/trade-exposed sector. More promising is to provide rebates related to output levels for vulnerable firms, as this preserves their incentives to lower energy intensity – but again this uses up revenue, and it might violate World Trade Organization (WTO) rules (which prohibit export subsidies). Border tax adjustments (particularly charges for the carbon content of imports whose production is energy intensive) can address competitiveness and leakage concerns but they are the most politically controversial option and face significant legal and administrative challenges. The best outcome would be for US trading partners to simultaneously price carbon, thereby weakening pressure for these types of measures.

Complementing carbon taxes with technology policy (Chapter 10)

The development and deployment of cleaner energy technologies over the longer term is critical for containing future mitigation costs and expanding emission reduction opportunities. Yet investment in these technologies is insufficient if firms are not fully rewarded for emissions reductions, and also if innovators are not appropriately rewarded for the benefits of technological advances to other firms (though the latter is a generic problem for all new innovations). A carbon tax therefore needs to be complemented with technology policies.

Richard Newell recommends ramping up, by about $3 billion annually, federal support for basic research into critical, climate-related needs (based on plausible assumptions about opportunities like solar energy and energy storage and rates of return to research). Making the R&D tax credit permanent across all sectors (to bolster incentives from intellectual property rights) would cost a further $10 billion a year (a modest fraction of potential carbon tax revenues). Additional incentives for demonstration projects, and for the deployment of new technologies, may also have a role, but they need to be carefully evaluated on a case-by-case basis.

Opportunities for reforming power sector policies (Chapter 11)

Carbon taxes set at an appropriate level would reduce, and in some cases eliminate, the need for other climate-related policies at federal and state level affecting the power sector, including support for renewables (currently about $5 billion a year from the federal government), energy efficiency requirements, and regulation of power plants.

As Dallas Burtraw and Karen Palmer suggest, however, it may in some cases make sense, at least initially, to preserve regulatory structures. For one thing, in practice the carbon tax may be set insufficiently high from an environmental perspective (though designating the Environmental Protection Agency to recommend tax levels based on scientific evidence may help). For another, various market barriers may justify a role for complementary measures, though the appropriate stringency of these measures needs to be carefully evaluated and cost containment safeguards should be built in. And a carbon tax should not necessarily preclude additional mitigation efforts at a sub-national level, though these need to be compatible with (rather than undermined by) the federal policy.

Opportunities for reforming transportation policies (Chapter 12)

Even if a carbon tax were introduced, according to Ian Parry and Kenneth Small, motor fuels in the United States would still be undercharged for the adverse side effects of vehicle use, which include highway congestion, accidents, and local pollution. Ideally, however, these other problems would be more effectively reduced through per-mile tolls, with fuel taxes retained only to reflect carbon damages and perhaps local pollution. Ideally these tolls would also vary, for example, across time of day and region to address congestion. Even with a carbon tax, other policies (e.g., air pollution regulations and transit fare subsidies) have a role, though the case for aggressive fuel economy standards is questionable given that they do not deter automobile use.

While earmarking can be problematic, revenues from carbon charges on motor fuels could go a long way toward covering current and growing shortfalls in highway funding, though they would not fully fund perceived infrastructure needs. But mileage taxes would be more stable as a source of highway funding than carbon charges on motor fuels, as they avoid the erosion of the tax base due to rising fuel economy and, by encouraging more efficient road use, they would also improve the productivity of highway investments.

Promoting international mitigation efforts (Chapter 13)

Ideally, carbon pricing in the United States would enhance momentum for similar pricing efforts elsewhere (not least, as already mentioned, to assuage concerns about competitiveness) and to facilitate this, metrics for comparing mitigation efforts across countries would be especially useful.

According to Joseph Aldy and William Pizer, no single metric is perfect, however. Comparing carbon prices alone does not account for other measures (e.g., energy efficiency mandates or tax reliefs for energy companies) that may reinforce or undermine mitigation incentives from the carbon price. Nor is comparing emission levels (relative to those in a baseline year) entirely satisfactory, as emissions depend on other factors (e.g., business-cycle-related shifts in energy demand), rather than just policy effort. Countries will therefore need to explore – perhaps based on analytical work – a suite of metrics for assessing whether they have undertaken appropriately comparable mitigation efforts.

Summing up, although there are many challenges to consider in crafting and implementing carbon tax legislation, they are largely manageable. What is important is not to lose sight of the basics – carbon tax design is not rocket science, but rather a straightforward extension of long-established, and easily administered, fuel excises. Any policy that deviates from common sense principles – whether that be excluding a significant portion of emissions, not exploiting the fiscal dividend, or failing to establish a credible and reasonably scaled price – will not realize its full potential for addressing environmental and fiscal challenges.

Notes

1. See www.cbo.gov/publication/44521.
2. See www.thepmr.org.
3. There are no previous books on a US carbon tax. Parry et al. (2012) provide guidelines on the design of fiscal policy to reduce greenhouse gases but for countries in general (without specifics for the US context), while Environmental Tax Policy Institute (2009) reviews carbon taxes in different parts of the world. A number of journal articles have been written on carbon tax design, but without getting into the detail considered in this volume (e.g., Aldy et al. 2008, Metcalf 2009, Metcalf and Weisbach 2009).
4. The papers were presented at a November 2012 conference hosted by the American Enterprise Institute (AEI), Brookings, the IMF, and Resources for the Future. AEI's publication rules preclude its inclusion as a co-publisher of this volume.

References

Aldy, Joseph, Eduardo Ley, and Ian Parry, 2008. "A Tax-Based Approach to Slowing Global Climate Change." *National Tax Journal* LXI, 493–518.

Bosetti, Valentina, Sergey Paltsev, John Reilly, and Carlo Carraro, 2012. "Emissions Pricing to Stabilize Global Climate". In I. Parry, R. de Mooij, and M. Keen (eds.), *Fiscal Policy to Mitigate Climate Change: A Guide for Policymakers*. Washington, DC, International Monetary Fund.

EIA, 2013a. *US Carbon Dioxide Emissions, 1990–2009*. Energy Information Administration, US Department of Energy, Washington, DC. Available at: www.eia.gov/environment/data.cfm#summary.

EIA, 2013b. *Annual Energy Outlook 2013 Early Release*. Energy Information Administration, US Department of Energy, Washington, DC. Available at: www.eia.gov/oiaf/aeo/tablebrowser/.

Environmental Tax Policy Institute, 2009. *The Reality of Carbon Taxes in the 21st Century*. Vermont Law School, South Royalton, VT. Available at: www.vermontlaw.edu/Documents/020309-carbonTaxPaper(0).pdf.

Griffiths, Charles, Elizabeth Kopits, Alex Marten, Chris Moore, Steve Newbold, and Ann Wolverton, 2012. "The Social Cost of Carbon: Valuing Carbon Reductions in Policy Analysis." In I. Parry, R. de Mooij, and M. Keen (eds.), *Fiscal Policy to Mitigate Climate Change: A Guide for Policymakers*. Washington, DC, International Monetary Fund.

IMF, 2013. *Fiscal Monitor: Taxing Times*. International Monetary Fund, Washington, DC. Available at: www.imf.org/external/pubs/ft/fm/2013/02/fmindex.htm.

IPCC, 2013. *Climate Change 2013: The Physical Science Basis*. Working Group I Contribution to the Fifth Assessment Report of the Intergovernmental Panel on Climate Change. Available at: www.ipcc.ch/report/ar5/wg1/.

Metcalf, Gilbert E., 2009. "Designing a Carbon Tax to Reduce US Greenhouse Gas Emissions." *Review of Environmental Economics and Policy* 3, 63–74.

Metcalf, Gilbert E. and David Weisbach, 2009. "The Design of a Carbon Tax." *Harvard Environmental Law Review* 3, 499–556.

Nordhaus, William D., 2013. *The Climate Casino: Risks, Uncertainty, and Economics for a Warming World*. New Haven, CT, Yale University Press.

Parry, Ian W. H., Ruud A. de Mooij, and Michael Keen (eds.), 2012. *Fiscal Policy to Mitigate Climate Change: A Guide for Policymakers*. Washington, DC, International Monetary Fund.

US IAWG, 2013. *Technical Support Document: Technical Update of the Social Cost of Carbon for Regulatory Impact Analysis under Executive Order 12866*. Interagency Working Group on Social Cost of Carbon, United States Government, Washington, DC. Available at: www.whitehouse.gov/sites/default/files/omb/inforeg/social_cost_of_carbon_for_ria_2013_update.pdf.

US IAWG, 2010. *Technical Support Document: Social Cost of Carbon for Regulatory Impact Analysis under Executive Order 12866*. Interagency Working Group on Social Cost of Carbon, United States Government, Washington, DC. Available at: www.epa.gov/oms/climate/regulations/scc-tsd.pdf.

GLOSSARY OF TECHNICAL TERMS AND ABBREVIATIONS

American Clean Energy Security (ACES) Act. A 2010 bill (introduced by Congressmen Waxman and Markey) with an ETS as its centerpiece. The bill passed the House, but the Senate version (Kerry-Lieberman's American Power Act) was never put to the vote.

Applied research. Research to gain knowledge or understanding to meet a specific, recognized need (e.g., with respect to commercial objectives for products or technologies).

Basic research. Research to gain improved knowledge or understanding of the subject under study without specific applications in mind (though it can be directed to areas or technologies of potential interest). Universities, other non-profits, and federal labs perform about 80 percent of basic research, more than half of which is funded by the federal government.

Biomass. Biological material derived from living, or recently living, organisms, particularly plants or plant-derived materials. As a renewable energy source, biomass can be combusted directly to produce heat, or used or indirectly after converting it to various forms of biofuel (e.g., ethanol).

Border tax adjustments (BTAs) or border carbon adjustments (BCAs). These measures impose charges on the embodied carbon content of certain products imported in the United States and (in some proposals) provide relief for domestic carbon taxes that might be reflected in production costs for categories of exports.

Bunker fuels. Fuel oil used to power ships.

Burden or incidence. Refers to whose economic welfare is reduced by this tax, and by how much. It is quite different from the formal or legal incidence – fuel suppliers, for example, may be responsible for remitting tax payments to the Internal

Revenue Service, but they may bear little economic incidence if they can charge higher prices.

Business as usual (BAU). Economic outcomes that would occur in the absence of a new policy or policy change.

Carbon budget. Specifies a maximum allowable amount of CO_2 emissions that a country can emit over a period of time (e.g., 10 years).

Carbon capture and storage (CCS). An (as yet unproven) technology for extracting CO_2 emissions from smokestacks and transporting them via pipelines to underground geological storage sites.

Carbon dioxide (CO_2). The main GHG. To convert tons of CO_2 into tons of carbon, divide by 3.67. To convert a price per ton of CO_2 into a price per ton of carbon, multiply by 3.67.

Carbon sequestration. The process by which carbon sinks remove CO_2 from the atmosphere.

Carbon sink. A natural or artificial reservoir that accumulates and stores, indefinitely, a carbon-containing compound. Natural sinks include the oceans, forests, and plants. Artificial sinks include CCS technologies.

Carbon tax. A tax imposed on CO_2 releases emitted largely through the combustion of carbon-based fossil fuels. Administratively, the easiest way to implement the tax is through taxing the fossil fuels – coal, oil, and natural gas – in proportion to their carbon content.

Clean Air Act (CAA). A federal law that requires the EPA to develop and enforce regulations to protect the public from pollutants known to be hazardous to human health. EPA has been regulating GHGs since 2011, following a US Supreme Court decision in 2007 affirming EPA's authority to do so under the CAA.

Clean energy standard (CES) and tradable performance standard. Policies to lower the CO_2 intensity of the power generation (or other) sectors. The CES does this through requiring a shift toward clean, or relatively clean, fuels, while the performance standard provides similar incentives through maximum allowable emission rates. Proposals typically include trading provisions to contain compliance costs.

CO_2 equivalent. The warming potential of a GHG over a long time period expressed in terms of the amount of CO_2 that would yield the same amount of warming.

Common but differentiated responsibilities (CBDR). A principle of the UNFCCC calling for developed countries to bear a disproportionately larger burden of mitigation costs (e.g., by funding emissions reduction projects in developing countries), given that they are relatively wealthy and contributed most to historical atmospheric GHG accumulations.

Corporate Average Fuel Economy (CAFE) standards. These impose requirements on the average fuel economy of new passenger vehicle sales. By 2016, the standards will average 35.5 miles per gallon and 54.5 miles per gallon by 2025, if applied to the current fleet mix with full compliance. Actual fuel economy may fall short of these levels, for example, if manufactures opt to pay out-of-compliance fines due to insufficient technological opportunities.

Corporate tax integration. Integration of the individual and corporate income tax systems such that corporate income would be taxed only once.

Corrective tax. A charge levied on a source of environmental harm and set at a level to reflect, or correct, for environmental damages.

Credit trading. In ETSs, credit trading allows firms with high pollution abatement costs to do less mitigation by purchasing allowances from relatively clean firms with low abatement costs. Similarly, in regulatory systems credit trading allows firms with high compliance costs to fall short of an emissions (or other) standard by purchasing credits from other firms that exceed the standard.

Cyclical unemployment. Unemployment caused by downturns in the business cycle.

Demonstration. Seeks to prove the viability of new technologies (from R&D) at commercial scale (prior to deployment).

Discount rate. In the present context, this refers to the rate at which damages in the future from climate change caused by current emissions are discounted back to the present.

Distribution-neutral policy. A policy that imposes the same burden as a proportion of income (or some other measure of household well-being) on all different income groups.

Double dividend. Refers to the idea that imposing an environmental tax and using the revenues, for example, to cut other taxes could produce two dividends: first, a reduction in pollution emissions and second, a boost in GDP and/or economic efficiency.

Downstream carbon tax. Refers to a tax imposed at the point where CO_2 emissions are released from stationary sources (primarily at coal plants and other industrial facilities). It could be levied on metered emissions out of the smokestack or on embodied carbon in fuel inputs.

Earned Income Tax Credit (EITC). A refundable tax credit available to low-income wage earners.

Economic welfare. A notion of household well-being encompassing everything that individuals value – market goods and services measured by GDP plus non-market items (e.g., childcare at home, leisure time).

Emissions leakage. Refers to a possible increase in emissions in other regions in response to an emissions reduction in one country or region. Leakage could result from the relocation of economic activity, for example, the migration of energy-intensive firms away from countries whose energy prices are increased by climate policy. Alternatively, it could result from price changes, for example, or increased demand for fossil fuels in other countries as world fuel prices fall in response to reduced fuel demand in countries taking mitigation actions.

Emissions trading system or scheme (ETS). A market-based policy to reduce emissions (sometimes referred to as cap-and-trade). Covered sources are required to hold allowances for each ton of their emissions or (in an upstream program) embodied emissions content in fuels. The total quantity of allowances is fixed and market trading of allowances establishes a market price for emissions. Auctioning the allowances provides a valuable source of government revenue.

Energy-efficiency gap or energy paradox. The observation that consumers fail to adopt energy-efficient technologies that appear to more than pay for their upfront investment costs in terms of the expected savings in energy costs over the life of the technology.

Energy-intensive and trade-exposed (EITE) firms. Firms (e.g., cement, aluminum, and chemicals producers) exposed to international trade whose production costs would increase significantly in response to higher energy prices.

Energy Modeling Forum (EMF). Based at Stanford University, the EMF provides a forum for discussing modeling results related to energy and environmental policy.

Enteric fermentation. A digestive process of ruminant animals, such as cows and sheep, by which microorganisms break down carbohydrates into simple molecules for absorption into the bloodstream. It accounts for about 20 percent of US methane emissions.

Environmental Protection Agency (EPA). The federal agency responsible for regulating GHGs (and other emissions) at a national level under provisions of the Clean Air Act.

Environmental (or green) tax shift. Increasing environmental taxes and lowering other taxes (e.g., on labor or capital income) in a way that has no effect on total tax revenues.

Exajoule (EJ). Unit of energy equal to one quintillion (10^{18}) joules.

Extenders. Various business tax breaks that have recently been extended through the end of 2013.

Externality. A cost imposed by the actions of individuals or firms on other individuals or firms (possibly in the future, as in the case of climate change) that the former do not take into account.

Feebate. This policy would impose a fee on firms with emission rates (e.g., CO_2 per kWh) above a 'pivot point' level and provide a corresponding subsidy for firms with emission rates below the pivot point. Alternatively, the feebate might be applied to energy consumption rates (e.g., gasoline per km) rather than emission rates. Feebates are the pricing analog of an emissions (or energy) standard, but they circumvent the need for credit trading (across firms and across time periods) to contain policy costs.

Fiscal consolidation. A reduction in the underlying fiscal deficit through tax increases and/or spending reductions.

Fiscal cushioning. In the present context, this refers to adjustments to broader energy taxes or subsidies that offset some of the environmental effectiveness of a formal carbon tax. It would be potentially important to monitor (and perhaps penalize) fiscal cushioning in an international carbon tax agreement.

Fiscal gap. The immediate and permanent increase in taxes and/or reduction in spending that would be needed to keep the long-term debt-to-GDP ratio at its current level.

Global warming potential (GWP). A measure of how much heat a ton of a non-CO_2 GHG traps in the atmosphere over a given period of time relative to the amount of heat trapped per ton of CO_2.

Great Recession. An ongoing global economic decline that began in December 2007 and took a particularly sharp downward turn in September 2008.

Greenhouse gas (GHG). A gas in the atmosphere that is transparent to incoming solar radiation but traps and absorbs heat radiated from the earth. CO_2 is easily the most predominant GHG.

Highway Trust Fund. A transportation fund financed from federal motor fuel taxes (and other vehicle excises) to pay for highway and transit projects.

Hybrid electric vehicle (HEV). A vehicle combining a conventional internal combustion engine propulsion system with an electric propulsion system.

Hydrofluorocarbons (HFCs). A source of GHGs, with especially high GWPs, used in refrigeration and air conditioning.

Inducement prizes. These can provide financial rewards for specific (high-priority) technological advances.

Input/output table. This provides detailed information on the value of output and value of various categories of input (including fuels and electricity use) by industries producing both intermediate goods and final consumer goods. These tables can be used to trace through the effects of carbon taxes (via higher energy prices) on the price of final goods (assuming all of the carbon tax is ultimately reflected in higher consumer prices).

Inter-agency working group on the SCC. A group of representatives from US executive branch agencies and offices tasked with developing consistent estimates of the SCC for use in regulatory analysis.

Intergovernmental Panel on Climate Change (IPCC). The IPCC assesses the scientific, technical, and socio-economic information relevant for understanding climate change. Its Fifth Assessment Report (AR5) is available at: www.ipcc.ch/index.htm.

Joint Committee on Taxation (JCT). Committee of the US Congress established under the Internal Revenue Code consisting of members from the Senate Finance Committee and the House Ways and Means Committee. The Committee monitors, reviews, and makes recommendations to improve the operation and administration of domestic federal taxes.

Kyoto gases. This refers to the six gases for which emission reduction pledges were made under the Kyoto Protocol. They include carbon dioxide (CO_2), methane (CH_4), nitrous oxide (N_2O), sulphur hexafluoride (SF_6), and two groups of gases hydrofluorocarbons (HFCs), and perfluorocarbons (PFCs).

Kyoto Protocol. Under this Protocol, 37 'Annex 1' or developed countries committed themselves to reducing CO_2 and five other GHGs to about 5 percent below 1990 levels by 2012 (China was not part of the Protocol and the United States never ratified it). The second Kyoto commitment period (2013 to 2020) replicates previously negotiated emissions targets for a subset of developed countries and a range of mitigation actions from a broader range of countries.

Low-Income Home Energy Assistance Program (LIHEAP). This program helps low-income families pay their energy costs for space heating and cooling.

Marginal effective tax rate (METR). The percentage reduction in the after-tax rate of return on a new investment due to corporate income taxes.

Market failure. A situation where the private sector by itself would not make production and consumption decisions that would be efficient from society's perspective, for example, excessive generation of GHG emissions due to the failure to price them for environmental damages.

National Energy Modeling System (NEMS). A modeling system used by the Energy Information Administration to project energy-related variables (as reported, for example, in the *Annual Energy Outlook*) and conduct policy simulations. NEMS contains considerable, up-to-date detail on a wide range of existing and emerging technologies across the energy system (e.g., fuel-saving technologies that might be incorporated into different vehicle classes), while also balancing supply and demand in all – energy and other, regional and national – markets of the economy.

Negative emissions technology. A technology that on net reduces atmospheric concentrations of GHGs (e.g., co-firing biomass in power plants that have installed CCS technologies).

Non-CO_2 GHGs. These gases account for about 16 percent of total US GHGs. These gases, which include methane, nitrous oxide, and HFCs, have relatively high GWPs.

Output-based rebate (OBR). In the context of a carbon tax, this is a payment per unit of output to compensate firms whose production costs rise significantly in response to higher energy prices.

Passed backwards. The extent to which a (carbon) tax lowers prices for energy suppliers.

Passed forward. The extent to which a (carbon) tax raises prices for energy consumers.

Per mile tolls. Charges levied in proportion to distance driven by a vehicle where the rate might vary, for example, with the amount of congestion on the road (charges might be metered electronically through Global Positioning Systems).

Plug-in hybrid electric vehicle (PHEV). A hybrid vehicle which utilizes rechargeable batteries, or similar energy storage device, that can be restored to full charge by connecting a plug to an external electric power source.

Polluter pays principle. The idea that those responsible for producing pollution should be charged the damages done to other parties.

Progressive policy. A policy that imposes a larger burden as a proportion of income (or some other measure of household well-being) on higher income groups and a smaller burden on lower income groups.

Public good. A good whose use by one person or firm does not reduce the availability of it for others and whose use by others is not easily prevented. Knowledge from research is a classic example.

R&D or RD&D. The former refers to research and development and the latter to R&D plus demonstration.

R&D tax credit. These credits allow businesses to apply for a dollar-for-dollar reduction of tax liability for qualified expenditures on research activities conducted in the United States. The tax credit is expired in December 2013 but is expected to be renewed for 2014.

Refundable tax credit. When a tax credit exceeds an individual's or business's tax liability and the difference is paid back to the individual or business.

Regressive policy. A policy that imposes a larger burden as a proportion of income (or some other measure of household well-being) on lower income groups and a smaller burden on higher income groups.

Renewable portfolio standard (RPS). A regulation requiring utilities to use renewable sources to generate a specific minimum amount of the power they produce and sell, where generators (with relatively emissions intensive portfolios) usually have the flexibility to fall short of the standard by purchasing credits from other (relatively clean generators) that exceed the standard. Renewable portfolio standards have been introduced in 29 US states and there have been several proposals to introduce them at the federal level.

Revenue-recycling. Use of (carbon) tax revenues to, for example, lower other taxes.

Sequester. Cuts in federal spending mandated by the Budget Control Act of 2011 that went into effect in March 2013. The cuts amount to $85 billion (about 2.3 percent of total federal spending) in fiscal 2013 and similar cuts are supposed to occur in every future year until 2021. The cuts are split evenly in dollar amounts between defense and non-defense sectors – Medicaid, Social Security, and a few small categories are exempt.

Shale gas. Natural gas that is found trapped within shale formations (i.e., fine grained, organic-rich, sedimentary rock).

Social cost of carbon (SCC). Refers to the net present value of damages (e.g., to agriculture, human health) due to the change in future global climate resulting from an additional ton of CO_2 emissions in a given year. It is expressed in monetary units and usually reflects worldwide damages (rather than damages to a particular country).

Starve the beast hypothesis. According to this hypothesis, keeping down government revenues is an effective approach to curtailing government spending.

Supplemental Nutrition Action Payments (SNAP). Formerly named Food Stamps, these payments help low-income households purchase food.

Supplemental Security Income (SSI) payments. Stipends provided to low-income people who are 65 or older, blind, or disabled.

Tax credit. An amount deducted from an individual's or business's tax liabilities.

Tax expenditure or preference. Revenue losses attributable to provisions of tax laws which allow special exclusion, exemption, or deduction from gross income on which tax is assessed or which allow a special credit, preferential rate of tax, or a deferral of liability. Prominent examples include the favorable tax treatment of employer-provided medical insurance and mortgage interest for owner-occupied housing.

Ton. Taken here to mean a metric tonne (1,000 kilograms). This is the standard unit for measuring CO_2 emissions (rather than a short ton, which is 2,000 pounds or 907 kilograms).

United Nations Framework Convention on Climate Change (UNFCCC). This is an international environmental treaty produced at the 1992 Earth Summit. The treaty's objective is to stabilize atmospheric GHG concentrations at a level that would prevent 'dangerous interference with the climate system'. The treaty itself sets no mandatory emissions limits for individual countries and contains no enforcement mechanisms. Instead, it provides for updates (called 'protocols') that would set mandatory emission limits.

Upstream carbon tax. In the present context, this refers to an emissions tax imposed at some point in the fossil fuel supply chain prior to fuel combustion, for example, it could be levied on petroleum refineries or coal processors. In extractive industries, 'upstream' refers to exploration and fuel production only (i.e., prior to fuel processing).

Waxman-Markey bill. See American Clean Energy Security Act.

World Trade Organization (WTO). An organization that seeks to promote and liberalize international trade.

1

CARBON TAXES AS PART OF THE FISCAL SOLUTION

William G. Gale, Samuel Brown, and Fernando Saltiel

KEY MESSAGES FOR POLICYMAKERS

- The United States faces substantial and unsustainable medium- and long-term budget deficits, which will require a combination of tax increases and spending cuts to resolve.
- A carbon tax could raise significant revenues, with several additional positive effects: it would improve environmental outcomes, increase economic efficiency, and allow the elimination of selected other tax subsidies and spending programs.
- While a carbon tax imposes a disproportionately larger burden on lower-income households, the opposite applies for many of the other options like scaling back tax expenditures. A long-term deficit reduction package that included a reduction in income tax expenditures as well as a carbon tax and offsetting payments could in principle provide a balanced distributional effect.

I. Introduction

The United States faces large federal fiscal deficits in the immediate future, the next 10 years, and the longer term. Although the current and recent deficits are thought to be helping the economic recovery, the deficits in the medium term and long term are more troubling because of their potential impact on national saving, economic growth, and financial markets. Addressing these medium- and long-term challenges will likely require a combination of spending cuts and revenue increases. None of the relevant options (some of which will need to be implemented sooner or later) are particularly attractive from a political perspective.

In this chapter, we consider the fiscal outlook, how new taxes on carbon could not only help address the fiscal problem but also bring about benefits on economic and environmental grounds, and how these taxes compare with some other revenue options. Section II discusses issues related to the fiscal outlook. In section III, we highlight the revenue, efficiency, and equity effects of taxes on carbon emissions and/or a higher tax on gasoline. Section VI provides a brief comparison of a carbon tax to other revenue options – including a VAT and income tax expenditure reform. Section V offers a short conclusion.

II. Fiscal outlook and implications

This section summarizes the fiscal outlook, discusses why both revenue increases and spending cuts will need to be considered as part of the solution, and examines the long-term impact of tax-financed deficit reduction policies.

A. Fiscal outlook

Figure 1.1 shows historical budget deficits and deficits projected under different future policy scenarios. Under the current-law baseline produced by the Congressional Budget Office (CBO) assumptions the deficit falls from 5.3 percent of GDP in 2013 to 2.9 percent in 2018, before rising to 3.8 percent by 2023.

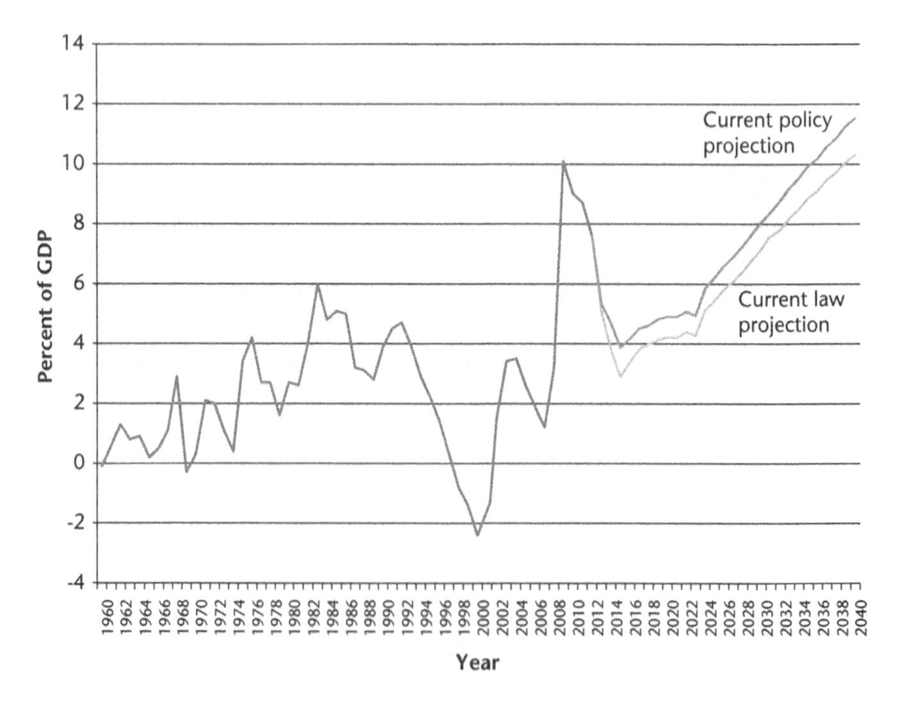

FIGURE 1.1 Deficits past and future, 1960–2040

Source: Auerbach and Gale (2013a).

TABLE 1.1 Major differences between current law and current policy baseline

Policy option	*Effect on the budget deficit in 2020 (percent of GDP)*
Extend expiring tax provisions	0.5
Cancel sequester	0.5
Institute "doc fix"	0.1
Drawdown in defense spending	–0.3
Remove disaster relief funding	–0.2

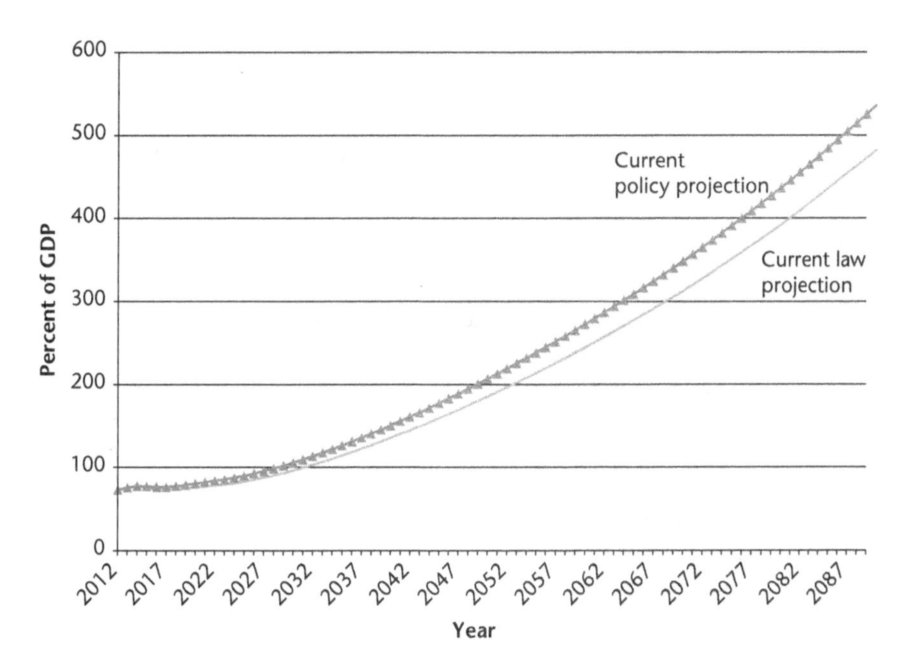

FIGURE 1.2 Alternative projections of the national debt, 2012–2090

Auerbach and Gale (2013a).

Auerbach and Gale (2013a), however, show that under a current policy baseline (reflective of more realistic policies), the federal deficit under current policies will hover around 3.5 percent of GDP between 2015 and 2019, before rising to 4.5 percent by 2023 (Figure 1.1). The policy differences between current law and current policy baseline are shown in Table 1.1.

Moreover, after 2022, projected deficits are poised to rise further under both scenarios (Figure 1.1), reaching 10 percent of GDP by 2036 under the current policy baseline and continuing to rise thereafter.

As for the federal debt-to-GDP ratio, after averaging 37 percent of GDP in the 50 years prior to the Great Recession that started in 2007 and attaining a value of 36.3 percent of GDP in 2007, the ratio is now projected to pass its 1946 high of 108.6 percent in 2035 under current policy baseline (Figure 1.2). Unlike the aftermath of World War II, however, the debt-to-GDP ratio will continue to rise after surpassing the previous peak. Expenditures are expected to rise significantly as the aging of the populace and excess cost growth of health care cause Medicare and Medicaid outlays to grow rapidly. Current estimates place the fiscal gap – the immediate and permanent increase in taxes or reduction in spending that would keep the long-term debt-to-GDP ratio at its 2012 level – or 70.1 percent of GDP – at 4–7 percent of GDP through 2089 and 5–7 percent on a permanent basis (Auerbach and Gale 2013b).

In contrast to the U.S. projected fiscal trajectory, many organizations place the desired debt/GDP ratio between 40 percent and 60 percent.[1] It is not entirely clear how an optimal debt/GDP ratio can be derived from theoretical first principles. What is clear, however, is the current trajectory for U.S. debt is not sustainable.

Although delayed implementation of deficit-reducing policies may be preferable given the current state of the economy, the longer it takes to put in place deficit-reducing policies, the larger will be the required spending cuts or tax increases in order to address the long-term fiscal gap. For example, if the adjustments are delayed until 2018, when the CBO projects the economy will reach potential GDP, the fiscal gap increases by up to 0.3 percentage points of GDP.

Budget projections (especially for the long term) embody considerable uncertainty, and deficit projections are particularly uncertain as relatively small percentage changes in outlays and revenues can lead to relatively large percentage changes in deficits. In the current environment, economic projections also may be more uncertain than usual, given uncertainty about the effects of the recent recession on the long-term growth rate. The other major uncertainty is the rate of growth of health care spending, which can have enormous impacts on the projected budget outlook. Despite this uncertainty, it is hard to paint an optimistic picture of the fiscal outlook. Indeed, the projections above are based on a series of economic and political assumptions that could be viewed as optimistic.

B. The need for spending cuts and revenue increases

Since projected spending is slated to rise faster than GDP for the indefinite future, it is clear that spending cuts must be part of the solution, in particular for government health care programs, which have been rising as a share of GDP for several decades and are projected to continue to rise.

There are several reasons to consider tax increases (beyond those already included in the January 2013 budget deal), however, as well as spending cuts, as part of the fiscal solution. First, the sheer magnitude of the fiscal gap suggests that a spending-only solution would need to impose very substantial reductions on spending that might not be seen as equitable. At 5–7 percent of GDP, the fiscal gap is several times larger than the savings that were generated in budget deals in the past. The 1983 Social Security Reform reduced deficits by about 1 percent of GDP in the four years after passage while the 1990 and 1993 budget deals reduced deficits by about 1.4 percent of GDP and 1.2 percent of GDP, respectively, over the 5 years after passage.[2] The recently enacted tax bill only raised 0.3 percent of GDP in revenue over the next decade. In addition, Americans seem particularly reluctant to cut government spending on Social Security and Medicare, two of the key drivers of long-term spending, than on other forms of spending. For instance, a 2011 Gallup poll showed that over 60 percent of Americans were unwilling to cut Social Security and/or Medicare, and this was true across the political spectrum.

Second, as a political equilibrium, it seems likely that a sustainable budget deal would draw from both sides of the ledger. Indeed, in the past, major deals have included both tax increases and spending cuts. With the 1983 Social Security reforms, the 1990 bipartisan budget deal, and 1993 budget deals, Congress both slashed spending and raised taxes. For example, in the 1990 budget deal, 49 percent of the reductions came from higher tax receipts, 34 percent from reduced defense spending, and 17 percent from other cuts in spending (Steuerle 2004).

Third, as a matter of equity, the only way that high-income households will share significantly in the burden of fixing the deficit is through revenue increases since spending cuts typically do not have a large impact on high-income households.

Fourth, spending appears to be controlled more effectively by requiring that it be paid for with current taxes, rather than allowing deficits to grow. In contrast, the "starve the beast" hypothesis argues that keeping revenues down is an effective approach to curtailing spending. However, the hypothesis does not appear to be consistent with recent experience.[3] And evidence in Romer and Romer (2009), for example, suggests that tax cuts designed to spur long-run growth do not in fact lead to lower government spending; if anything, they find that tax cuts lead to *higher* spending. This finding is consistent with Gale and Orszag (2004a), who argue that the experience of the last 30 years is more consistent with a "coordinated fiscal discipline" view, in which tax cuts were coupled with increased spending (as in the 1980s and 2000s) and tax increases were coupled with contemporaneous spending reductions (as in the 1990s).

C. Long-term growth effects of tax-financed deficit reductions

An increase in taxes will not necessarily slow long-term economic growth. Tax changes have two broad sets of long-term effects on the economy.[4] The first set operates through direct changes in relative prices, incentives, and after-tax income. These changes affect the degree to which households are willing to work, save, invest in education and training, and so on, and to which firms invest and hire; these effects are known as income and substitution effects. Thus, for example, increases in marginal tax rates, holding other factors constant, can reduce the size of the economy and reduce economic growth.

However, other factors are not constant. The second broad effect is on national saving. A reduction in the deficit tends to raise public saving, which typically results in higher national saving (national saving is the sum of household, corporate, and government saving). This effect is often ignored in discussions of tax policy and economic growth, but it can be quite important.

Containing deficits matters for several reasons.

Sustained deficits may enhance the risk of a financial crisis. Even in the absence of precipitating a financial crisis, however, sustained deficits have deleterious long-term effects, as they translate into lower national savings, higher interest rates, and increased indebtedness to foreign investors, all of which reduce future national income. In addition to the growth impacts, sustained deficits may impose unfair burdens on future generations and may constrain U.S. foreign policy or defense positions, especially as they relate to creditor nations.

Gale and Orszag (2004b) estimate that a 1 percent of GDP increase in the deficit will raise interest rates by 25 to 35 basis points in the United States and reduce national saving by 0.5 to 0.8 percentage points. Engen and Hubbard (2004) obtain similar results with respect to interest rates. Thus, relative to a balanced budget, this study suggests a deficit equal to 6 percent of GDP would raise interest rates by at least 150 basis points and reduce the national saving rate by at least 3 percent of GDP. The IMF (2010) estimates that, in advanced economies, an increase of 10 percentage points in the initial debt/GDP ratio reduces future GDP growth rates by 0.15 percentage points. Hence (if this result is extrapolated linearly, and we do so with caution, since it would be easy to think of reasons that would make a larger debt change have more-than-proportional or less-than-proportional effects), the increase in the debt-to-GDP ratio from about 40 percent earlier in the decade to 85 percent by 2022 (Auerbach and Gale 2012) would be expected to reduce the growth rate by a whopping 0.675 percentage points. Thus a deficit reduction plan that included tax increases (at least one that did not primarily rely on raising taxes on savings and investment) could, on balance, help spur economic growth in contrast to continuing policy as normal.

The net long-term effect of a tax change is the result of the two effects outlined above, which are sometimes offsetting and sometimes mutually reinforcing. Stokey and Rebelo (1995), for example, show that even the very large tax increases associated with World War II – on the order of 10 percent of GDP – apparently had no discernible impact on the long-term economic growth rate. Likewise, the 1981 tax cuts, which cut the top rate from 70 percent to 50 percent, accounted for only a very small share of the growth of the economy between 1981 and 1986, according to Feldstein and Elmendorf (1989). Auerbach and Slemrod (1997) also document tepid economic growth responses to the 1986 tax act. Gale and Potter (2002) find that the impact of the 2001 tax cuts on the deficit and national saving outweighed its impact on incentives, so that the net effect on growth was negative. This suggests that raising taxes by undoing the 2001 tax cuts would raise long-term economic growth (due to the beneficial effect of lower deficits).

III. Carbon taxes

The discovery and exploitation of natural resources by humans gave rise to the advanced civilization in which we live today. Coal, petroleum, and natural gas fueled industrialization, raising living standards and life expectancy for most. Energy use continues to fuel economic growth and development today. But along with the benefits of energy consumption come substantial societal costs – including those associated with air and water pollution, road congestion, and climate change. Many of these costs are not directly borne by the businesses and individuals that use fossil fuels and thus are ignored when energy production and consumption choices are made. As a result, there is too much consumption and production of fossil fuels.

Economists have long recommended specific taxes on fossil-fuel energy sources as a way to address these problems. That recommendation has gained additional urgency in recent years in light of the fiscal situation outlined above. New revenue from energy taxes could be used to reduce the debt or finance reform or reductions in other taxes.

Throughout this chapter we use the phrase "carbon tax" to refer to a tax on carbon dioxide. Although a carbon tax would be a new policy for the federal government, the tax has been implemented in several other countries (though – as discussed in the introduction to this volume – not always in a way that conforms to the design principles advocated by economists). Finland, Norway, Sweden, and Denmark instituted carbon taxes in the early 1990s, followed by the Netherlands and Germany in the latter part of the 1990s. The United Kingdom followed suit in 2001. Australia introduced a carbon tax in 2011. North American jurisdictions have also implemented carbon taxes. The town of Boulder, Colorado, adopted a carbon tax in 2006, and Montgomery County, Maryland, did so in 2010. The Canadian provinces of Alberta and Quebec adopted carbon taxes in 2007, followed by British Columbia in 2008.

A. Revenue

Carbon taxes can raise significant amounts of revenue. For instance, in 2007 the tax raised revenue equivalent to about 0.3 percent of GDP in Finland and Denmark, and 0.8 percent in Sweden. A well-designed tax in the United States could raise similar amounts. As shown in Table 1.2, a number of studies have estimated the net revenue effects of carbon taxes – accounting for the reduction in revenues from broader taxes that would occur – with estimates (for the year 2015) ranging from 0.5 percent of GDP for a $15 per ton tax (McKibbin et al. 2012) to 0.8 percent of GDP for a $31 per ton tax (Metcalf 2010) with intermediate estimates including CBO (2011) and Rausch and Reilly (2012).[5]

TABLE 1.2 Carbon tax revenue, 2015

Author	*Level of tax ($/tCO$_2$)*	*Net revenue (percent of GDP)*	*Net revenue after distributional offsets (percent of GDP)**
McKibbin et al. (2012)	15	0.51	0.32
CBO (2011)	22	0.60	0.37
Rausch and Reilly (2012)	23	0.66	0.41
Metcalf (2010)	31	0.79	0.49

*For discussion of the notion and nature of distributional offsets, see the text.

Based on analysis in Dinan (2012) discussed below, we assume 38 percent of net carbon tax revenues would need to be used to offset distributional effects – as noted later, this might be viewed as a generous estimate if a carbon tax is part of a broader package of measures to reduce the deficit, and other measures are progressive (i.e., they impose a disproportionately larger burden on higher income households). Our assumption leaves the net revenue yield after distributional compensations at between 0.32 percent and 0.49 percent of GDP. In terms of gauging how large these taxes are in practical terms, a tax of $25 per ton of carbon dioxide would raise gasoline prices by 25 cents a gallon (Bauman 2010).

B. Efficiency

In principle, carbon taxation receives high marks on efficiency criteria. Indeed, the basic rationale for a carbon tax is that it makes good economic sense: unlike most taxes, carbon taxation can improve the efficient allocation of resources by accounting for externalities in the market price. Externalities can be severe. Stavins (2007) notes that the efficiency benefits of a carbon tax are often understated since the largest efficiency gains come in the form of internationally shared reduced greenhouse gas emissions. While the United States is the largest per capita emitter of carbon dioxide, China is the largest overall emitter, and the European Union makes a significant contribution as well. Therefore, enacting a program that would lead to better cooperation with other countries, and reduce emissions across the world, would be better suited to deal with the well-known problems brought about by global warming, such as rising sea levels, more frequency in extreme temperatures, among others.

Taxes on energy can address these externalities. Not surprisingly, most analyses find that a carbon tax could significantly reduce emissions. Metcalf (2008) estimates that a $15 per ton tax on CO_2 emissions that rises over time would reduce greenhouse gas emissions by 14.0 percent, while Sumner et al. (2009) estimate that the European countries' carbon taxes have had a significant effect on emissions reductions, attributing reductions of up to 15 percent to the carbon tax. Furthermore, the University of Ottawa (2012) found that the carbon tax implemented in British Columbia led to a 9.9 percent reduction in greenhouse gas emissions in the province, compared to just 4.6 percent for the rest of Canada, where comprehensive carbon taxes were not applied.

In addition to reducing emissions, a carbon tax could improve other economic incentives by reducing other tax rates or paying down the deficit (Parry and Williams 2011). A carbon tax could have other benefits too. It would reduce the U.S. economy's dependence on foreign sources of energy, and would create better market incentives for energy conservation, the use of renewable energy sources, and the production of energy-efficient goods. The permanent change in price signals from enacting a carbon tax would stimulate new private sector research and innovation in developing new ways of harnessing renewable energy and energy-saving technologies. The implementation of a carbon tax also offers opportunities to reform and simplify other climate-related policies affecting the transportation sector.

C. Distribution

The net effects of a carbon tax will depend, of course, not only on the magnitude of the tax and the behavioral response by consumers and firms, as the studies above consider, but on how the funding is used. To be clear, all uses of carbon tax revenues (or of other revenues for that matter) involve some form of giving the money back to taxpayers. What varies is which taxpayers receive the funds, during what time period, and under what conditions. Providing a rebate to consumers obviously returns the revenue to citizens. But so do all other uses of the funds. For instance, paying down the deficit implicitly gives the money to future citizens by reducing the extent to which they have to pay higher taxes or bear the burden of spending cuts. Likewise, using the funds to provide corporate tax cuts reduces burdens for whichever individuals ultimately bear the burden of corporate taxation.

In many instances to date, carbon tax revenues have not been used for deficit reduction. Norway and Sweden do include carbon tax revenue as part of general government receipts, which suggests a possible effect on deficit reduction. But carbon tax revenue in Denmark is returned to industry and directed toward environmental subsidies. Several nations have used carbon tax revenue to reduce other taxes (Sumner et al. 2009). Australia coupled its carbon tax with a substantial increase in the tax-free level of income (and other tax changes).[6] The Netherlands and Sweden have exempted a large portion of the industrial sector from the tax, as well as helping low-income households offset the burden of the tax (the latter measure was also implemented by Germany) (Johansson 2001). Quebec deposits carbon tax revenues into a fund devoted to public transportation and environmental initiatives, while British Columbia makes its carbon tax revenue-neutral by reducing corporate and personal income tax rates and providing an annual credit of $100 per adult and $30 per child to lower-income citizens (British Columbia Finance Ministry 2008).

Distributional concerns over carbon taxes stem from the observation that low-income households devote a higher proportion of their income to consumption and will thus bear a higher burden of the tax relative to high-income households. The distributional effects of carbon taxation have been well-studied (Bull et al. 1994, Hassett et al. 2009, Metcalf 1999, Metcalf 2007). The regressivity finding is consistent across studies, but varies in magnitude. Metcalf (2008) analyzes the distributional effects of a carbon tax and finds that it would reduce the after-tax income of taxpayers in the first decile by 3.7 percent, compared to just an 0.8 percent reduction for the wealthiest decile. Findings are dependent on whether incidence is measured on a current income versus lifetime basis, with the tax being more regressive when measured on a current income basis relative to lifetime income basis. For example, Hassett et al. (2009) find that the indirect component of a carbon tax (i.e., higher prices due to higher costs of production) is significantly more progressive, whereas the direct component, which focuses on the changes in the cost of gas and electricity, is regressive. Lastly, the incidence varies with timing: the carbon tax can either fall forward in the form of higher consumer prices or backward in the form of lower returns to factor inputs. Bovenberg and Goulder (2001) and Paltsev et al. (2007) find that the short- and medium-term incidence falls primarily on consumer prices.

Importantly, the regressive impact of a carbon tax could be offset in any of a number of ways, similar to offsets for distributional effects of the VAT, as will be discussed in the next section. Most prominent among these options would be refundable income tax credits (Dinan 2012) or payroll tax refunds (Metcalf 2007). Dinan (2012) notes that CBO analysis suggests that fully offsetting the effects of carbon taxes for households in the lowest quintile would require about 12 percent of gross revenues, while fully offsetting the effects for households in the second quintile would require 27 percent of gross revenues. These figures do not account for added government costs (of indexing transfers or higher payments for inputs, for example). Nor do they account for the reduction revenues from other taxes noted above. As a rough approximation, for now we assume that 38 percent of net carbon tax revenues would have to be used for offset purposes. This is not inconsistent with Dinan's estimates and is similar to the calculations derived by Toder and Rosenberg (2010) for a VAT. Thus, while the regressivity of a carbon tax is clearly a concern, it should not be considered an obstacle to the implementation of carbon taxes.

D. Motor fuel taxes

Raising taxes on gasoline and (motor) diesel is another option. While modest excise taxes on these fuels already exist in the United States, they are substantially lower than in other industrialized nations.

For example, in the United States, federal excise taxes on gasoline amount to 18.4 cents per gallon, with local tax rates typically taxing gasoline at additional 20–30 cents per gallon in 2010. The OECD average for gasoline excise taxes is approximately $3.39 per gallon, about 7 times the rate of the U.S. tax.[7] OECD taxation of gasoline ranged from $0.34 per gallon (Mexico) to $5.14 per gallon (Turkey); the United States has the second-lowest rate of gasoline taxation among OECD countries (OECD 2011). In addition, per-mile fuel taxes in the United States are low by historical standards, falling by 40 percent in real terms since 1960 (Parry et al. 2007). Moreover, fuel taxes at least three times as high as current levels (and perhaps higher still) appear to be justified by the adverse side effects of motor vehicles – pollution, congestion, and so on (Parry et al. 2007).

Higher excise taxes on motor fuels could raise significant amounts of revenue. For example, Parry (2011) estimates that raising gasoline and diesel fuel taxes to their corrective levels would increase revenue by around 0.8 percent of GDP, while CBO (2009) estimates that a 50-cent increase in the gasoline excise tax alone would raise about 0.3 percent of GDP. Raising the gas tax by 25 cents per year for 10 years would raise substantially more in revenues, but would still leave U.S. gas tax rates well below those of European countries.

Although higher fuel taxes would have some impact on reducing carbon emissions, they are much less effective than a carbon tax at reducing carbon emissions, since the former covers a much narrower range of externality-producing goods.[8] Davis and Kilian find (2009) find that a 10 cent per gallon increase in the U.S. gasoline excise tax would reduce total carbon emissions by 0.5 percent overall and by 1.5 percent from vehicles. Like carbon taxes, gasoline taxes will fall disproportionately on low-income households, especially in the short-run when households have difficulty adjusting their behavior to avoid the tax (Poterba 1989 and 1991).

IV. Other revenue options

A carbon tax can be compared to other tax options – not necessarily because the ultimate choice will be one of those options versus another, as the country will probably need several ways to raise revenue, but rather to discuss the relative revenue-generating potential, efficiency, and equity effects of the different taxes. A full-scale comparison is beyond the scope of this chapter (see Gale and Brown 2012 for a more comprehensive discussion of the options). We do briefly describe options relating to the value-added tax (VAT) and to income tax expenditure reform, however, to provide some sense of the trade-offs, and possible complementarities, between carbon taxes and broader fiscal options.

A. Value-added tax

Under a VAT, businesses would pay taxes on the difference between their revenues from total sales to other businesses and households and their purchases of inputs from other businesses. That difference represents the value-added by the firm to the product or service in question.[9] The sum of value added at each stage of production (including extraction of the raw materials) is the retail sales price, so the VAT simply replicates the tax patterns created by a retail sales tax and is like other taxes on aggregate consumption. The key distinction is that VATs are collected at each stage of production, whereas retail sales taxes are collected only at point of final sale. Furthermore, the VAT is easier to enforce and is widely regarded as having a superior administrative structure to a retail sales tax. Although it would be new to the United States, the VAT is in place in about 150 countries worldwide and in every OECD country other than the United States. Experience suggests that the VAT can raise substantial revenue, is administrable, and minimally harmful to economic growth. Toder and Rosenberg (2010) show that a 5 percent VAT with a relatively broad base could raise revenue equal to 1 percent of GDP in the United States, even after accounting for distributional issues via rebates and adjusting for revenue losses from other taxes (Table 1.3).

The distributional burden of the VAT is regressive relative to current income (though not relative to current consumption). Concerns about the regressivity of the VAT are valid, but they should not obstruct the creation of a VAT for two reasons. First, while we accept the validity of distributional considerations, what matters is the progressivity of the overall tax and transfer system, not the distribution of any individual component of that system. Clearly, the VAT can be one component

TABLE 1.3 Revenue effects of various tax reform options

Proposal	*Year*	*Net revenue (percent of GDP)*	*Net revenue after distributional offsets (percent of GDP)**
5 percent broad-based VAT (Toder and Rosenberg 2010)	2012	1.64	1.02
Cap itemized deductions at $50,000**	2015	0.33	N/A
Cap the tax value of tax expenditures for high-income households at 2 percent of AGI (Baneman et al. 2011)	2015	0.26	N/A

*For discussion of the notion and nature of distributional offsets, see the text.

** http://www.taxpolicycenter.org/numbers/displayatab.cfm?DocID=3587

of a progressive system. Second, it is straightforward to introduce policies that can offset the impact of the VAT on low-income households. The most efficient way to do this is simply to provide households either refundable income tax credits, adjustments to cash-transfer benefits, or outright payments.[10] In contrast, many OECD governments and U.S. state governments offer preferential or zero rates on certain items like health care or food to increase progressivity. This approach is largely ineffective because the products in question are consumed in greater quantities by middle-income and wealthy taxpayers than they are by low-income households.[11]

B. *Tax expenditure reform*[12]

A third alternative is reform of income tax expenditures. In formal terms, tax expenditures are "revenue losses attributable to provisions of the Federal tax laws which allow a special exclusion, exemption, or deduction from gross income or which allow a special credit, preferential rate of tax or a deferral of liability" (Congressional Budget and Impoundment Control Act of 1974 (P.L. 93–344)). The canonical focus for income tax reform is to create a system with a broad base that taxes all sources and uses of income at the same rate so as to generate lower statutory rates. Tax expenditure reform would be essential to achieving these goals. Broadening the base entails restricting the use of exclusions and deductions. Taxing all sources and uses of income, at the same effective rate, entails restricting the use of preferential rates, credits, and deferrals. This would reduce distortions between the taxation of different sources and uses of income and therefore could be efficiency improving.

Many major tax expenditures act essentially as government spending programs that happen to be embedded in the tax code rather than in outlays (Batchelder and Toder 2010; Marron 2012; Marron and Toder 2012). Tax expenditure reform in many cases can be thought of as reducing effective government spending.

The value of most tax expenditures, other than credits, rises with the marginal tax rate. A deduction or exclusion of $1,000 would reduce tax liability by $150 for an individual in the 15 percent bracket but $330 to one in the 33 percent bracket.

Although different types of tax expenditures are distributed differently, the aggregate distribution of tax expenditures tends to be tilted toward high-income households because they itemize their deductions, receive a substantial share of the income in the form of returns to investment, which is often subject to preferential rates, they have more tax to offset, and they receive a higher benefit per dollar of deduction or exclusion due to higher marginal tax rates.

Tax expenditure reform can raise significant amounts of revenue. Although precise estimates are difficult to compute, illustrative calculations indicate the potential for revenue-raising. The FY2013 Budget lists 173 individual and business tax expenditures, the total value of which would approach 7.5 percent of GDP in the 2015 fiscal year (relative to current law) and about 80 percent coming from individual income receipts (Office of Management and Budget 2012; Marron 2012). Interaction effects increase the revenue loss: Toder and Baneman (2012) estimated that interaction effects increased lost revenue from non-business individual income tax expenditures by 9.6 percent in 2011.

Yet potential revenue raised from a realistic tax expenditure reform would be much less for administrative and political reasons. Some expenditures are difficult to eliminate for various administrative reasons. Many of the largest tax expenditures (e.g., mortgage interest deduction, employer-sponsored health insurance) are broadly popular because they benefit middle-income, as well as high-income, taxpayers. Recent proposals have focused on capping overall tax expenditures for a tax filer rather than eliminating individual policies to ease the political constraints to tax expenditure reform. Such proposals still can raise revenue and increase the progressivity of the tax system.

One recent proposal would cap itemized deductions at $50,000. The Tax Policy Center estimates that, relative to current policy, a $50,000 cap would raise 0.33 percent of GDP in 2015 (Table 1.3). The policy will have a small effect on households in the bottom 90 percent of the income distribution. Households in the 90th to ninety-ninth percentiles would see their after-tax income decrease by between 0.3 and 0.5 percent. After-tax income would decrease by 3 percent in the top 1 percent of the income distribution.

Feldstein et al. (2011) propose a cap on the tax value of certain tax expenditures to 2 percent of the earner's AGI.[13] Baneman et al. (2011) applied the cap to earners making more than $250,000 (married) or $200,000 (single) and estimated that it could raise 0.26 percent of GDP relative to current policy (Table 1.3). The cap would not affect taxpayers below the ninety-fifth income percentile. It would decrease the after-tax income by 0.9 percent of income for filers between the ninety-fifth to ninety-ninth percentiles and by 3 percent for filers in the top 1 percent (Baneman et al. 2011).

V. Conclusion

The United States faces substantial and unsustainable medium- and long-term budget deficits, which will require a combination of tax increases and spending cuts to resolve. On the tax side, one relatively attractive option for raising revenue would be to impose a carbon tax. Besides its impact on revenues, the tax would improve environmental outcomes, increase economic efficiency, and allow the elimination of selected other tax subsidies and spending programs. The distributional effects would be regressive but could be offset by other policy changes. As policymakers search for solutions to the fiscal problem and for ways to improve the tax system, carbon taxation could play a positive role in addressing each situation.

Notes

1. The IMF (2012) suggested 60 percent as an appropriate ratio of gross general debt-to-GDP for advanced countries while emerging and low-income countries had a lower ratio of 40 percent. The Peterson-Pew Commission on Budget Reform (2011), the Bipartisan Policy Center's Debt Reduction Task Force (2010), and the President's National Commission on Fiscal Responsibility and Reform (2010) all had medium-term goals of a 60 percent of debt-to-GDP ratio by 2020 or 2021. Johnson and Kwak (2012) suggested a goal of 50 percent to err on the side of caution to account for the fact that the U.S. workforce is growing more slowly and for the fears that the U.S. may lose its global reserve currency status or face another financial crisis.
2. Authors' calculations based on Steuerle (2004) and CBO (1983, 1991, 1993).
3. Bartlett (2007) outlines the development of the "starve the beast" theory and shows how it failed to apply during the George W. Bush administration.
4. Short-term economic effects of tax-financed deficit reductions often differ from long-term effects. Consequently, the relative benefits of a tax-financed deficit reduction policy depend on the time frame of the analysis. Since this chapter is concerned with a long-term fiscal solution, we focus on the long-term economic effects.
5. The creation of carbon taxes will cause a partial, automatic reduction in other tax revenues. As one simple example of how this might work, a firm that pays $100 in carbon taxes would, in the absence of any other changes, have $100 less in corporate profits and so would owe less in corporate taxes. Studies estimate overall automatic tax offsets between 25 percent and 31 percent of the gross revenue levels, thus resulting in the net revenue levels reported in the text and shown in Table 1.2.
6. Australia really has an emissions trading system. However, because most of the allowances are auctioned, and there is a price collar (at least until 2015) it looks more like a tax.
7. Authors' calculations based on OECD (2011).
8. Sterner (2007), for example, estimates that fuel demand in Europe would be twice as high if European countries had faced U.S. gas tax rates.
9. There are several options for administering the tax which we do not go into here. See Bickley (2006) and Cnossen (2009) for some discussion of these options.
10. Toder et al. (2011) propose a two-pronged rebate. The rebate would be a credit equal to the VAT rate multiplied by a base of $12,000 for single households and $24,000 for married households (in 2012); the base could not exceed employment income. In addition, they propose an upward adjustment to Social Security payments to offset the reduction in real wages over time.
11. Congressional Budget Office (CBO; 1992, xv) finds that "excluding necessities such as food, housing, utilities, and health care would lessen the VAT's regressivity only slightly." Toder and Rosenberg (2010) find that excluding housing, food consumed at home, and private health expenditures from the consumption tax base can somewhat increase progressivity, but not as much as a per-person payment would.
12. All of the revenue estimates here refer to pre-ATRA baselines. Since ATRA raised tax rates, post-ATRA revenue estimates of tax expenditure reform would yield somewhat larger revenue estimates than indicated here.
13. The Feldstein-Feenberg-MacGuineas proposal limited the tax value of itemized deductions, the health insurance exclusion, and the child tax credit, dependent care credit, and

general business credit. For deductions and exemptions, the tax value is equal to the face value of the deduction or exclusion multiplied by the filer's marginal tax rate. The tax value of a tax credit is equal to the credit.

References

Auerbach, Alan J. and William G. Gale. 2012. "The Federal Budget Outlook: No News Is Bad News." Available at: <http://emlab.berkeley.edu/~auerbach/Auerbach-Gale%20 2012-08-27.pdf>

Auerbach, Alan J. and William G. Gale. 2013a. "Fiscal Fatigue: Tracking the Budget Outlook as Political leaders Lurch from one Artificial Crisis to Another." Available at: <www.brookings.edu/research/papers/2013/02/28-fiscal-fatigue-budget-outlook-gale#_ftn9>

Auerbach, Alan J. and William G. Gale. 2013b. "Fiscal Myopia." Available at: <www.brookings.edu/~/media/research/files/papers/2013/09/30%20fiscal%20myopia%20auerbach%20 gale/30%20fiscal%20myopia%20auerbach%20gale.pdf>

Auerbach, Alan J. and Joel Slemrod. 1997. "The Economic Effects of the Tax Reform Act of 1986." *Journal of Economic Literature* 35(2): 589–632.

Baneman, Daniel, Jim Nunns, Jeff Rohaly, Eric Toder, and Roberton Williams. 2011. "Options to Limit the Benefit of Tax Expenditures for High-Income Households." Washington, D.C.: Tax Policy Center.

Bartlett, Bruce. 2007. "Starve the Beast: Origins and Development of a Budgetary Metaphor." *The Independent Review* 12(1): 5–26.

Batchelder, Lily and Eric Toder. 2010. "Government Spending Undercover: Spending Programs Administered by the IRS." Washington, D.C.: Center for American Progress.

Bauman, Yoram. 2010. "Comments on Nordhaus: Carbon Tax Calculations." *The Economists' Voice* 7(4), ISSN 1553-3832, DOI: 10.2202/1553-3832.1796, October 2010.

Bickley, James M. 2006. "Value-Added Tax: A New U.S. Revenue Source?" Congressional Research Service Report to Congress RL33619.

Bipartisan Policy Center Debt Reduction Task Force. 2010. "Restoring America's Future: Reviving the Economy, Cutting Spending and Debt, and Creating a Simple, Pro-Growth Tax System." November.

Bovenberg, A. Lans and Lawrence H. Goulder. 2001. "Environmental Taxation and Regulation." In Alan J. Auerbach and Martin Feldstein (eds), *Handbook of Public Economics*, Vol. 3. Amsterdam: North Holland Publishing, 1471–1545.

British Columbia Finance Ministry. 2008. "Balanced Budget 2008." Available at: <www.bcbudget.gov.bc.ca/2008/backgrounders/backgrounder_carbon_tax.htm>

Bull, Nicholas, Kevin A. Hassett, and Gilbert E. Metcalf. 1994. "Who Pays Broad-Based Energy Taxes? Computing Lifetime and Regional Incidence." *The Energy Journal* 15(3): 145–164.

Cnossen, Sijbren. 2009. "A VAT Primer for Lawyers, Economists, and Accountants." *Tax Notes* 124(7): 687–698. August 17.

Congressional Budget and Impoundment Control Act of 1974. (*P.L.* 93–344, 88 *Stat.* 297, *2 U.S.C. §§ 601–688*).

Congressional Budget Office. 1983. "The Economic and Budget Outlook: An Update." Washington, D.C.: Congressional Budget Office.

Congressional Budget Office. 1991. "The Economic and Budget Outlook: An Update." Washington, D.C.: Congressional Budget Office.

Congressional Budget Office. 1992. "Effects of Adopting a Value-Added Tax." February 1992. Washington, D.C.: Congressional Budget Office.

Congressional Budget Office. 1993. "The Economic and Budget Outlook: An Update." Washington, D.C.: Congressional Budget Office.

Congressional Budget Office. 2009. "Budget Options." Washington, D.C.: Congressional Budget Office.

Congressional Budget Office. 2011. "Reducing the Deficit: Spending and Revenue Options." Washington, D.C.: Congressional Budget Office.

Davis, Lucas W. and Lutz Kilian. 2009. "Estimating an Effect of a Gasoline Tax on Carbon Emissions." NBER Working Paper No. 14685, National Bureau of Economic Research.

Dinan, Terry. 2012. "Offsetting a Carbon Tax's Costs on Low-Income Households: Working Paper 2012–16." Congressional Budget Office Working Paper.

Engen, Eric M. and R. Glenn Hubbard. 2004. "Federal Government Debt and Interest Rates." *NBER Macroeconomics Annual* 19: 83–138.

Feldstein, Martin and Douglas W. Elmendorf. 1989. "Budget Deficits, Tax Incentives and Inflation: A Surprising Lesson from the 1983–84 Recovery." NBER Working Paper No. 2819, National Bureau of Economic Research.

Feldstein, Martin, Daniel Feenberg, and Maya MacGuineas. 2011. "Capping Individual Tax Expenditure Benefits." *Tax Notes* 131(5): 505–509.

Gale, William G. and Samuel Brown. 2012. "Tax Reform for Growth, Equity, and Revenue." Brookings Institution. Available at: <www.brookings.edu/research/papers/2012/11/30-tax-reform-brown-gale>

Gale, William G. and Peter R. Orszag. 2004a. "Bush Administration Tax Policy: Starving the Beast?" *Tax Notes* 105(8): 999–1002.

Gale, William G. and Peter R. Orszag. 2004b. "Budget Deficits, National Saving, and Interest Rates." *Brookings Papers on Economic Activity* 2: 101–187.

Gale, William G. and Samara Potter. 2002. "An Economic Evaluation of the Economic Growth and Tax Relief Reconciliation Act." *National Tax Journal* 55(1): 133–186.

Hassett, Kevin, Aparna Mathur, and Gilbert E. Metcalf. 2009. "The Incidence of a U.S. Carbon Tax: A Lifetime and Regional Analysis." *The Energy Journal* 30(2): 155–177.

International Monetary Fund. 2010. *Fiscal Monitor (May 2010): Navigating the Challenges Ahead*. Washington, D.C.: International Monetary Fund.

International Monetary Fund. 2012. *Fiscal Monitor (April 2012): Balancing Fiscal Policy Risks.* Washington, D.C.: International Monetary Fund.

Johansson, Bent. 2001. *Economic Instruments in Practice: Carbon Tax in Sweden.* Swedish Environmental Protection Agency. Available at: <www.oecd.org/dataoecd/25/0/2108273.pdf>

Johnson, Simon and James Kwak. 2012. *White House Burning: The Founding Fathers, Our National Debt, and Why It Matters to You.* New York, NY: Random House.

Marron, Donald. 2012. "How Large Are Tax Expenditures? A 2012 Update." *Tax Notes.* Tax Policy Center, April 9, 235.

Marron, Donald and Eric Toder. 2012. "How Big Is the Federal Government?" Discussion Paper, Tax Policy Center, Urban Institute and Brookings Institution.

McKibbin, Warwick J., Adele C. Morris, Peter J. Wilcoxen, and YiYong Cai. 2012. "The Potential Role of a Carbon Tax in U.S. Fiscal Reform." Available at: <www.brookings.edu/~/media/research/files/papers/2012/7/carbon%20tax%20mckibbin%20morris%20wilcoxen/carbon%20tax%20mckibbin%20morris%20wilcoxen.pdf>

Metcalf, Gilbert E. 1999. "A Distributional Analysis of Green Tax Reforms." *National Tax Journal* 52(4): 655–682.

Metcalf, Gilbert E. 2007. "A Proposal for a U.S. Carbon Tax Swap: An Equitable Tax Reform to Address Global Climate Change." *The Hamilton Project, Brookings Institution.* October.

Metcalf, Gilbert E. 2008. "Designing a Carbon Tax to Reduce U.S. Greenhouse Gas Emissions." NBER Working Paper No. 14375, National Bureau of Economic Research.

Metcalf, Gilbert E. 2010. "Submission on the Use of Carbon Fees to Achieve Fiscal Sustainability in the Federal Budget." Available at: <http://works.bepress.com/gilbert_metcalf/86>

National Commission on Fiscal Responsibility and Reform. 2010. "The Moment of Truth." December.

OECD. 2011. "Energy Prices and Taxes: Quarterly Statistics, Fourth Quarter 2010." OECD: Paris, France.

Office of Management and Budget. 2012. *Budget of the U.S. Government: Fiscal Year 2013*. Washington, D.C.: Government Printing Office.

Paltsev, Sergey, John M. Reilly, Henry D. Jacoby, Angelo C. Gurgle, Gilbert E. Metcalf, Andrei P. Sokolov, and Jennifer F. Holak. 2007. "Assessment of U.S. Cap-and-Trade Proposals." MIT Joint Program on the Science and Policy of Global Change Report 146. April.

Parry, Ian W. G. 2011. "How Much Should Highway Fuels Be Taxed?" In Gilbert E. Metcalf (ed.), *U.S. Energy Tax Policy*. Cambridge: Cambridge University Press, 269–297.

Parry, Ian, Margaret Walls, and Winston Harrington. 2007. "Automobile Externalities and Policies." *Journal of Economic Literature* 65: 373–399.

Parry, Ian W. H. and Roberton C. Williams III. 2011. "Moving US Climate Policy Forward: Are Carbon Taxes the Only Good Alternative?" Resources for the Future Discussion Paper 11–02.

Peterson-Pew Commission on Budget Reform. 2011. "Paths to Debt Stabilization: A Comparison of Debt Triggers." October 19.

Poterba, James M. 1989. "Lifetime Incidence and the Distributional Burden of Excise Taxes." *American Economic Review* 79(2): 325–330.

Poterba, James M. 1991. "Is the Gasoline Tax Regressive?" In David Bradford (ed.), *Tax Policy and the Economy*, Vol. 5. Cambridge, MA: National Bureau of Economic Research, 145–164.

Rausch, Sebastian and John Reilly. 2012. "Carbon Tax Revenue and the Budget Deficit: A Win-Win Solution?" MIT Joint Program on the Science and Policy of Global Change.

Romer, Christina D. and David H. Romer. 2009. "Do Tax Cuts Starve the Beast? The Effect of Tax Changes on Government Spending." *Brookings Papers on Economic Activity* 2009(Spring): 139–200.

Stavins, Robert N. 2007. "A U.S. Cap-and-Trade System to Address Global Climate Change." Hamilton Project Discussion Paper 2007–13.

Sterner, Thomas. 2007. "Fuel Taxes: An Important Instrument for Climate Policy." *Energy Policy* 35: 3194–3202.

Stokey, Nancy L. and Sergio Rebelo. 1995. "Growth Effects of Flat-Rate Taxes." *Journal of Political Economy* 103(3): 510–550.

Steuerle, C. Eugene, 2004. "Contemporary U.S. Tax Policy." Urban institute, Washington, DC.

Sumner, Jenny, Lori Bird, and Hillary Smith. 2009. "Carbon Taxes: A Review of Experience and Policy Design Considerations." National Renewable Energy Laboratory Technical Report NREL/TP-6A2–47312.

Toder, Eric and Daniel Baneman. 2012. "Distributional Effects of Individual Income Tax Expenditures: An Update." Discussion Paper, Tax Policy Center, Urban Institute and Brookings Institution, February 3.

Toder, Eric, Jim Nunns, and Joseph Rosenberg. 2011. "Methodology for Distributing a VAT." Discussion Paper, Tax Policy Center, Urban Institute and Brookings Institution, April 12.

Toder, Eric and Joseph Rosenberg. 2010. "Effects of Imposing a Value-Added Tax to Replace Payroll Taxes or Corporate Taxes." Discussion Paper, Tax Policy Center, Urban Institute and Brookings Institution, April 7.

University of Ottawa. 2012. "British Columbia's Carbon Tax Shift: The First Four Years." Sustainable Prosperity. Available at: www.sustainableprosperity.ca/dl872&display.

2

CHOOSING AMONG MITIGATION INSTRUMENTS

How strong is the case for a US carbon tax?

Ian Parry

KEY MESSAGES FOR POLICYMAKERS

- Carbon pricing policies are substantially more effective at reducing energy-related CO_2 emissions than most other climate-related mitigation policies: for example, incentives for renewables, energy efficiency mandates, and motor fuel taxes by themselves achieve only a small fraction of the potential emissions reductions from carbon pricing. Even broad combinations of regulations will at best miss about a third of the emissions reductions opportunities exploited by carbon pricing (and extensive credit trading is needed to contain their costs).
- Making productive use of carbon tax revenues (e.g., to lower the burden on the economy from broader taxes) keeps the costs of carbon taxes very low or even negative. In contrast, if this fiscal dividend is not exploited (e.g., revenues are dissipated in low-value spending) the case for carbon taxes on cost effectiveness grounds is seriously undermined.
- Although some compensation for vulnerable groups may be desirable (and necessary politically) this should not be excessive (as this reduces revenue available for other purposes). Ideally compensation would be in the form of schemes that increase economic efficiency, such as incentives for low-income households to increase work effort.
- While emissions trading systems can be designed to look like carbon taxes, the latter (building off long-established fuel excises) are a more natural and simpler way to achieve the dual objectives of raising revenue and price stability.

1. Introduction

Numerous options are being debated for strengthening the US fiscal position, such as scaling back preferences and altering rate schedules in the personal income, payroll, and corporate income tax systems, and cutting any number of entitlement or discretionary government spending programs (Chapter 1). Yet a carbon tax stands out, from an economic perspective, as an especially attractive source of new revenue. Unlike most other major potential revenue sources, carbon taxes correct for a market failure, namely CO_2 emissions that would otherwise be excessive,[1] and therefore potentially involve net economic benefits, rather than net costs – benefits that would be magnified if US action galvanized climate mitigation initiatives in other countries.

The focus of this chapter, however, is on the case for a carbon tax from an environmental, rather than fiscal, perspective – it is important to understand whether a carbon tax should be the main emissions mitigation instrument on the table, or whether there are reasonable alternatives, such as taxes on fuels and electricity, subsidies to clean products, energy efficiency standards, renewable fuel mandates, emissions trading systems, and so on. Section 2 compares the effectiveness of different mitigation policies at reducing energy-related CO_2 emissions. Section 3 discusses their fiscal implications, overall economic costs, and some policy implications. Section 4 touches on issues in the choice of tax and emissions trading approaches.[2] Section 5 offers concluding remarks.

2. How much more effective are carbon pricing policies than other mitigation instruments?

This section starts with a very quick summary of (existing and prospective) sources of energy-related CO_2 emissions in the United States and then considers where the CO_2 emissions reductions would come from under a carbon tax and realistic possibilities for alternative policies.

Brief overview of energy-related CO_2 emissions[3]

In 2011 total, energy-related CO_2 emissions in the United States amounted to 5,632 million tons. Power generation accounted for 39.8 percent of these emissions, transportation 32.7 percent, industry 17.3 percent, residences 6.2 percent, and the commercial sector 4.1 percent.

In the power sector, coal accounted for 42 percent of generation in 2011, reflecting its low cost relative to other fuels, the ease of dispatching it at any time of day, and the failure to charge for carbon and local pollution damages in its price. Nuclear power made up 19 percent of generation, though barriers to expanding this power source include high construction costs, and concerns over risks from accidents, terrorism, and waste disposal. In the past natural gas, which accounted for a further 25 percent of generation in 2011, has had trouble competing with coal because of limitations in pipeline capacity and its higher price and price volatility, though the rapid expansion of shale gas has alleviated the latter problems. Renewables accounted for 13 percent of generation, with non-hydro renewables accounting for 4 percent. The latter have been, and are expected to continue, expanding rapidly (albeit from a small base), though intermittency and the mismatch between their location and urban centers pose limitations to their growth.

In the transportation sector, the average fuel economy of passenger vehicles on the road was 20.6 miles per gallon in 2011, while that for new vehicles was 25.5 miles per gallon. Hybrid vehicles currently account for 2 percent of new light-duty vehicle sales, while plug-in electrics have not yet penetrated the fleet in significant numbers (BTS 2013, Tables 1–17, 1–19).

With current policies, economy-wide CO_2 emissions are projected to remain approximately flat out to 2035 (with the shares from different sectors approximately constant). Although energy demand is growing across all sectors (due to rising population and living standards), this is being roughly offset by improved energy efficiency and reduced emissions per unit of energy. In the power sector, this reflects efficiency improvements (e.g., in buildings) and a progressive shift away from (carbon-intensive) coal-fired plants towards natural gas and renewables (the latter's projected generation shares increase to 31 and 15 percent, respectively, by 2035 while that for coal falls to 36 percent). In the transport sector, the projected reduction in emissions intensity reflects improvements in the average fuel economy of (light- and heavy-duty) vehicle fleets in response to regulation (see below) and higher expected fuel prices (though the latter are notoriously difficult to project).

For our purposes, it is helpful to classify the potential sources of emission reductions from mitigation policies into the following four categories:

- *Reducing emissions intensity in the power sector.* This refers to reducing average CO_2 emissions per kilowatt hour (kWh) of power generation, primarily through fuel switching – that is, switching from (high-carbon) coal plants to (intermediate carbon) natural gas plants, and from both of these fuels towards (zero-carbon) renewables (mainly wind, but also solar, geo-thermal, and biomass) and nuclear power. Emissions intensity can also be reduced through technologies that improve plant efficiency (i.e., that reduce average fuel requirements per kWh of generation).[4]
- *Reducing electricity demand.* Residential and industrial (including commercial) electricity demand can be reduced through adoption of electricity-saving technologies (e.g., more efficient lighting) and reduced overall usage of products requiring electricity (e.g., economizing on the use of space heating and cooling).
- *Reducing transportation fuels.* Use of transportation fuels can be scaled back by reducing the overall amount that vehicles are driven and by improving the average fuel economy of vehicle fleets (via adoption of fuel-saving technologies in conventional vehicles and shifting the fleet mix towards smaller vehicles, hybrids, etc.).
- *Reducing (non-electricity) fuel use in industry and homes.* Similarly, direct industrial and residential use of natural gas (and to a limited extent oil) can be reduced through adoption of fuel-saving technologies and other measures to conserve on use of these fuels.

Where would emissions reductions come from under a carbon tax?

This and the next subsection discusses a selection of policies simulated in Krupnick et al. (2010) which used a variant of the Energy Information Administration's National Energy Modeling System (NEMS), incorporating energy projections as of 2009 (Box 2.1 provides a brief discussion of this model).[5] The results are a little outdated, given the major reduction in the price of natural gas due to the expansion of shale gas. However, they are still very useful in providing some broad sense of the relative effectiveness of different policies (a couple of more recent policy simulations using a 2012 baseline are noted below).

Krupnick et al. (2010) focus on a carbon pricing policy covering all energy-related emissions, where the emissions price (in year 2007) rises to $33 per ton of CO_2 in 2020 and $66 per ton in 2030. For now the carbon pricing policy is referred

BOX 2.1 METHODOLOGY FOR ASSESSING THE EFFECTIVENESS OF CLIMATE MITIGATION POLICIES

Krupnick et al. (2010) simulate the effect on projected US CO_2 emissions from a diverse range of climate-related policies, and various combinations of those policies, using a variant of the Energy Information Administration's National Energy Modeling System (NEMS) model. Energy projections from NEMS are widely used by other teams seeking to model the effect of US energy policies.

NEMS contains considerable, up-to-date detail on a wide range of existing and emerging technologies across the energy system (e.g., fuel-saving technologies that might be incorporated into different vehicle classes), while also balancing supply and demand in all – energy and other, regional and national – markets of the economy. The model incorporates existing regulations, taxes, and subsidies at the federal level (e.g., production subsidies for renewable generation, vehicle fuel economy standards), and some state-level policies (e.g., renewable portfolio standards, regulated electricity prices). All of these policies are updated regularly. Based on the recommendations of outside experts, Krupnick et al. (2010) made certain adjustments to the NEMS model to improve the realism of the policy simulations (e.g., the responsiveness of demand for oil products beyond the passenger vehicle sector to emissions pricing was increased, while the responsiveness of nuclear generation capacity to carbon pricing was reduced).

The policy simulations should be viewed with some caution. In particular, NEMS tends to be on the conservative side in terms of projecting the emissions impacts of alternative policy interventions (that is, some other models project larger emissions impacts).

Source: Author.

to as a carbon tax though an emissions trading system establishing the same price on emissions would have equivalent effects (the distinction between these instruments is discussed in Section 4).

The carbon tax reduces projected CO_2 emissions by 8.5 percent in 2020 and by 22.2 percent in 2030 below levels for those years that would otherwise occur. Over the 2010–2030 period cumulative CO_2 reductions are about 12.2 billion tons (or 10 percent).

Figure 2.1 summarizes (for the 2009 baseline) the contribution from individual sources to the emissions reductions in 2020 and 2030. The striking finding is that

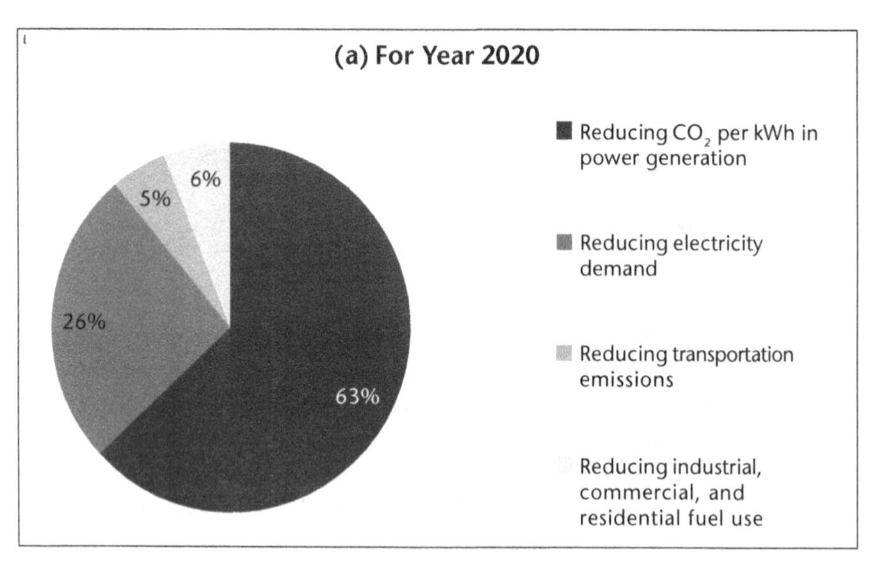

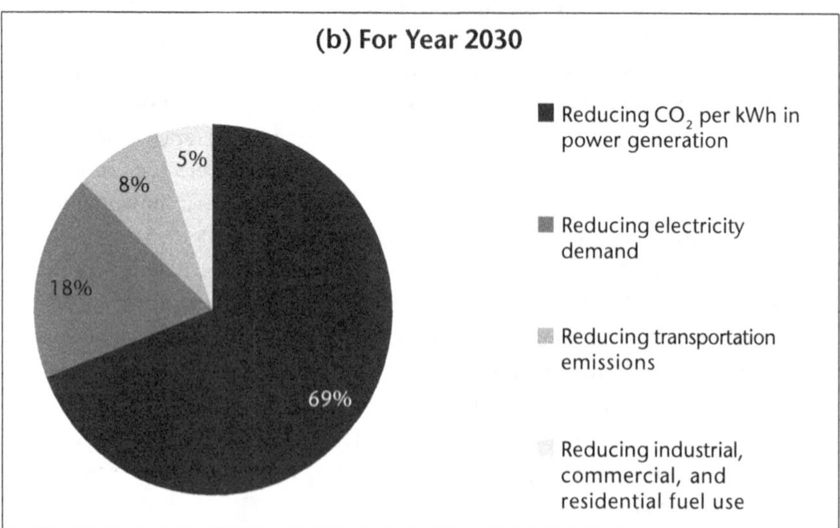

FIGURE 2.1 Breakdown of CO_2 emissions reductions under a US carbon tax

Source: Krupnick et al. (2010), p. 73.

the huge bulk – 63 to 69 percent – of the emissions reductions come from just one channel, reducing the emissions intensity of power generation, mostly by an expansion of renewables and nuclear at the expense of coal.[6] The other major source of emissions reductions is reduced demand for electricity, which accounts for 18 to 26 percent of the CO_2 reductions. The contributions from transportation and reduced industrial/residential fuel use are relatively modest at 5–8 and 5–6 percent, respectively.

The much smaller (and perhaps surprising) emissions response in the transport sector compared with the power sector reflects several factors. In particular, the carbon tax has a relatively modest impact on motor fuel prices (which increase by 24 cents per gallon or about 7 percent in 2020) compared with coal prices (which increase by about 100 percent). Moreover, there is a wide array of possibilities for substituting away from coal to other fuels in the power sector, while alternatives to gasoline and diesel vehicles in the transport sector (such as electrics and natural gas vehicles) have a way to go before becoming more competitive. Carbon pricing also has a relatively large impact on electricity prices, which increase by 1.7 cents per kWh or 19 percent in 2020, encouraging significant conservation measures. On the other hand, rises in motor fuel prices (besides being smaller in proportionate terms) have only limited effects on encouraging adoption of fuel-saving technologies in vehicles, given that many of these technologies are already being incorporated in response to rising fuel economy standards.

While some of the low-cost options for reducing CO_2 in power generation induced by carbon pricing in Krupnick et al. (2010) are now occurring in practice without this policy (due to the expansion of natural gas),[7] the main point here is that how any policy (or combination of policies) performs on effectiveness grounds relative to a (comprehensive) carbon pricing policy will largely hinge on whether it exploits the full range of possibilities for reducing CO_2 emissions per kWh in power generation. This point is now fleshed out in a bit more detail by looking at emissions reductions under alternative policies.

Other mitigation policies

Policies affecting the power and transportation sectors are briefly discussed in turn. The effectiveness of the different policies are summarized in Figure 2.2, which shows cumulative CO_2 reductions over the 2010–2030 period, expressed relative to those reductions under the above CO_2 pricing policy.

Power sector policies

Policies for renewable generation fuels. Renewable portfolio standards have been introduced in 29 states and there have been proposals to introduce them at the federal level. These policies require utilities to generate a minimum amount of the power they produce and sell from renewable sources, where generators (with relatively emissions-intensive portfolios) usually have the flexibility to fall short of the

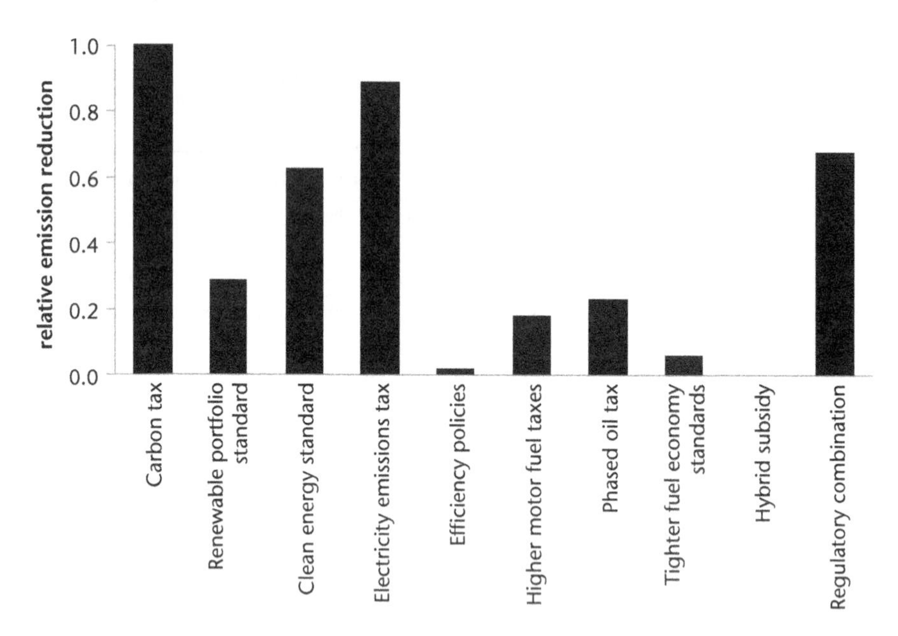

FIGURE 2.2 Cumulative CO_2 reductions under policies relative to those from carbon pricing, 2010–2030

Source: Krupnick et al. (2010) and author's calculations.

standard by purchasing credits from other (relatively clean generators) that exceed the standard. The effectiveness of these policies is somewhat limited however: Krupnick et al. (2010) studied a progressively rising federal requirement resulting in a doubling of the share of renewables in generation in 2030 to 20.5 percent. The policy reduces cumulative CO_2 emissions during 2010–2030 by 29 percent of the reductions under the carbon tax.[8]

One reason for this limited effectiveness is that the policy does not promote a switch from coal to natural gas, nor from these fuels to nuclear. The policy also reduces electricity demand by less than under the carbon tax, as there is no pass through of tax revenue in higher electricity prices.[9]

Clean energy standard. A related, but more promising, approach is a clean energy standard. Although it can take a variety of forms, basically this policy requires that a portion of generation comes from 'clean' sources, usually defined broadly to include nuclear power, and allowing a partial credit for natural gas, given its lower carbon intensity compared with coal. Krupnick et al. (2010) evaluated a policy designed to achieve a similar generation mix over time as under the carbon pricing policy (by 2030, 52 percent of generation comes from carbon-free sources). Cumulative CO_2 reductions under this policy are 63 percent of those under the carbon tax (Figure 2.2), the difference reflecting the weaker impact of the clean energy standard on electricity demand and the failure to exploit mitigation opportunities outside of the power sector.

Tax on power sector emissions. On the other hand, a tax on CO_2 emissions levied at the same rate as the comprehensive carbon tax, though applied to power sector emissions only, does raise revenue and achieves an estimated 89 percent of the cumulative emissions reductions as under the comprehensive tax. The policy is more effective than the clean energy standard, as it has greater impact on reducing electricity demand (emissions tax revenues are generally passed forward into higher electricity prices).

Energy efficiency policies. These policies take several forms – such as tax credits, and information programs (e.g., the Energy Star labeling program), to encourage voluntary purchase of more energy-efficient equipment – but potentially the most effective are national standards for the energy efficiency of buildings, household appliances (e.g., air conditioners), and other energy-using capital. Some standards already apply at the federal level, notably those introduced in 2007 to raise the efficiency of light bulbs by about 25 percent by 2014 (effectively banning most incandescent light bulbs) and many states already impose energy efficiency requirements on buildings.

Krupnick et al. (2010) evaluate energy efficiency provisions in the Waxman-Markey bill which included national energy standards for new buildings, measures to improve efficiency of existing buildings, some retrofit policies, and some relatively modest improvements in lighting and appliance efficiency (all beyond those achieved by any existing federal and state policies). As indicated in Figure 2.2, however, the effectiveness of these combined measures is very limited – they reduce cumulative CO_2 emissions by only 2 percent of those under the carbon tax.

One reason for this is that standards mainly apply to new buildings, which replace old ones only very slowly (and whose emissions reductions would continue beyond 2030). Another is that efficiency improvements would be occurring anyway, even in the absence of new standards, due to ongoing technological advances and pre-existing policy initiatives. Moreover, the efficiency provisions considered are not that comprehensive as they omit a range of energy-consuming equipment used by firms and households (e.g., small appliances, audio and entertainment equipment, industrial assembly lines).

Transportation policies

The two main options for reducing CO_2 emissions from the road transport sector (which accounts for almost 80 percent of emissions from all forms of transport) are fuel taxes and fuel economy (or, almost equivalently, emissions per mile) standards for new vehicles, or some pricing analog of the latter like 'feebates' (see Chapter 12 and Krupnick and Parry 2012). While both policies increase fuel economy, fuel taxes are more effective because (by increasing fuel costs per mile driven) they also discourage vehicle use. Fuel economy standards in fact have the opposite effect (because they reduce fuel costs per mile driven), though the extra mileage as a result of this 'rebound effect' is estimated to offset only moderately (by about 10 percent) the fuel saving from better fuel efficiency. While federal taxes on gasoline and diesel (18.4 and 24.4 cents per gallon, respectively) have been fixed in nominal terms (and hence have fallen substantially in real terms) since 1993, in contrast fuel economy standards are now being ramped up aggressively (Chapter 12).

Fuel taxes. Krupnick et al. (2010) considered an increase of $1.27 per gallon in gasoline and diesel taxes in 2010 and maintained in real terms thereafter (in 2020, for example, this policy increases gasoline prices by about four times the increase under the carbon pricing policy). But even this (quite large and politically rather unrealistic) tax increase reduces cumulative CO_2 emissions by 'only' 18 percent of the reductions under the carbon tax (Figure 2.2), given that the former forgoes the low-cost mitigation opportunities in the power sector, and the difficulty of reducing emissions in transport (see above).

Oil taxes. Taxing all oil uses would exploit more margins of behavior for reducing emissions including, for example, opportunities to conserve air travel, truck freight, and industrial oil uses (gasoline accounts for slightly less than half of total US petroleum consumption). But the prospects for a large, comprehensive, and immediate oil tax in the United States are extremely remote. And even if an oil tax were to be implemented it would probably be phased in gradually. Krupnick et al. (2010) simulated a tax on all oil products of initially 8 cents per gallon, rising by 8 cents per gallon each year to reach $1.73 per gallon (for all oil products) by 2030. As indicated in Figure 2.2, this policy reduces cumulative CO_2 emissions by 23 percent of those under the carbon tax – more than under higher motor fuel taxes (because the tax base is larger), but only moderately so, given the tax increase is progressive rather than immediate.

Fuel economy standards. Fuel economy standards are being raised and integrated with new targets for limiting CO_2 emissions per mile. By 2016 standards will average 35.5 miles per gallon across light-duty vehicles and 54.5 miles per gallon by 2025, if applied to the current fleet mix with full compliance (Chapter 12). However, actual fuel economy may fall far short of these levels if manufactures shift their production mix to larger-size vehicles (with larger 'footprints') for which fuel economy requirements are lower, or if they choose to pay out-of-compliance fines (instead of incorporating enough fuel-saving technologies to meet the standards).

Figure 2.2 reports the effectiveness of a (slightly less aggressive) policy (than the one recently enacted) that raises nominal standards to 52.2 miles per gallon by 2030 (though as just noted actual fuel economy may fall well short of this standard). Cumulative reductions in CO_2 emissions are only 6 percent of those under carbon pricing, or one-third of those under higher fuel taxes. This reflects, for example, the failure, even within the transport sector, to exploit emissions reductions through discouraging vehicle use.

Hybrid subsidies. Hybrid electric vehicles (HEVs) and plug-in hybrid electric vehicles (PHEVS) are a potentially promising alternative to traditional gasoline vehicles, though some of the savings from reduced gasoline consumption in PHEVS is offset by increased emissions from power generation.[10] HEVs currently account for about 2.5 percent of new automobile sales, though this sales share is projected to expand rapidly to roughly a quarter by 2030, while PHEVs are just being introduced (their future costs, which critically hinge on battery costs, remain highly uncertain). Tax credits for HEVS have now been phased out and those for PHEVs phase out once manufactures have sold 200,000 vehicles.

Krupnick et al. (2010) considered vehicle purchase subsidies for HEVs and PHEVs (varying between $2,000 and $3,500 per vehicle depending on its type). Although these policies give a further boost to hybrid sales, they essentially have no overall effect on CO_2 emissions – manufactures selling more hybrids can lower the average fuel economy of other vehicles in their fleet and still meet existing fuel economy standards.

Summary

Most existing and prospective climate-related policies (e.g., incentives for renewables or energy efficiency) by themselves provide at best only a minor fraction of the incentives for emissions reductions that can be provided under comprehensive carbon pricing policies. The two exceptions are the tax on electricity emissions, and the clean energy standard, as they both fully exploit the biggest source of low-cost options for emissions reductions, namely fuel switching possibilities in power generation. These policies can be combined with other measures to further close the gap between their effectiveness and that of carbon pricing. However, there are limits to this – a regulatory package combining the clean energy standard, fuel economy policy, and energy efficiency standards yields emissions reductions that are a third lower than that of carbon pricing (Figure 2.2) as the policy fails to encourage reductions in the use of vehicles and other energy-using goods. And as climate policies become more aggressive over the longer term it will become increasingly important to exploit the entire range of mitigation opportunities, once the 'low-hanging fruit' in power generation has been exploited.

3. Fiscal impacts and costs of alternative policies

This section considers the future revenue consequences of different climate mitigation policies and their costs, accounting for any fiscal dividend. Some broad guidelines for the use of revenues, and the balance between carbon and other taxes in the government's budget, are also discussed.

Revenue impacts

The concern here is with the revenue potential (or revenue loss) of other policies *relative* to that for a carbon tax, or in other words, how much of the revenue potential is forgone by using other policy instruments in place of a carbon tax, leaving aside (for the moment) the possibility that some carbon tax revenues may be used up in compensation schemes.

Again, to give some broad sense for this the simulations in Krupnick et al. (2010) are used. Specifically, Figure 2.3 shows the revenue gains/losses, in present value over the 2010–2030 period from the policies described above, expressed as a ratio of the present value of revenues under the carbon tax ($2,276 billion). These revenue effects reflect the direct impacts from carbon pricing, oil taxes, and so on, net of any erosion of revenues from pre-existing (federal and state) taxes on gasoline and diesel fuel.[11]

The main points from Figure 2.3 are as follows.

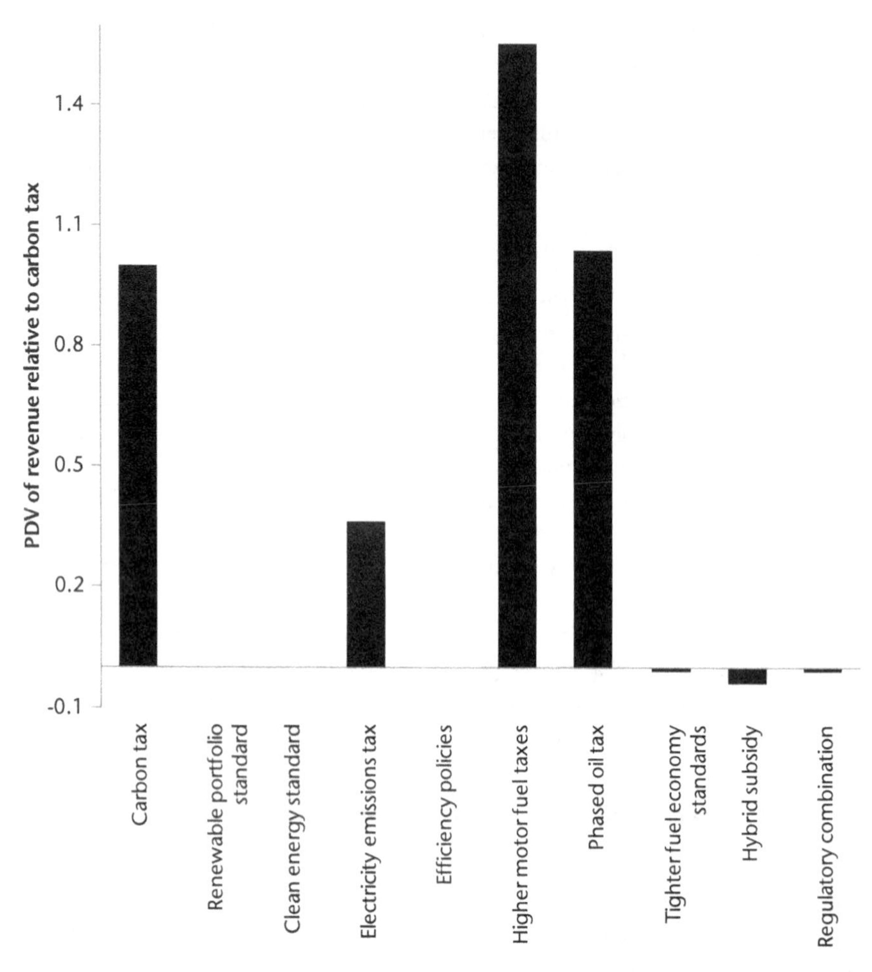

FIGURE 2.3 Revenue gains (or losses) from policies relative to carbon tax, 2010–2030

Source: Krupnick et al. (2010) and author's calculations.

First, (stating the obvious) regulatory policies like renewable fuel and clean energy standards forgo all the revenue that would be raised under a carbon tax (in fact tighter fuel economy standards reduce revenue a little by eroding the base of pre-existing fuel taxes).

Second, the electricity emissions tax forgoes about two-thirds of the discounted revenue raised by the broader carbon tax. Therefore, although the former policy performs reasonably well on effectiveness grounds (because it exploits low-cost abatement opportunities in the power sector) it is a relatively poor substitute for a carbon tax from a fiscal perspective (because the base of the tax is much narrower than that for the carbon tax).

Third, on the other hand fuel tax increases (of about $1.25 per gallon) would raise about 50 percent more revenue over 20 years than the carbon tax. And even the tax applied to all oil products, but phased in progressively, raises about the same amount of revenue as under the carbon tax.

Finally, while on net the hybrid incentive policy does not generate any fuel savings (in the presence of fuel economy standards) the revenue losses from this policy are actually significant, almost 4 percent of the revenue gains from the carbon tax.

Comparing policy costs

Achieving a given emissions reduction with the lowest burden on firms and overall economy is important, not only for its own sake, but also to sustain support for the policy over time. And these costs depend critically on how the any revenues from mitigation policies are used.

Here the costs of policies within the energy sector, and then linkages with the broader fiscal system, are discussed. The focus is purely on *economic costs*, which are generally accepted as the appropriate metric for regulatory cost assessment[12] – other economic impacts (e.g., on national output) are discussed in Chapter 6.

Costs within the energy sector

Climate mitigation policies cause several types of economic costs within the energy sector. One is the costs to power generators of producing with a cleaner, but more expensive, fuel mix than would otherwise be efficient. Others include the net costs from adopting more energy efficient equipment, vehicles, or products – that is, the upfront cost, less savings in energy expenditures over the lifetime of the investment.[13] Costs also include those from other measures to conserve on fuel, such as the value to motorists of trips forgone in response to higher fuel prices, or from people heating their homes less than they would otherwise prefer.[14]

Figure 2.4 reports estimates of the average costs within the energy sector per ton of CO_2 reduced from the policies discussed above – or more precisely, the present value of costs out to 2030, divided by cumulative emissions reductions due to policy impacts over the period (see Krupnick et al. 2010, Appendix C, for more details on these calculations).

The carbon tax, renewable portfolio standard, clean energy standard, and electricity emissions tax are all reasonably cost effective (for the emissions reductions each achieves) in the sense that the average costs within the energy sector are around $12–15 per ton of CO_2 reduced.[15] This follows because all of these policies exploit the low-cost opportunities for reducing emissions through fuel switching in the power sector. On the other hand, the transportation policies, energy efficiency polices, and oil tax involve higher average costs – between $17 and $34 per ton of CO_2 reduced, as these policies are pushing on emissions reduction opportunities that are relatively more difficult or costly, though account is not being taken here of ancillary benefits such as reduced incidence of traffic congestion and highway collisions, which may easily dominate these costs (see Chapter 12 of this volume).

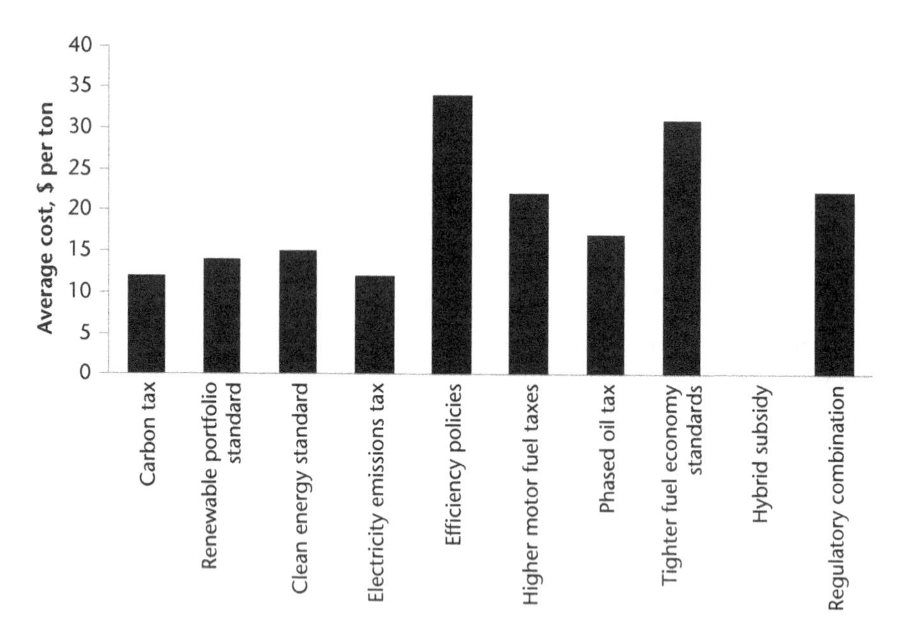

FIGURE 2.4 Comparing average costs within the energy sector of mitigation policies (cost per ton of CO_2 reduced over 2010–2030 in year 2007)

Source: Krupnick et al. (2010) and author's calculations.

Note: Average costs are calculated by the present discounted value of costs for a policy over 2010–2030, divided by the cumulative emissions reductions from the policy (including emissions reductions beyond 2030 associated with policy-induced investments up to 2030). See Krupnick et al. (2010), Appendix C.

Costs due to linkages with the broader fiscal system

The broader fiscal system distorts economic activity in a variety of ways, and to the extent that climate policies impinge upon these distortions, there are important implications for the overall costs of such policies.

One major distortion arises from the impact of taxes on labor income – particularly personal income and payroll taxes – which reduce take-home pay and discourage work effort. For example, these taxes may discourage some people from participating in the labor force, others from putting in extra effort and hours on the job, and still others from investments in education and training to raise their earnings potential. Personal taxes on the income from savings, and corporate taxes on the return to business investments, cause further distortions by reducing the overall level of capital accumulation. And the tax system induces a variety of tax-sheltering behavior that distorts the composition of economic activity, such as shifting of activity into the (less productive) informal sector, and excessive spending on employer medical insurance and other fringe benefits, home ownership, and so on, at the expense of other spending that does not receive favorable tax treatment.

These broader economic distortions affect the costs of climate mitigation policies in two main ways. First, using the revenue from carbon taxes, fuel taxes, and so on, to reduce the rates of personal, payroll, or corporate taxes produces economic benefits, by improving incentives for work effort and investment and alleviating some of the artificial biases for tax-sheltering. As a rough rule of thumb, cutting broader taxes might produce a net economic gain to the US economy of about 25 cents or more per dollar of revenue used to cut broader taxes.[16] Second, on the other hand, to the extent that climate mitigation policies push up production costs by increasing the costs of energy, transportation, and so on, they tend to (slightly) contract the overall level of economic activity, which in turn reduces employment and investment, thereby exacerbating economic distortions from the broader fiscal system.

Figure 2.5, which is from Parry and Williams (2012), provides a very rough snapshot, for the year 2020, of how estimated costs for certain climate mitigation policies change when linkages with the broader fiscal system are taken into account. The policies include the same carbon tax as simulated by Krupnick et al. (2010) ($33 per ton in 2020), along with other policies that are scaled to achieve the same CO_2 reductions as the tax (8.5 percent below levels that would otherwise occur) in 2020.

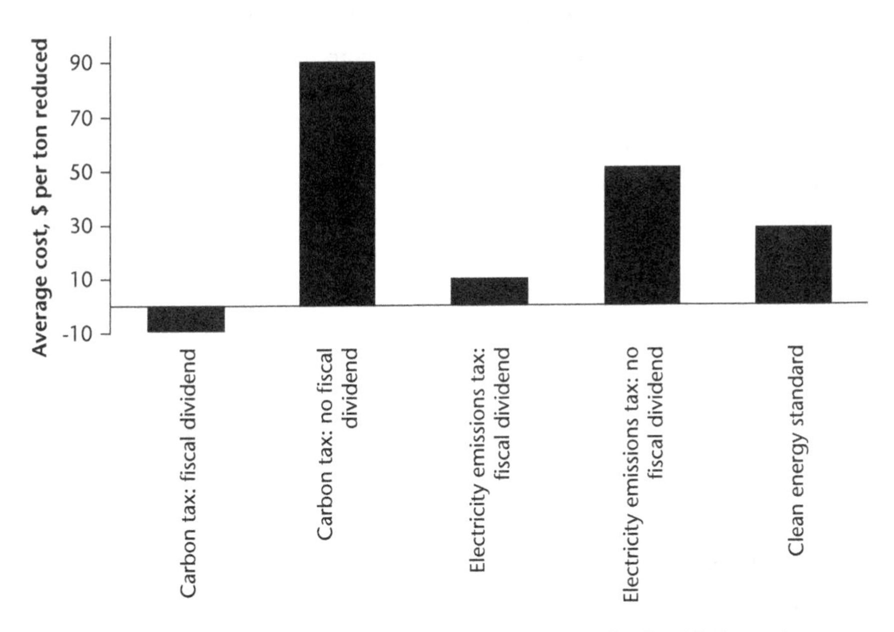

FIGURE 2.5 Comparing average costs of policies accounting for fiscal linkages (costs per ton of CO_2 reduced in year 2020 in year 2007)

Source: Parry and Williams (2012).

Note. These estimates reflect the average costs of policies within the energy sector and very rough estimates of (a) additional costs due to the impact of higher energy prices on reducing work effort and investment less, in some cases (b) benefits from recycling revenues in broader income tax reductions. All policies are scaled to achieve the same emissions reduction (8.5 percent in year 2020).

According to these estimates, the average costs in year 2020 of a \$33 per ton CO_2 tax are very high at \$90 (in year 2007) per ton of CO_2 reduced, when the adverse effect of higher energy prices on economy-wide employment, investment, and so on are taken into account. But this assumes revenues do not generate any net economic benefits (e.g., they are used to finance projects whose benefits exactly outweigh the costs, or revenues are not raised in the first place as in an emissions trading system with free allowance allocations).

In contrast, if the fiscal dividend is fully exploited (in this case by financing a general reduction in personal income tax rates) the resulting benefit to the economy dramatically lowers the overall cost. In fact overall the estimated costs of the policy are slightly negative at -\$9 per ton (without counting environmental benefits) – that is, the benefits from cutting other taxes outweigh the costs in the energy sector, and from the adverse effect of higher energy prices on the broader economy. In effect, the tax reform (partly) replaces a very distortionary tax – an income tax with significant exemptions and deductions – with an idealized tax on (the carbon content of) fuels with no exemptions. More generally, the costs of a carbon pricing policy where some, but not all, of the fiscal dividend is exploited lie somewhere between the two extreme cases shown by the first two bars in Figure 2.5.

Average costs are higher (\$10 per ton of CO_2 reduced) under a tax on electricity emissions compared with the carbon tax, when both fully exploit the fiscal dividend. Mostly, this difference reflects the smaller amount of revenue collected and recycled under the electricity emissions tax, given its much smaller tax base. On the other hand, if the fiscal dividend is not exploited, the average costs of the electricity emissions tax are lower than that for the economy-wide carbon tax, as the latter has a greater adverse effect on the overall economy – besides raising electricity prices, the carbon tax also raises costs in the transportation and other sectors. Finally, the clean energy standard has lower average costs (\$30 per ton) than either the carbon or electricity emissions tax with no fiscal dividend, because it has the lowest impact on energy prices and hence overall economic activity (prior to any use of revenues from other policies). Unlike the electricity emissions tax, for example, the policy does not involve the pass through of emissions tax revenue in higher electricity prices.

Some policy implications

Broad guidelines for revenue use

Other chapters in this volume go into details on the possible use of carbon tax revenues for compensating vulnerable households and firms (Chapters 7 and 9), reforming corporate taxes (Chapter 8), funding technology programs (Chapter 10) and highway spending (Chapter 12). Here some general principles are noted for the use of revenues that follow from the points just emphasized, namely the potentially large economic benefits from using carbon tax revenues to offset broader tax burdens. These principles include:

- *Be wary of earmarking carbon tax revenues,* such as for clean energy programs, climate adaptation. Ideally, any earmarked spending should generate comparable net economic benefits from alternative uses of the revenue. That is, the benefits of the program should exceed the costs by roughly 25 percent or more – if not, then the money is better spent (at least from a pure economic perspective) on reducing other distortionary taxes. A more general issue related to earmarking is that there is no relation between the economically appropriate amount of spending on clean energy or related programs and the amount of revenues raised by an appropriately scaled carbon tax. Furthermore, even if initially meant to be transitory, once earmarked spending programs are established, they tend to gain constituencies that can retard appropriate phasing out of the program.
- *Provide appropriate compensation for vulnerable or influential groups.* It is sometimes suggested that emissions trading systems are easier to implement than carbon taxes because the allowances can be given away for free to build political support for the program. But Figure 2.5 underscores the problem with this argument: if allowances are given away rather than auctioned, the fiscal dividend is lost, and the costs of the carbon pricing program rise dramatically – in fact regulatory approaches (that have weaker impacts on energy prices) can be more cost effective than trading systems with allowance giveaways. The more general point here is that, although appropriate compensation for vulnerable groups may be needed, compensation should strike the necessary balance and be neither inadequate (which risks compromising standards for the vulnerable) nor excessive (which leaves less revenue available for other purposes). In other words, policymakers should bear in mind the trade-offs – more compensation for those at risk from higher energy prices means, by definition, that taxes must be higher, or spending lower, elsewhere in the government's budget.
- *Look for compensation schemes that improve economic efficiency.* Following from the last point, the harsh trade-offs between compensation and cost effectiveness might be ameliorated to some extent if the compensation scheme can be designed in a way that improves economic efficiency. Most obviously, providing targeted relief to low-income households through reductions in their tax burdens (e.g., extending the earned income tax credit, rebating payroll taxes) improves their incentives for work effort while providing them relief in the form of lump-sum cash transfers does not (see Chapter 7).
- *Raising revenue by itself is not enough* – the revenue needs to be used productively. Before a carbon tax with revenue going to the government can be evaluated relative to an emissions trading system where allowances are given away for free, the accompanying legislation specifying how the revenues will be used needs to be considered. Specifying reductions in other taxes to offset most of the revenue

gain from the carbon tax seems to be the best way to guard against the concern, among some, that carbon taxes might lead to increases in public spending that may not be socially desirable.

Balance between carbon and other taxes in the government's budget

The Introduction to this volume discuss why a tax in the ballpark of $25 per ton of CO_2 or more might be a reasonable level to aim for in the United States in the near to medium term, from an environmental perspective. But when account is taken of the fiscal dividend from carbon taxes, is a higher level of taxation appropriate?

In principle a higher level of taxation (than justified on pure environmental grounds) can be warranted when the fiscal dividend is fully exploited (while a much lower tax level is appropriate if the fiscal dividend is not fully exploited).[17] However, fine tuning these adjustments to the economically efficient tax level is difficult (given imprecision in the data). For practical purposes, the carbon tax level should be largely set on environmental grounds with revenues largely contributing to the general budget, leaving the remainder of the government's revenue needs to be met through broader fiscal instruments.

4. Taxes versus trading: A quick look

This section briefly touches on issues in the choice of carbon taxes and emissions trading systems (ETS).

In principle, the choice between these instruments systems is less important than implementing one of them, but getting the design details right. This includes comprehensively covering emissions, pricing them at an appropriate level, exploiting the fiscal dividend through, for example, auctioning allowances in ETS, and – in the case of ETS – including provisions to limit year-to-year volatility in prices. Price volatility raises policy costs,[18] deters investments in clean technologies with high upfront costs and long-range payoffs, and creates uncertainty about potential revenues from allowance auctions.

In practice the carbon tax, at least if administered as an extension of existing motor fuel excises, might be the more likely of the two instruments to exploit the fiscal dividend, given that the revenues would be collected by the Internal Revenue Service. The Environmental Protection Agency, were it to be charged with administering a trading program, might be reluctant to hand over 100 percent of the revenues from allowance auctions to the Treasury (given pressure on the agency for using some of the revenues for environmentally related spending). The Waxman-Markey ETS bill was not designed to provide a large new source of federal government revenue – at least initially, only 7 percent of the allowances were to be auctioned for the purpose of deficit reduction (most of the allowances were to be set aside for compensation schemes or to fund energy-related spending).

Possibly another strike against cap-and-trade systems is that they are less compatible with other climate-related initiatives at the federal and state level. In the presence of a federal ETS, other programs designed to reduce emissions might affect the distribution of emissions reductions across sectors, or across states, but they cannot affect the economy-wide level of emissions, which are fixed by the federal cap. In contrast, under a federal carbon tax, individual states with preferences for more aggressive action on climate would be free to implement additional measures without their emissions impacts being neutralized at the national level.

Finally, it is often suggested that a trading system is the most natural instrument for channeling flows of climate finance to developing countries, though emissions offset provisions. However, there is no reason why such provisions could not be included under a tax regime (as reliable capability for verifying these emissions reductions is developed) – domestic tax credits could be awarded to fuel suppliers purchasing emission reduction credits in developing countries.

In short, while in principle ETS can be well designed, and carbon taxes can be badly designed – it all depends on legislative specifics – the risks of key design problems (most obviously failure to exploit the fiscal dividend and price volatility) might be greater in the case of ETS.

5. Conclusion

The need to address the twin long-term challenges of global warming and the US federal budget deficit has heightened interest in the possibility of a US carbon tax. Carbon pricing instruments are the most effective policies for exploiting emissions reduction opportunities across the economy and they can raise substantial amounts of revenue.

While the introduction of a carbon tax may not be imminent, there are reasons for believing that its appeal may only grow over time as extreme events (like Hurricane Sandy) become more common, the limitations of other emissions mitigation policies are realized, other countries (like China, Brazil, South Africa) introduce carbon pricing policies, and the political difficulties of broader deficit control measures become apparent.

But getting the design details right is critical, not only for environmental, economic, and fiscal reasons, but also to enhance the prospects that policy will be sustained and strengthened over time, and copied by other countries.

Notes

1. These emissions currently account for about 80 percent of total US GHG emissions (Chapter 3).
2. For a more in-depth discussion of general (rather than US-specific) issues in the choice of climate mitigation instruments see Krupnick and Parry (2012). The appropriate role of complementary policies to address other possible market inefficiencies in the context of technological innovation and the transport and power sectors are discussed in Chapters 10, 11, and 12 of this volume, respectively.
3. All figures in this subsection are from EIA (2011).

4. Another mitigation possibility is to build new coal plants, or retro-fit existing plants, with carbon capture and storage technologies. However, to encourage significant penetration of this technology, emissions prices would need to rise beyond levels considered in this chapter.
5. Ideally it would be better to compare results from a selection of models rather than just one, but no other study evaluates such a broad range of policies on a consistent basis.
6. Although carbon pricing makes natural gas more attractive relative to coal, this is approximately offset because gas becomes less attractive relative to renewables and nuclear.
7. An updated policy simulation (for the 2012 baseline) still suggests a broadly similar pattern of emissions reductions across the economy as described above, though with natural gas being more of a beneficiary from the contraction of coal generation, and electricity demand not falling quite as much (Palmer et al. 2011).
8. Renewables have also been subsidized through production tax credits, which amounted to 2.1 cents per kWh for wind, geothermal, and biomass (for the first 10 years of plant operation) and a 30 percent tax credit for upfront investments in solar. Extending these incentives out to 2030 has only a minor impact on economy-wide emissions, however (Krupnick et al. 2010).
9. Furthermore, some of the expansion of renewables comes at the expense of natural gas rather than coal (the most carbon-intensive fuel). This happens less in the more recent simulation in Palmer et al. (2011) with lower natural gas prices, which increases the effectiveness of renewable portfolio standards, but only moderately so.
10. All-electric vehicles are not discussed here as their projected market penetration is relatively small in the policy simulations considered.
11. As discussed below (and in Chapter 1), further revenue losses will occur to the extent that carbon policies indirectly reduce the base of broader taxes on labor and capital.
12. The notion of economic cost has been endorsed by governments around the world for purposes of evaluating regulations, government investments, taxes, and other policies. In the United States, a series of executive orders since the 1970s has required government agencies to perform hundreds of cost-benefit analyses a year, using economic costs (and benefits) to determine whether regulations are warranted from society's perspective.
13. There remains much dispute about whether some of these investments might involve negative net costs – that is, the energy-saving benefits might exceed the upfront costs – there being several possible market failures that might explain why such technologies are not already adopted. Chapter 10 of this volume elaborates on this issue. The cost estimates discussed below are from Krupnick et al. (2010) who take an intermediate position on the possibility of market failures.
14. Revenue transfers between government and the private sector are not a component of these costs. How government revenues are spent or raised, however, matters a lot for policy costs, as discussed below.
15. These cost estimates assume that, for the regulatory policies, there are credit trading provisions allowing firms the flexibility to fall short of a standard by purchasing credits from other firms that exceed the standard (without any slippage in aggregate from an environmental perspective). Without these flexibility provisions, costs for these policies would be higher, perhaps substantially so.
16. That is, the private sector receives the dollar of returned revenue plus 25 cents in additional benefits through reducing distortions from the tax system.
17. There is some confusion on this issue in the literature. Some early studies suggested that, even if carbon tax revenues were used to cut other taxes, the efficient level of carbon tax is still significantly below environmental damages per ton. However, those studies substantially understate revenue recycling benefits as they do not capture the full range of distortions created by the broader fiscal system (see, for example, Parry and Bento 2000 for more explanation on this).
18. Studies suggest that for the United States, an ETS scaled to achieve the same cumulative emissions reductions over the longer term as a carbon tax will be about 15 percent more

costly than the tax if the former lacks any provisions to contain price stability (e.g., Fell et al. 2012). Price volatility raises policy costs by causing sizable discrepancies in incremental abatement costs at different points of time.

References

BTS, 2013. "National Transportation Statistics." Bureau of Transportation Statistics, US Department of Transportation, Washington, DC.

EIA, 2011. *International Energy Outlook 2011*. Energy Information Administration, US Department of Energy, Washington, DC. Available at: www.eia.gov/forecasts/ieo/index.cfm

Fell, Harrison, Ian A. MacKenzie, and William A. Pizer, 2012. "Prices versus Quantities versus Bankable Quantities." *Resource and Energy Economics* 34: 607–623.

Krupnick, Alan J. and Ian W.H. Parry, 2012. "What is the Best Policy Instrument for Reducing CO_2 Emissions?" In I. Parry, R. de Mooij, and M. Keen (eds.), *Fiscal Policy to Mitigate Climate Change: A Guide for Policymakers*. Washington, DC: International Monetary Fund, 1–26.

Krupnick, Alan J., Margaret Walls, Tony Knowles, and Kristin Hayes, 2010. *Toward a New National Energy Policy: Assessing the Options*. Washington, DC: Resources for the Future and National Energy Policy Institute.

Palmer, Karen, Anthony Paul, Matt Woerman, and Daniel C. Steinberg, 2011. "Federal Policies for Renewable Electricity: Impacts and Interactions." *Energy Policy* 39: 3,975–3,991.

Parry, Ian W. H. and Antonio M. Bento, 2000. "Tax Deductions, Environmental Policy, and the 'Double Dividend' Hypothesis." *Journal of Environmental Economics and Management* 39: 67–96.

Parry, Ian W. H. and Roberton C. Williams, 2012. "Moving US Climate Policy Forward: Are Carbon Tax Shifts the Only Good Alternative?" In Robert Hahn and Alistair Ulph (eds.), *Climate Change and Common Sense: Essays in Honor of Tom Schelling*. Oxford: Oxford University Press, 173–202.

3

ADMINISTRATION OF A US CARBON TAX

*Jack Calder**

KEY MESSAGES FOR POLICYMAKERS

- A clear and preferably simple policy objective is important for both design and effective administration of a carbon tax.
- Emissions from fossil fuel combustion are simpler to tax than other greenhouse gas emissions, and are by far the largest source of emissions. It is best, therefore, to concentrate on taxing these first.
- If the policy objective is comprehensive taxation of those emissions, upstream taxation based on fossil fuel sales has significant administrative advantages over downstream taxation based on measured emissions.
- It may be difficult to identify a single common point of upstream taxation for all fossil fuel sales, but the Internal Revenue Service can develop practical regulatory solutions, using its experience of administering excise tax on petroleum products and coal.
- There is no need to re-invent the wheel for administration of an upstream carbon tax. It should use the established procedural framework for existing taxes, particularly excise.
- If carbon capture and storage and measurement of carbon "sunk" in forestry become practicable, a simple system of government payments for carbon stored and sunk will be easier to administer than carbon tax repayments and credits.
- If the policy objective is to exempt energy-intensive trade-exposed businesses, administration will be more challenging. Upstream taxation may be less advantageous.
- Taxation of non-fossil fuel emissions will be more complex, and in some cases impractical, to administer. It will require development of new

measurement systems, in some cases requiring specialist expertise, and special forms of tax or regulation. These could be introduced more gradually as capacity is developed.

- If the policy objective is to tax emissions generated to produce goods and services for US consumption, border tax adjustments (to charge for embodied carbon in imports and rebate it for exports) will be required, but the administrative difficulties and complexities these would pose should not be underestimated.

Introduction

This chapter considers the administration of a US carbon tax (CT). It does not consider administration of other methods of controlling greenhouse gas (GHG) emissions, such as emissions trading systems (ETS) or regulation.

The chapter starts with a brief illustration of US GHG emissions and of the revenues that would potentially arise from taxing them. It then briefly outlines alternative policy options for taxing them, not with a view to evaluating those options but merely with a view to considering their administrative implications. The options considered are first, comprehensive taxation of US GHGs so far as is practicable; second, taxation of US GHGs with exemptions for energy-intensive trade-exposed (EITE) or other businesses; and third, taxation of GHG emitted in the US or elsewhere to produce goods and services consumed in the US (described here in short as taxing GHG *consumption*).

For fossil fuel combustion emissions, the chapter discusses the practical advantages and disadvantages, administrative and otherwise, of "upstream" taxation (taxation at points where fossil fuels enter the economy) and "downstream" taxation (taxation at points of combustion or energy consumption). It considers the administrative implications of carbon capture and storage (CCS) and carbon "sinks." It looks at the practicalities of extending CT to non-fossil fuel emissions. It considers administration of international transactions, and in particular the issues that would arise if it were intended to apply carbon pricing to carbon intensive goods and services imported into the US from countries that do not price carbon. Finally it outlines how a US CT might be administered in practice.

US GHG emissions and potential CT revenues

Table 3.1 shows 2010 US emissions of CO_2 and other GHG measured in terms of their CO_2 equivalence (CO_2e). Total US GHG emissions vary from year to year (reflecting changes in energy demand, fuel prices, fuel efficiency, etc.). Various factors (such as expanding shale gas production) may increase relative contributions from some sources and reduce them from others. These figures are therefore just a snapshot, but can be used for illustration since the broad picture is unlikely to change significantly in the near future. Ignoring the role of carbon sinks, US GHGs were

TABLE 3.1 Sources of US GHG emissions and potential revenue from a carbon tax, 2010

Gas/source	*CO_2 or CO_2e emissions, MMT*	*Percent of total gross US GHGs*	*Revenue from $20 per ton CO_2 tax, $billion*
CO_2	**5706.4**	83.6	
Fossil fuel – combustion	5387.8	79.0	107.8
Electricity	2258.4	33.1	
Transportation	1745.5	25.6	
Industrial	777.8	11.4	
Residential	340.2	5.0	
Commercial	224.2	3.3	
US Territories	41.6	0.6	
Fossil fuel – non-energy use	125.1	1.8	2.5
Industrial processes	193.5	2.8	3.9
Iron and steel	54.3	0.8	
Natural gas systems	32.3	0.5	
Cement	30.5	0.5	
Other	76.4	1.1	
Methane	**666.5**	9.8	13.3
Natural gas systems	215.4	3.2	
Enteric fermentation	141.3	2.1	
Landfill	107.8	1.6	
Coal bed	72.6	1.1	
Manure	52.0	0.7	
Petroleum systems	31.0	0.4	
Other	46.4	0.6	
Nitrous oxide	**306.2**	4.5	6.1
Soil management	207.8	3.0	
Stationary combustion	22.6	0.3	
Mobile combustion	20.6	0.3	
Other	55.2	0.8	
HFCs	**123.0**	1.8	2.5
Other non-CO_2 GHGs	**19.7**	0.3	0.4
Total Gross GHGs	**6821.8**	100	136.4
Land use (sinks) CO_2	**(1074.7)**		
Net total GHGs	**5387.8**		

Source: Author's calculations based on EPA, 2012.

6,821.8 million metric tons (MMT) in 2010. Fossil fuel combustion accounted for 70.0 percent of these emissions, with the major contributors including power generation (33.1 percent), transport (25.6 percent), industrial fuel use (11.4 percent), residential use (5.0 percent), and commercial uses (3.3 percent). Non-CO_2 GHGs include methane (which accounted for 9.8 percent of GHG emissions), nitrous oxide (which accounted for 4.5 percent of emissions), and hydro-fluorocarbons (HFCs) (which accounted for 1.8 percent).

The table also shows government revenues from a CT applied to 2010 emissions at an illustrative rate of $20 per ton. These would have amounted to around $136 billion, but it should be remembered that a CT could itself be expected to alter emissions levels and the relative contributions from different sources, perhaps significantly. As will be discussed, it would also be impracticable to tax all GHG emissions. If a CT reduced the illustrative 2010 emissions by, say, 15 percent, that would reduce revenues to around $116 billion; each percent of GHG taxed would produce $1.16 billion. And if, say, only 90 percent of those reduced emissions could be taxed in practice, that would reduce revenues further to around $104 billion. This "back-of-the-envelope" calculation[1] is useful as an indication of potential order of magnitude of a US CT. For comparison total US federal taxes in 2010 were $2,163 billion, of which corporate income tax was $191 billion and excise taxes $67 billion. A CT would therefore be a substantial tax, comparable to other major business taxes, but, in the context of a $700 billion deficit, would not by itself transform US government finances or the overall US tax burden.

Policy and administration

Effective administration of a CT, as of any other tax, requires a well-defined tax base, with a clear underlying policy objective. Lack of clarity is often a major barrier to compliance and effective administration. The choice of tax base also needs to command broad public acceptance – without which compliance is inevitably problematic. Administration should not drive policy, but policy intentions must be capable of effective implementation in practice.

The *general* rationale for a CT is to internalize an externality (GHG emissions) that causes universal harm (global warming) to future humankind. If an all-seeing, all-powerful government representing future humankind were to impose a CT, it would probably impose it on those citizens whose consumption of fuels and products caused GHG to be emitted, and spend it to benefit humankind generally, particularly those who would suffer most from the consequences of global warming. This would transfer value from countries responsible for higher-than-average per capita GHG emissions, such as the US and Australia, to countries responsible for lower-than-average per capita emissions (i.e., broadly speaking, from rich countries to poor countries, though of course there is not an exact correlation between wealth per capita and GHG emissions per capita).

But future humankind does not have a government, and taxes are generally imposed and collected by national governments. Conceptually a *national* CT might seem an odd mechanism for tackling a *global* harm. But it is hard to imagine the US, or any other government for that matter, accepting an international CT imposed disproportionately on its own citizens and spent disproportionately for the benefit of citizens of other countries. A national tax is also a practical approach: it allows countries to use established tax collection and expenditure mechanisms to tax transactions or events within their own borders. It is hard to imagine the US ceding tax collection and expenditure to an international body like the United Nations Framework Convention on Climate Change (UNFCCC). Governments that have introduced carbon pricing elsewhere (e.g., in Australia, Sweden, British Columbia) have generally collected carbon taxes or auction fees on their own national emissions, and spent them primarily for the benefit of their own citizens.[2]

A US CT would almost certainly likewise be collected by the US government and spent primarily for the benefit of US citizens. It would thus be broadly neutral for the US as a nation, transferring value from US citizens affected by the CT to US citizens who benefitted from the disposition of its revenues (obviously those classes would overlap). This would probably be necessary (though perhaps not sufficient) to secure US government and public acceptance of a CT. The policy objective might be as much to increase government revenue (to reduce the deficit, lower other taxes, or increase spending) as to curb GHG emissions. But a US CT on US emissions, even if collected and spent within the US, would still benefit humankind globally to the extent it reduced global emissions and spurred emissions pricing initiatives in other countries. Indeed if other countries similarly taxed their own emissions, the combination of national CTs would have the same impact on global emissions as an international CT.

Unfortunately, however, it cannot be assumed that all countries will tax their GHG emissions. In that case a US CT on US emissions would reduce global emissions so far as it applied to non-mobile sources of emissions. But if it applied to mobile sources (i.e., emissions from activities that could be carried out abroad), there is a risk that these would gravitate to countries that did not price emissions. This would negate the benefit of taxing those sources, and at the extreme could even result in an increase in global emissions, if the activities generated higher GHGs abroad then they would have done in the US. A number of policy responses to this are possible. For example:

- Tax all US GHG emissions anyway. Treat the risks of business moving as not significant (for example, because the advantages of cheap energy resulting from shale gas might compensate for the disadvantage of a CT); or manage those risks in some other way, for example, by output-based rebates.
- Tax US GHG emissions but with exceptions or exemptions for particular industries, especially energy-intensive trade-exposed (EITE) industries.
- Tax GHG emitted in the US or elsewhere to produce goods and services *consumed* in the US by means of border tax adjustments.

These different options (and no doubt there are other possibilities) would have implications for CT design, which would in turn have implications for administration.

The first policy option is very clear and straightforward. Implementing it would be largely a question of administrative practicality. To internalize the cost of US GHG emissions accurately, a CT should ideally apply to them comprehensively and at a uniform rate based on their global warming potentials (GWP). In practice administrative difficulty or cost may make this difficult to achieve. Gaps in coverage and non-uniformity of rates would not necessarily make a CT ineffective, just less effective than it might be. The aim should be to find the best possible balance between coverage/uniformity and administrative practicality.

The second option is more complex. Exemptions may be difficult to define and monitor. As will be discussed, administrative options that are suitable for comprehensive GHG taxation – for example, collection based on "upstream" sales of fossil fuels – may be less suitable for selective GHG taxation.

Taxation of US GHG consumption would be more complex still. This is discussed in more detail later under International Considerations.

It is important to define policy objectives clearly. If a CT is planned and designed on the basis that its objective is to tax US GHG emissions, and it emerges later that its true objective is to tax non-mobile GHG emissions or US GHG consumption, it may be difficult to re-structure it to meet that revised objective in practice.

CO_2 from fossil fuel combustion

CO_2 from fossil fuel combustion accounted for 79 percent of 2010 US GHG (see Table 3.2), and so is the main source of GHG to which a CT would need to apply. (Non-energy use of fossil fuels, discussed later, accounted for a further 1.8 percent.)

Distinct stages can be identified in the use of fossil fuels:

- Production (discovery and extraction)
- Processing
- Combustion
- Consumption (of heating, transport, goods, etc., that use energy from combustion).

Between stages there will often, though not always, be a transmission or distribution stage. For example, natural gas is distributed from gas processing plants to users via a pipeline network; electricity generated by fuel combustion is distributed

TABLE 3.2 2010 CO_2 fossil fuel combustion emissions, and revenue potential, by fuel type

	MMT CO_2 or CO_2e emissions	*Percent of gross US GHG emissions*	*Revenue potential from \$20 per ton CO_2 tax*
CO_2 from fossil fuel combustion	5387.8	79.0	\$107.85bn
Coal	1933.2	28.3	\$38.7bn
Natural gas	1261.6	18.5	\$25.2bn
Petroleum	2192.6	32.1	\$43.9bn
Geothermal	0.4	Neg.	

Source: Author's calculations based on EPA, 2012.

to consumers via an electricity "grid"; in other cases (such as petroleum-powered transport) combustion and consumption happen at the same point.

Most fossil fuel undergoes processing after production. Crude oil is distilled and subjected to other processes at refineries to produce a range of usable petroleum products, most but not all of which are combusted as fuel. Natural gas is normally processed at gas processing plants to separate gas liquids and to remove water and impurities. Coal is normally "washed" and graded at a coal preparation plant to remove rocks and dirt and other unwanted content. Processing may be carried out as a separate business – most common in petroleum refining – or as part of a production business.

CO_2 emissions from fossil fuels almost all occur at the combustion stage. But a CT can be imposed *indirectly* at earlier "upstream" stages. "Upstream" is traditionally used in the extractive industries to mean exploration and production (and "downstream" to mean other stages), but in the carbon pricing literature (and this chapter) it just means imposed at a stage prior to combustion/consumption, that is, it may include processing.

CT could be imposed at different stages for different fuels or uses, with an obvious requirement that it should apply only once. For example, producers could be required to pay CT only on sales but not sales to approved processing or refining companies, who would then pay CT on their sales instead; refiners could pay CT on their sales of petroleum products, but not on gasoline sales to wholesalers, who could be taxed on their sales instead.

The choice of which stage should form the CT base will depend on various factors. Administrative considerations – finding the best balance between coverage and accuracy of measurement of carbon content on the one hand, and administrative cost, convenience, and practicality on the other – may be important, but non-administrative political and economic considerations may be at least as important.

Advantages and disadvantages of upstream CT[3]

Administrative advantages

- *Limited number of measurement points/taxpayers.* Tax is paid by taxpayers, not points, but the number of points at which measurement is required affects taxpayers' compliance costs and burdens, and the ease of government monitoring.
    - CT on *petroleum* could be applied to output from refineries, of which there are around 150, and to imports, which occur at a small number of points (diminishing as tanker sizes increase).[4] The number of refining firms who would measure output from their own (multiple) refineries and pay the tax would be considerably smaller. CT based on production measured at the well-head would involve a larger number of taxpayers (oil producing firms), and a much larger number of measuring points, and would have no obvious advantages.
    - CT on *coal* could be applied to output from mines measured at the mine mouth (and again to imports measured at import points). There are around

1,300 US coal mines, but a much smaller number of producer firms who would measure the output from their mines and pay the tax. In 2010, 26 producers accounted for more than 85 percent of total US coal production, and the total number of producers was fewer than 500. CT could alternatively be applied to output from coal preparation plants – likely to be fewer in number than mines – where coal is routed through these.

- For *gas* there are over 450,000 gas wells, but a much smaller number of gas producing firms (around 8,000, of whom no more than 500 are of significant size). CT could therefore be paid by producer firms, based on total well output. A better option might again be to tax output from gas processing plants – there are around 500 significant ones (though again also some small ones) and again the number of processor firms that measured and paid the tax would be smaller than the number of plants. Again imported natural gas would have to be taxed at import points (of which there are around 50).[5]

So possibly not more than, say, 1,250 to 1,500 taxpayers could account for CT on the vast bulk of fossil fuels if taxed upstream. Even if that number were increased 20-fold (for example, because producing firms rather than processing firms were taxed) it would still be a small number for a tax yielding $100 billion. Contrast company income tax, for example, where in 2010 there were more than 1.8m US C corporations paying $191 billion. Furthermore, if it was decided to impose CT on methane emissions from gas, coal, and petroleum production (see later discussion) – which produce up to 4.5 percent of GHG emissions – some of that CT might also be collected from those same taxpayers.

- *Clearly identifiable taxpayers.* Upstream CT would apply to taxpayers who sell coal, gas, or petroleum. Those are recognized business activities, making CT taxpayers easier to identify. Contrast combustion, which, unlike refining, is not a business, but a business by-product. Almost all businesses combust fossil fuel to some extent, so defining a manageable number of taxpayers requires identification of those emitting CO_2 above a chosen threshold. This may be difficult at the margin since emissions (or fuel inputs) are not measured and reported for normal business purposes, and may present avoidance opportunities.
- *Coverage.* An upstream tax could achieve comprehensive coverage whereas by definition a tax on emissions above a chosen threshold implies less than perfect coverage. Of course if de minimis limits were applied to exclude small upstream businesses from CT, then an upstream CT would likewise not be comprehensive, but these often complicate rather than simplify administration, for example, by creating avoidance opportunities, and would probably not be necessary or worth the trouble, given the small total number of upstream taxpayers.
- *Measurability.* The CT base must be visible and measurable, and must provide an accurate measure of GHG emissions, even if indirect. This reduces taxpayer costs, and avoids disputes. The technical difficulty of measurement is also clearly relevant to administration. It is an advantage if the base is measured for normal business purposes, since again this reduces costs and makes measurement easier to

monitor. The volumes of output from fuel producers and processors are relatively easy to classify and measure, and the CO_2 output is in turn broadly measurable on the basis of those measurements, limiting technical difficulty and scope for dispute. It is worth noting, however, that likely CO_2 output can generally be more accurately measured from processed fuel outputs than from unprocessed outputs, which may contain varied amounts of impurities (which lower the carbon content per unit of the fuel). For petroleum, measurement of refinery outputs also has the advantage of allowing different distillates to be taxed at rates reflecting their different CO_2 content or different likelihood of combustion. Coal has four different grades (anthracite, bituminous, sub-bituminous, and lignite) with even more significant variations in CO_2 content, and again it is likely to be easier to measure CO_2 content after coal has been washed and graded.

- *Limited scope for avoidance.* Note that avoidance by reducing GHG emissions is acceptable – indeed is one of the prime objectives of a CT. By-passing the point of measurement to avoid tax on US GHG emissions would, on the other hand, be unacceptable, as it would frustrate the policy intention. There would be limited opportunities for fuels to by-pass upstream measuring points, since they are generally measured at those points as part of normal business processes. The fossil fuel industries are closely regulated for environmental and other reasons, further reducing opportunities for avoiding measurement points. Avoidance of CT by relocation of GHG emissions abroad is a more difficult issue, as discussed later under International Considerations, but the location of upstream activities would not be relevant to this – imports of fuels destined to create emissions from combustion in the US would be taxed, and exports destined to be combusted abroad would be exempted.
- *Use of existing tax mechanisms.* This (rather than the limited number of taxpayers) is probably the key advantage of upstream taxation. An upstream CT on fossil fuels would be broadly similar in operation to an excise tax. The US Internal Revenue Service (IRS) already collects excise[6] on petroleum products and coal (though not on natural gas unless compressed for use as a transport fuel). CT would involve some refinements that do not apply to current excise – for example, different rates for different fuel types, possibly credits or refunds for non-combustion uses. But broadly speaking, key procedures developed for excise purposes would be similar to those required for CT: for example, procedures to ensure that fuels do not by-pass excise measuring points and are taxed only once, to provide refunds where fuels are used for excise-exempt purposes, and to impose excise on imports and exempt exports. Extending this kind of regime to natural gas should raise no major new issues of principle. The existing excise legal framework for taxpayer registration, returns, payments, audit, and dispute resolution could probably be adapted for CT without too much alteration, and would have the advantage of familiarity to most CT taxpayers. The training and skills of IRS staff working in this field would be well suited to applying a similar regime to CT and making any necessary modifications. Excise is encrusted with the idiosyncrasies of any

long-established tax, and full integration with CT might therefore be difficult, but prospects for rationalization of their relationship would likely be stronger if they were administered together rather than separately.

So in short, an upstream CT on fossil fuels appears likely to provide a good balance of coverage and administrative cost and convenience. It would cover 80.8 percent of GHG emissions (including non-combusted fossil fuels), the vast bulk of which would be collected from no more than, say, 1,500 taxpayers, with the possibility of further methane emissions being taxed on broadly the same population.

Non-administrative advantages

- Opposition and pressure for compensation or exemption might be less widespread than if CT were imposed directly on downstream industries. (At any rate they are unlikely to be worse.) It would also make it slightly more difficult to grant special exemptions in practice, which proponents of a universal CT would consider an advantage.
- The low visibility of an upstream CT to consumers and voters might make it more acceptable. An upstream CT might have the same price impact on consumers as a downstream CT, but it might have the same low visibility as a corporate income tax, which consumers tend to think of as a tax on someone else, even if passed on to them in higher prices. Taxing big mining and petroleum companies might even be popular with voters, increasing general public acceptance. There are competing considerations here. Politicians sometimes advocate transparency in taxation (particularly if opposed to the government imposing the taxes); and there is a special case for visibility of CT since it is meant not just to influence consumer behavior but raise consumer awareness. But in practice governments hoping for re-election might find low visibility attractive.

Administrative disadvantages/problems

- As discussed, it may be difficult to measure CO_2 content accurately if production rather than processing is used as the CT base. But it may be possible to use processed fuels in most cases. In any case minor inaccuracies in measurement of CO_2 content, while undesirable, would not make a CT ineffective.
- In some cases fuels may by-pass processing facilities. For example, dry gas may be fed directly into transmission pipelines after limited well-head processing. It has been estimated that CT based on outputs from gas processing plants plus imports would catch only 70 percent of emissions from natural gas combustion. Similarly coal may be delivered direct from mine mouth to end user. But it should be possible to tax fuels at an alternative upstream point where they are not routed through processing plants. Similar problems are successfully managed in applying excise taxes.

- A small proportion of gas and petroleum outputs may not be combusted (e.g., asphalt, chemical feedstocks), or CO_2 from combustion may be captured and stored. CT should not apply in those circumstances. But practical ways of dealing with them could be developed, as discussed later.
- Taxation of non-fossil fuel emissions (e.g., nitrous oxide, CO_2 from industrial processes) would have to be administered separately, whereas it could be combined with downstream taxation of fossil fuel emissions.

Non-administrative disadvantages

- Upstream CT may be seen as inconsistent with the "polluter pays" principle. This principle may partly explain why most countries tax combustion emissions. (Other anti-pollution programs, for example on SO_2, have tended to focus on emissions.) Upstream CT may therefore be seen as less "fair." But a case can be made for allocating some GHG responsibility to producers and processors – they let the genie out the bottle.
- There is likely to be industrial opposition. This will happen however CT is imposed, and, as discussed earlier, the lower visibility of upstream taxation might actually reduce opposition. But if policymakers wished to respond to it, for example by giving reliefs or exemptions to industry, or more generally by grandfathering existing CO_2 production levels, it might be more difficult if fuels were comprehensively taxed upstream. This would depend on the type of exemption proposed. For example, if the aim was to exempt EITE industry from CT on power consumption but not transport costs, upstream taxation might make this more difficult.
- As discussed above, it might be argued that the low visibility of an upstream CT to consumers would lessen its impact on consumer behavior and choice (although economists tend to be skeptical of this argument).
- In countries with well-established downstream fuel and energy taxes, governments may be reluctant to replace them with a new upstream tax that is likely to be controversial, however much preferred by economists. (An old tax is a good tax.) This, however, is of limited relevance to the US, which does not currently impose any significant downstream federal fuel and energy taxes.
- There is a treaty bar on taxing air fuel used for international flights. But a CT on refinery output might not fall foul of this.

Are there advantages to applying CT at the point of combustion?

Most other countries apply carbon pricing to large emitters at the point of combustion. The reasons for this may be mainly non-administrative. As discussed, it may be seen as more consistent with the "polluter pays" principle. It may also be seen as more visible to consumers. (It may, however, be difficult for business reasons for CT taxpayers to make CT visible to consumers even if applied at the point of combustion – for example, a power company with power stations using different fuels might find it awkward to discriminate between customers and prefer just to pass on CT as an average price increase.) Another (political) advantage of taxing at the downstream level is that it facilitates targeted "fiscal cushioning" (such as special CT exemptions) to placate powerful industrial lobbies (as has happened in various countries).

One administrative advantage of a combustion-based CT is that it avoids the need under an upstream CT for a mechanism to exempt non-combusted fossil fuels (asphalt, etc.). But this is a relatively minor issue, because only a small proportion of fossil fuels are not combusted and, as discussed later, exempting them under an upstream CT should not be a major problem.

One claimed administrative disadvantage of taxing at the point of combustion is lower coverage (most countries that tax on this basis cover less than 50 percent of their GHG emissions). It is not practical, for example, to tax transportation fuel at the point of combustion. But again low coverage may to some extent reflect political rather than administrative considerations, for example, the desire to protect certain industries; or the fact that many countries already have high taxes on transport fuels, and may be reluctant to impose carbon charges on top of them. There is no reason in principle why CT at the point of combustion for some fuel uses (such as electricity generation) should not be combined with CT at the point of wholesale distribution for other uses (such as transportation).

But there may be other reasons why it is inherently difficult to make a combustion-based CT comprehensive. There are a larger number of taxpayers (and measuring points) where CT applies at this point rather than upstream. It has been estimated that the number within the European ETS is 12,000 and that a similar number would be subject to a US combustion-based CT, whereas under an upstream system broad coverage could probably be achieved with around 1,500 taxpayers. But 12,000 is not a large number of taxpayers for a major tax (again contrast the numbers for US corporate income tax).

A more important issue may be the technical difficulty of measuring CO_2 emissions at this point. For example, the EU rules on measurement for the purposes of its ETS are highly technical and complex.[7] It imposes major bureaucratic burdens on businesses, such as measuring fuel inputs and applying emissions coefficients, which it would not do for normal business purposes. It is often considered impractical to impose such burdens on any but the largest businesses (for example, the Australian CT applies only to only around 300 companies emitting more than 25,000 tons p.a., and is estimated to cover only around 60 percent of CO_2 emissions). Restriction to large companies thus reduces coverage, and may distort markets and create opportunities for avoidance and perceptions of unfairness. Combustion emissions need special technical expertise to monitor – they are generally not monitored by normal tax administrations. So the increase in administrative cost and complexity from taxing at the point of combustion may be much greater than the mere increase in taxpayer numbers would suggest.

CT at point of distribution to final consumer?

It would be impractical to impose CT at the point of consumption if by this is meant taxing millions of end consumers. But where consumers buy fuel directly for heating or transport purposes, it would be relatively straightforward to impose CT on retail distributors in the same way as a retail sales tax. At the other extreme it would be impracticable to impose CT on retail distributors of manufactured goods, since it would be extremely difficult and administratively onerous to calculate the GHG emissions involved in the manufacture of complex goods such as cars, which have hundreds of components with different origins, manufacturing histories, power inputs, and so on. In between, there are cases where consumers do not purchase fuel directly, but where it may be possible to quantify the GHG input reasonably accurately (for example, electricity – but even there it could be difficult if the supplier provided electricity from a different kind of power plant).

For similar reasons, it is in general unlikely to be possible for CT to be passed on cumulatively through a supply chain in the same way as a VAT, where tax accumulating on each addition to value is eventually borne by the final consumer (though not collected by the government direct from the final consumer). The accumulation of GHG inputs and outputs would be far more complex and onerous to measure than sales inputs and outputs used to calculate VAT. (Again car manufacture is an obvious example where this would be impractical.) In many cases CT would for practical purposes have to be passed on as a price increase (as is normal for excise taxes) rather than as a tax (as is normal for sales taxes).

Point of taxation of fossil fuel emissions – general conclusions

On balance an upstream CT on fossil fuel sales seems likely to be an administratively better option than a downstream CT imposed directly on power and industrial plant emissions, providing a better balance of coverage/accuracy and administrative practicality; and it has no conclusive non-administrative disadvantages. But countries have in practice managed to administer downstream CT on power and industrial plant emissions, and if combined with CT on retail sales of fuel for transport and domestic consumption, this could achieve quite high coverage, even though not as high as an upstream CT. If CT is imposed upstream on fossil fuels there does not have to be a single point of taxation. The experience of collecting excise taxes is likely to be useful for working out the most practical points of measurement to ensure that they are comprehensively taxed in practice. If petroleum excise tax is applied to refinery output, and coal excise to mine output, these might be best points of measurement for CT purposes too. This chapter cannot claim to have identified and addressed every practical problem that might arise with an upstream CT. But in general it seems likely that such problems will be similar in kind and degree to those that arise with excise taxes, and should therefore be capable of being addressed in practice. British Columbia probably comes closest to an upstream CT, and seems to combine reasonable coverage (estimated at 77 percent of Canada's total GHG emissions[8]) with relatively simple administration.

Sequestered CO_2 from fossil fuels

Non-energy use of fossil fuels

Non-energy use produced 1.8 percent of GHG emissions in 2010, but if CT applied upstream, it would be necessary to ensure that it applied only to CO_2 emitted and not to CO_2 stored, for example, in asphalt (where 100 percent is stored), chemical feedstock (62), waxes (58), lubricants (9), others (25). If CT applied to refinery output it would be possible to identify and exempt asphalt at that stage, and similarly to identify and exempt (or partially exempt) other products destined for non-combustion use. If not possible, then purchasers would have to apply for refunds (as currently happens with excise where petroleum fuels are used for exempt purposes such as agriculture).

Carbon capture and storage (CCS)

CCS is not an immediate concern since there has been limited development of CCS technology in practice so far.[9] If it did become viable in future, steps would have to be taken to exempt captured CO_2 from CT. If CT applied upstream, it would not be practical to identify and exempt in advance fuels whose CO_2 would be captured by CSS. Retrospective linking of CCS to the relevant upstream supply, refund of CT on that supply, and revision of the price charged to the customer for it would be administratively complex, and if CCS ever became common, with CO_2 piped from numerous sources to CCS plants, it would probably negate some of the administrative advantage of upstream CT. The problem with CT credits or refunds in an upstream system is that upstream CT payers might not be involved in CCS projects, and downstream businesses who were involved would be unable to use them. One option would be to make CT credits tradable, but this would add complexity and might be open to fraud. A simpler option would be for the government just to pay an amount equal to CT per ton to a company that implemented CSS.[10] The government would monitor CCS facilities, and charge CT on any CO_2 that escaped. Monitoring CCS would be a specialist technical function, and monitoring and payment could be allocated to a specialist technical agency either in the IRS or outside it. International issues might have to be considered in due course – for example, import or export of liquefied CO_2 for CCS. Credit for export would need international cooperation (to ensure CO_2 was genuinely stored). The US would give no credit for imported CO_2, but would impose CT on any that escaped.

Land use (forestry, etc.) carbon sinks

Forestry provides the main potential carbon "sink" (though of course deforestation can *increase* GHG emissions). In 2010, forestry and other land use were estimated by the EPA to have reduced US CO_2 emissions by a substantial 1,075 MMT, primarily through increases in forest density or acreage. In principle there is a case for the government to provide some form of CT set-off for additional CO_2 sunk in forestry. As with CCS, it would probably be better for the government not to try and embed forestry within the CT regime through CT refunds, tradable CT credits, and the like, but instead simply to make payments equivalent to the rate of CT per ton direct to forestry businesses that could provide an acceptable measure of additional tons of CO_2 sunk, since this would be easier to administer and less open to fraud. The basic problem, however, would be how to establish additionality. It has been suggested that if potential qualifying companies were limited to large paper and forest product companies, who were required to measure sunk CO_2 by reference to their entire land stock, it might make additionality easier to establish. But there would be a host of other complex technical issues in measuring CO_2 sunk in forestry, even if additionality could be established. These issues are far too complex to do justice to in this short chapter. Measurement, if it did become feasible, would be a highly technical operation, and, as for CCS, administration would probably have to be carried out by a specialist technical agency either within the IRS or entirely separate from it (since the required expertise might already exist in other government departments).[11]

Extending CT to non-fossil fuel emissions

With non-fossil fuel emissions it is again normally possible to identify different stages at which a CT could theoretically be imposed, but it is generally harder to identify anything equivalent to an upstream taxing point; and a large proportion of these emissions are fugitive, not measured for normal business purposes, and in some cases very difficult to measure at all in relation to individual taxpayers. This means that administration is likely to be more complex, and coverage in some cases may be impractical. A factor to take into account is the relative cost or difficulty of reducing the emissions. If the prime purpose of a CT is seen as reducing emissions, it may be felt that if CT would result in a relatively large reduction in GHG (particularly those with high GWPs), it may be worth the administrative effort even if the emission is relatively small; and conversely that there is little point in applying CT to a substantial but hard-to-administer emission if no one can reduce it anyway. (An alternative point of view is that governments should apply CT to significant emissions even if they are difficult to reduce, primarily as a means of raising revenue.)

Non-fossil fuel CO_2

CO_2 emissions from industrial processes (over and above their energy input) were 2.8 percent of total GHG emissions in 2010. Natural gas systems accounted for 0.5 percent. CT on combusted fossil fuels would increase the competitiveness of natural gas against other fossil fuels, which might provide a special reason for trying to ensure that all emissions from natural gas industrial processes were fully taxed, so that it did not also enjoy an unfair advantage. In view of the numbers, small producers might need to be excluded. Iron and steel producers and cement producers between them accounted for 1.3 percent, and a further four industries – lime production, waste incineration, limestone and dolomite use, and ammonia production – for 0.6 percent. The numbers of companies engaged in those businesses are not large – for example, there are around 116 cement plants owned by 39 companies, and 23 steel mills (per Metcalf and Weisbach 2009) – and estimating the CO_2 output from those industrial processes should on the whole be feasible, though it would require special technical measurements and procedures, adding administrative complication. This would bring a further 2.4 percent of GHG emissions within CT (or around 85 percent of the CO_2 from industrial processes). Most of those businesses are, however, EITE, so leakage and international competitiveness issues would be particularly important with some of these taxpayers. There might therefore be strong political pressure not to tax CO_2 emissions from industrial processes, or alternatively to protect them from competition by border tax adjustments on imports and exports (as discussed later).

Non-CO_2 GHG emissions

Non-CO_2 GHG emissions accounted for 16.4 percent of 2010 emissions, methane being the most significant. These gases have high, sometimes very high, GWPs. A CT would have a very high impact on their price. They may therefore be a relatively low cost source of emissions reductions. Metcalf and Weisbach (2009) estimate that they could possibly account for one-third of the emissions reductions that might arise from carbon pricing, presenting a strong policy case for bringing them within CT as far as possible.

Methane

Methane accounted for 9.8 percent of 2010 GHG emissions. The main sources were natural gas systems (3.2 percent), enteric fermentation[12] (2.1), landfills (1.6), coal mining (1.1), and manure management (0.7).

- On natural gas systems, Metcalf and Weisbach (2009) estimate that around one-quarter of emissions come from field operations and are not practical to measure and tax, but suggest that it would be practicable to tax the remaining three-quarters emitted from processing, transmission and storage, and distribution, by monitoring inputs and outputs and taxing the difference. These proportions may have changed, since it is claimed that shale gas development produces higher emissions from field operations than conventional gas. For the reasons discussed earlier, there may be a special case for taxing standard *estimated* methane emissions from gas field operations, since failure to do so would unfairly increase the advantage CT gave to natural gas over other fossil fuels.

- Metcalf and Weisbach (2009) suggest that it could be cost effective to bring enteric fermentation within CT, by taxing cattle per head, with varied rates for five different cattle types (which emit different amounts of methane), and reduced rates where there was proof of emissions-reducing cattle diets. Although this may be feasible, there is no doubt that it would be complex to administer – the US has many farms and cattle herds, whose composition and diets might change regularly. (Other countries such as Australia have in general concluded that agricultural emissions are too difficult to tax and simply exempted them from CT.)
- There are around 1,800 landfills, which currently capture and combust 50 percent of their methane emissions (turning them into less harmful CO_2). It would be feasible to measure and tax those emissions, providing a worthwhile incentive to increase their capture.
- For coal bed methane it would be relatively straightforward to capture and tax the two-thirds emitted from underground mines, since companies have to control those emissions for safety reasons. But it would be less easy to capture and measure the one-third emitted by surface mining. It could be argued that it was inappropriate to tax emissions that were difficult to measure and control, but again it could be countered that there is a special case for taxing *all* methane emissions from production of fuels, even if this necessitates taxing on the basis of a standard estimate.
- Manure management, the least important source discussed here, is not practical to tax.

Nitrous oxide

This gas accounted for 4.5 percent of 2010 US GHG emissions. But three-quarters of this comes from agricultural management (soil, fertilizer practices, and so on) and is not practical to tax. (Alternative approaches such as taxing fertilizer could have unintended consequences, such as more cows and enteric fermentation.) The remaining quarter comes mainly from combustion emissions from cars and other machinery. Metcalf and Weisbach (2009) suggest that mobile combustion emissions could be taxed downstream by means of annual vehicle tests and calculations of emissions per gallon, miles per gallon, and annual mileage. The sheer number of vehicles and machines involved would make taxing this source a costly and difficult procedure. An alternative option would be to identify a limited number of businesses that emit nitrous oxide above a certain limit, and restrict CT to those. An alternative or additional approach might be to keep tightening nitrous oxide emissions standards for new vehicles and machinery.

HFCs (and PFCs) are synthetic chemicals that substitute for ozone-depleting substances in the manufacture of products such as refrigerators and air conditioning systems. They have very high GWPs, but are hard to tax because emissions are fugitive (resulting from leakage or incorrect disposal). Metcalf and Weisbach (2009) suggest use of a downstream deposit/refund mechanism (with refunds also

TABLE 3.3 Potential US carbon tax coverage

Source	*% US GHG*	*Practicable to tax?*	*Comments*
CO_2			
Fossil fuel emissions	79.0	79.0	Assumes comprehensive upstream taxation, no exemptions.
Fossil fuel non-energy	1.8	1.8	As above.
Industrial processes	2.8	2.5	Develop special measurement systems for main industries concerned. Will probably need de minimis limits. Opposition likely.
Methane	9.8	6.0	Mainly fossil fuel producers (possibly only large ones) and landfill.
Nitrous oxide, other GHG	6.6	0.7	Very difficult to tax – would probably need special fees, deposits, etc., outside mainstream CT system.
Total	**100.0**	**90.0**	

for "banked" gases in products already produced). The biggest problem would be how to charge the appropriate deposits on imports of HFCs and of goods containing HFCs, which would be difficult to quantify.

Summary

Altogether non-fossil fuel emissions accounted for 19.2 percent of 2010 US GHG emissions (Table 3.3). It might be possible to tax nearly half of those at reasonable administrative cost (though at considerably greater cost than upstream taxation of fossil fuel emissions). This would include most CO_2 emissions from industrial processes, and CO_2e from methane emitted by natural gas systems, landfills, and coal mining. Methane from enteric fermentation, nitrous oxide from mobile sources, HFC from selected products, and possibly various other minor emissions might also be considered, though are more difficult. Taxing those sources would require setting up new taxation procedures and mechanisms, and the development of new technical skills. It would probably be best to introduce this gradually over time.

International considerations

If the policy objective of a US CT was to tax only US GHG emissions, and it applied upstream to fossil fuels (for simplicity let us say at the processing stage), administration of international transactions would be relatively straightforward. Where fossil fuels were imported in unprocessed form, they would be taxed at the processing stage rather than on import. Where they were imported as processed products, they would be subject to a border tax adjustment (BTA). This would impose CT in exactly the same way as on domestically produced fuels, using the same simple volume measurements and product classifications. Fuel imports occur at a relatively small number of points, subject to regulation. Taxing imports at those points would not greatly complicate administration. The imposition of tax on imported refined fuels should not cause WTO problems because the taxes imposed would be calculated on exactly the same basis as on domestically produced products. Where fossil fuels were exported in unprocessed form, CT would not apply. Where they were exported in processed form, they would be exempted. This should not impose major administrative burdens – again the same taxpayers and measurement systems would be involved as for CT imposition. The exemption might in theory cause WTO problems, but it could be argued that CT on fossil fuels was essentially a destination-based tax on consumption, so that exemption should be allowed. It should be noted that WTO rules do not prevent zero-rating of exports for VAT purposes or exemption of exports from excise taxes, which would provide clear parallels for CT exemption of exported fuels. The taxation of emissions from imported fossil fuels and the exemption of exports would mean that US producers did not suffer competitive disadvantages relative to foreign producers in the same markets (though of course CT would affect competition between different fuel types within the US).

Countries could in theory *not* apply CT to imported refined fuels and *not* exempt exported ones, effectively applying an origin-based system. But it is hard to see why this would make sense on theoretical grounds, since it would mean that the country that collected CT was one where emissions were neither emitted nor consumed. Apart from that, realpolitik is likely to influence countries' decisions on how CT taxing rights should be allocated internationally. It is hard to see why the US would support the principle of allocating them primarily to fuel-exporting countries (which would include Saudi Arabia, Russia, and Iran, to mention a few at random).

Imports of other non-CO_2 sources of GHG would likewise be taxed and exports exempted. In practice the main issue would arise with imported HFCs, including those incorporated in manufactured goods, which, as discussed earlier, could be difficult to measure. Imports and exports of other items would, generally speaking, have no impact on US GHG emissions, though of course CT would apply in the normal way where US emissions occurred later as a result of importation (for example, if cattle were imported and enteric fermentation subsequently taxed).

With a CT based on emissions a problem arises with "bunker" fuels used for international shipping or air transport. These would not clearly fall within any particular country's jurisdiction. If an upstream CT were applied to fossil fuels, shipping and air companies might be able to avoid it by fuelling overseas. EU proposals to charge aviation fuel emissions on flights using EU airports have been controversial, and international agreement may be required to determine how CT on such fuels should be applied and administered.[13]

If there was a clear and simple policy to tax GHG emissions in the US, then international transactions would be unlikely to cause major problems (other than the problem of bunker fuels just discussed). The law would be easy to apply and interpret; disputes would be unlikely; avoidance opportunities would be limited. This policy would be consistent with the general international approach, which is to tax domestic GHG emissions and not consumption. There would be no need to distinguish the tax treatment of exports and imports on the basis of whether the other countries concerned were members of the carbon pricing "club."

But the position would change radically if the aim were to tax GHG consumption rather than GHG emissions. There are theoretical arguments for this, but practical arguments against. The theoretical arguments include:

- The demand of US consumers causes GHG to be produced.
- Foreign GHG resulting from US demand harm the US (and everyone else) just as much as US emissions.
- The US government is best placed to influence US consumer behavior through its tax system.
- If it does not tax GHG consumption, there may be leakage of GHG emissions from mobile businesses, because US firms that would suffer from CT might re-locate activities to countries that did not tax GHG, or might lose business to non-US competitors, negating the effect of CT either way (and possibly even resulting in an *increase* in global emissions).[14]
- Another way of putting this is that US firms might avoid CT by continuing to emit GHG but re-locating their emissions abroad.
- And if they did not they might be competitively disadvantaged in both domestic and export markets against firms from those countries.

Taxing US GHG consumption would mean applying CT to US imports of products whose production had involved untaxed foreign GHG emissions; and also exempting US exports from CT that they would otherwise, directly or indirectly, have suffered. Assuming effective implementation, this would have the following advantages:

- It would meet the theoretical argument for using GHG consumption as the CT tax base.
- It would prevent leakage – there would be no advantage in foreign re-location of production intended for consumption in the US.
- It would prevent avoidance – again there would be no tax saving from foreign re-location of production intended for the US market.
- It would not harm the competitiveness of EITE US firms. Imports would have no CT advantage over domestic production. Exports would be exempted from CT.
- Double taxation could be avoided by exempting imports from fellow members of the international carbon pricing "club," and *not* exempting exports to them

(this would be administratively simpler than taxing imports and exempting exports).

- It would provide an incentive for other countries to join the carbon pricing "club" – they would collect CT on their exports rather than letting the US collect it (though this would be counter-balanced if staying out of the "club" meant that imports were taxed that would otherwise be exempt).

A variant on this approach would be to tax carbon-intensive imports from countries not in the "club" but *not* to exempt carbon-intensive exports to those countries. EITE US industry would suffer competitive disadvantage in export markets, but would be protected from unfair competition in its home market. This might be sufficient to reduce leakage to acceptable levels. It would have the advantage of maximizing US government CT revenues and creating a clearer incentive for other countries to join the "club."

This is all fine in theory, but there are major practical and administrative problems in taxing GHG consumption.

1. *Measuring the GHG "content" of imports and exports.* As discussed earlier, this can be extremely difficult to measure, particularly for manufactured products – and the cumulative effect of CT cannot be measured in the same way as with a tax like VAT. Even if it were possible to measure GHG inputs with all the relevant information to hand, that information would be difficult to get hold of for imports. BTAs could be possibly be simplified by basing them on standard input/output tables reflecting average energy input per unit and average emissions intensity of energy production *in the US.* This would not be an accurate measurement of GHG input, since foreign GHG input might be higher or lower than US GHG input (because different fuels and/or processes were used), but arguably it would create a level playing field in the US and might therefore be enough to prevent leakage and avoidance. But even this would be very complex. Any application of CT to imports and exemption of exports would inevitably be incomplete and based on somewhat arbitrary distinctions and measurements. It would be onerous to apply, open to avoidance, and probably subject to frequent dispute.
2. *Reconciling BTAs with WTO/GATT legal rules.* Since it is hard to measure GHG content and the amount of CT imposed, it would be hard in many cases to square taxation of imports and exemption of exports with WTO rules. Where CT was passed on to consumers as a price increase and not explicitly as a tax (for example, on manufactured goods), then applying CT to such goods on import or exempting them on export would appear discriminatory. It might, however, be possible to apply BTAs in limited circumstances – for example, if CT applied downstream to CO_2 emissions from industrial processes like steel making, it might be possible under WTO rules to apply similar BTAs to tax imports and exempt exports of those products. (In general it might be easier to conform with WTO rules where CT was applied downstream.) But again

this would inevitably be an incomplete and somewhat arbitrary solution to the problem.

3. *Identifying which countries should be treated as carbon pricing club members.* This would involve numerous difficulties. For example:

   - How to define the coverage level and carbon pricing rate required for club membership.
   - How to define an acceptable limit of "fiscal cushioning" (by subsidies, CT exemptions or reduced rates, output-based rebates, etc.) and establish whether a country was within that limit at any particular stage.
   - If particular industries were favored in the other country, whether to exclude them from preferential treatment.
   - Whether to allow countries with alternative regulation-based approaches to join the club.
   - How to establish whether club members were implementing carbon pricing effectively in practice (failure could result from administrative inefficiency, lax enforcement, even corruption).
   - If other club members taxed emissions and not consumption (as at present), how to ensure that imports from them were not products they had in turn imported from non-club members.
   - Who would administer club membership rules – an international body would be necessary to avoid duplication and apply common standards, but the US might not be happy to rely on it.

These may not quite amount to three "impossible things before breakfast," but they would undoubtedly complicate CT administration very significantly. Instead of a clear underlying policy objective (CT applies to defined US GHG emissions), there would be a confused underlying policy objective (CT applies to defined US GHG emissions, except sometimes it applies to US GHG consumption). Disputes and uncertainties would have to be resolved according to complex and detailed rules rather than according to a clear principle. Rather than complicating CT administration in this way (contrary to normal international practice), it might be better to find simpler ways of addressing the policy and commercial objections to a CT based on US emissions – for example, by temporary grants or subsidies to affected industries. Obviously these should not completely cancel the effect of CT (which would make CT administration pointless altogether) but they might temporarily cushion the blow and reduce (though not eliminate) competitive disadvantages and the potential resulting leakage and avoidance.

Conclusion – practical administrative arrangements for a US carbon tax

It should be fairly straightforward for the IRS to administer an upstream CT tax on US emissions from fossil fuels. CT would be payable by a couple of thousand taxpayers or fewer, with a relatively easy-to-measure tax base, and should not require major addition to resources. There would be no need for a major new addition to technical skills – staff currently monitor and audit the outputs of the companies concerned. CT taxpayers might need to report and analyze their output in more detail for CT than for other tax purposes, but this analysis would often already be done for commercial purposes, since fuel products with different energy outputs have different commercial values. Border tax adjustments on fuel imports and exports would be relatively straightforward to administer, since they would be similar to those applying for excise tax purposes.

CT administration should be integrated into the normal tax administration of the companies concerned – with similar self-assessment–based procedures for taxpayer service, registration, returns filing, payment, audit, and dispute resolution, and similar risk-based strategies for encouraging compliance and detecting and dealing with non-compliance. It would be unnecessary and undesirable to re-invent the wheel for CT. On the contrary the aim should be to exploit established tried and tested mechanisms as far as possible.

Staff involved in taxing natural resource companies normally need to coordinate with and obtain information from industry regulatory agencies – this would equally be needed for CT.

Extending CT to other emissions would be more complex, and in some cases involve setting up new taxing arrangements. But so far as possible it would be best to develop those within existing tax arrangements. Only where exceptional specialist technical skills were required, for example, for monitoring CCS or for quantifying the GHG effects of land use, would it be necessary to set up specialist technical offices. It would be a matter of choice whether these should be attached to the IRS – it might be better to attach them to government agencies that already had the technical knowledge required.

These conclusions would have to be revisited if the intention was to tax US GHG *consumption* rather than emissions. Administration of complex BTAs, applied selectively depending on membership of the carbon pricing "club," would add significant complication. It would have a major impact on the resources and skills required by the IRS. It would also likely need the establishment of an international agency to administer the club membership rules, with whom the IRS would need to cooperate.

The IRS reports to the US Treasury, which is primarily responsible for revenue policy. But a CT would have dual revenue-raising and environmental protection objectives, and the EPA is responsible for the latter. It may therefore be appropriate for the EPA to have some oversight and input with regard to IRS administration of CT. Some technical functions relating to CT, for example, measurement of non-CO_2 emissions or of CO_2 sinks, might furthermore, as discussed, be better carried out by specialist agencies reporting to the EPA (or possibly some other department) rather than the IRS. There would therefore need to be close cooperation between the Treasury and IRS on the one hand and the EPA on the other, on both policy and administration. Formal arrangements for this would have to be put in place.

Notes

* I am grateful to Katherine Baer and Ian Parry for their helpful comments and suggestions.

1. See Chapters 1 and 3 for further discussion of revenue impacts.
2. Under the third phase of the EU ETS, individual member states will largely auction their own emissions permits.
3. See seminal paper by Metcalf and Weisbach (2009) on which this chapter draws heavily (but with occasionally different conclusions for which they have no responsibility).
4. According to the Energy Information Administration, the US imported 32 percent of its oil in 2013, down from 60 percent in 2005.
5. Bluestein (2008) estimates that charges levied on processors and importers would cover 70 percent of natural gas emissions.
6. See www.irs.gov/pub/irs-pdf/p510.pdf
7. http://eur-lex.europa.eu/LexUriServ/LexUriServ.do?uri=CELEX:32012R0601:EN:NOT
8. The BC tax does not, generally speaking, apply to emissions other than those from fossil fuel combustion.
9. For a discussion of the obstacles to this technology see Deutch et al. (2007), pp. 43–62.
10. One can imagine competitive market-based solutions being developed, with CCS providers competing to buy CO_2 from emitters at a price reflecting CT per ton less their CCS costs. But if separate firms carried out capture and storage (for example, a coal plant captured CO_2 which an oil producing company stored in a depleted oil reservoir), an issue to be addressed (whether CT applied upstream or downstream) would be how to apportion any CT refund or credit.
11. This chapter does not discuss the complicated issue of how anything equivalent to international ETS offsets might be built into a CT and administered. Again some separately administered mechanism might have to be developed.
12. This is a digestive process of ruminant animals, such as cows and sheep, by which microorganisms break down carbohydrates into simple molecules for absorption into the bloodstream.
13. See Keen et al. (2013) for further discussion.
14. Another form of leakage occurs if a US carbon tax reduces US energy demand and in turn reduces world energy prices, encouraging higher energy use elsewhere. It is not possible to address this through tax adjustments.

References

Bluestein, Joel, 2008. *Coverage of Natural Gas Emissions and Flows under a Greenhouse Gas Cap-and-Trade Program*. Available at: www.pewclimate.org/docUploads/NaturalGasPointof Regulation09.pdf

Deutch, John, et al., 2007. *The Future of Coal: Options for a Carbon-Constrained World*. Available at: http://web.mit.edu/coal/The_Future_of_Coal.pdf

EPA, 2012. *Inventory of US Greenhouse Gas Emissions and Sinks 1990 – 2010,* April 15, 2012. Available at: http://epa.gov/climatechange/ghgemissions/usinventoryreport.html

Keen, Michael, Ian Parry and Jon Strand, 2013. "Ships, Planes, and Taxes." *Economic Policy* 28, 701–749.

Metcalf, Gilbert E. and David Weisbach, 2009. "The Design of a Carbon Tax." *Harvard Environmental Law Review* 3, 499–556.

4

CARBON TAXES TO ACHIEVE EMISSIONS TARGETS

Insights from EMF 24

Allen A. Fawcett, Leon C. Clarke, and John P. Weyant*

KEY MESSAGES FOR POLICYMAKERS

- According to the most recent Energy Modeling Forum (EMF 24), U.S. CO_2 emissions are projected to be 2 percent below 2005 levels in 2020 (mean value across all reference case scenarios), though recent trends in energy markets that are not included in vintage of the models used in this exercise (e.g., the shale gas expansion) would tend to lower those projections.
- On average, the estimated 2020 CO_2 price consistent with the Administration's pledge to reduce emissions by 17 percent below 2005 levels in 2020 is $35 per ton.* However, estimated prices vary dramatically with different modeling assumptions, especially in regard to the future costs and availability of emissions-saving technologies and complementary policies – e.g., recently enacted fuel economy regulations when combined with carbon capture and storage (CCS) requirements for new coal-fired power plants and a renewable portfolio standard (RPS) lower the average price to about $20 per ton in 2020.
- Across all model runs, emissions reductions come from a broad range of sources including shifting away from fossil sources of primary energy through reduced energy consumption and replacement with non-emitting energy sources. In the latter regard, the largest reductions come from a shift away from coal, with smaller reductions coming from shifts away from oil, and in some cases from natural gas.
- Technology options such as nuclear, carbon capture and storage, renewables, and energy efficiency are, when taken together, extremely

important for cost-effective emissions mitigation, though no single technology is crucial. Instead, different combinations of technologies can lead to cost-effective decarbonization of the energy system.

- This underscores the need for a credible and sustained carbon price, complemented with measures to further encourage a broad portfolio of technology developments. Complementary measures to contain uncertainty over future emissions (e.g., carbon budgets specifying allowable cumulative emissions over an extended budget period) may also be desirable.

*Prices here are in year 2005 – to express in year 2012 multiply by 17.5 percent.

1. Introduction

In the one hundred eleventh Congress, considerable attention was given to cap-and-trade legislation. The American Clean Energy Security Act of 2010 (introduced by Congressmen Waxman and Markey) passed the House, though the Senate version of the bill (Kerry–Lieberman's American Power Act) was never put to the vote. With the prospects for cap-and-trade legislation stalled (and not helped by depressed prices in the European Trading System), and the need to strengthen the U.S. fiscal position, increasing attention is being paid to a domestic carbon tax.[1]

Although both of these policies would place a price on greenhouse gas (GHG) emissions, they behave differently under uncertainty (e.g., about future fuel prices and technology costs). A cap-and-trade policy (at least in its pure form without price stability provisions) specifies an emissions quantity target, leaving the carbon price uncertain, while a carbon tax specifies a price, leaving the quantity of GHGs uncertain. Economists generally recommend that emissions prices are stable (though rising) over time, to help equalize incremental abatement costs in different periods, and create a stable environment for clean technology investments (e.g., Pizer 2005). However, international negotiations generally focus on specific emissions reductions pledges, and the quantity of GHGs realized under a policy may be of significant domestic concern, so it is important to understand what emissions prices might be consistent with alternative emission reduction targets.

However, as the shale gas boom reminds us, the future, even the near future, evolution of the energy system is very difficult to predict. What is needed is not one model, with one future scenario, but rather a broad range of models and scenarios, to give some sense of what carbon price trajectories might be consistent with specific U.S. emissions reductions goals, and what the resulting transformation of the U.S. energy system might look like. This chapter therefore uses results from the Energy Modeling Forum (EMF) 24 (Weyant et al. 2014),[2] which encompasses a broad range of future scenarios (e.g., for economic growth, fuel resources and prices, and the costs and availability of energy technologies) and expert opinion.

The discussion is organized as follows. Section 2 gives an overview of the motivation for and the design of the EMF 24 workgroup. Section 3 examines the results from a range of scenarios that all achieve emissions reductions of 17 percent below 2005 levels by 2020 and 50 percent below 2005 levels by 2050, under a variety of different technology and policy assumptions, focusing on emissions, carbon price, and primary energy impacts. Section 4 explores the results for alternative emissions targets ranging from constant 2005 emissions to 80 percent below 2005 levels by 2050. Finally, section 5 concludes.

2. Overview of the study design

The Stanford Energy Modeling Forum – which has engaged in 28 studies since it was established in 1977 – brings together leading energy and economic model builders and model users representing academic, corporate, and government perspectives. The primary goals of the Forum are to utilize the collective expertise of the participants to improve understanding of an important energy and environmental problem; identify policy-relevant insights and analyses that are robust across a wide range of models; explore and explain the strengths and limitations of alternative modeling approaches; and help identify high-priority areas of future research.

The EMF 24 exercise began with three motivating questions. What would the U.S. energy system transition look like under pricing policies needed to meet U.S. emissions goals that are roughly consistent with an ambitious international climate goal, such as containing mean projected global warming above pre-industrial levels to 2°C or 3°C? What are the potential implications of transportation and electric sector regulatory approaches to emissions reductions in meeting this goal? And, how might the technological improvements and technological availability influence the answers to both of the above questions?

To address these questions, each of the modeling teams involved in EMF 24 (insofar as possible) used a matrix of 42 scenarios, covering 8 different technology, and 15 different policy, assumptions, though only a subset of these runs is reported here. Table 4.1 summarizes these scenarios and those considered here (indicated by the shaded cells).

The technology assumptions consist of optimistic and pessimistic assumptions for end-use technology (e.g., energy efficiency), carbon capture and storage (CCS), nuclear energy, wind and solar energy, and bioenergy – the full list of technology assumptions used in EMF 24 is given in Table 4.A.1 in the Appendix.[3] The policy assumptions include baseline or reference scenarios with no policy; emissions target scenarios of varying stringency, that all linearly reduce emissions over time to different percentages below 2005 levels; electricity and transportation regulatory scenarios combined with a 50 percent emissions target policy (by 2050); and various regulatory policies in isolation. The policy assumptions used in EMF 24 are further described in Table 4.A.2 in the Appendix.

TABLE 4.1 EMF 24 scenario matrix

Technology dimension								
	Optimistic	Single technology sensitivities				Combined sensitivities		Pessimistic
End Use Technology	Optimistic	Pessimistic	Optimistic	Optimistic	Optimistic	Optimistic	Pessimistic	Pessimistic
CCS	Optimistic	Optimistic	Pessimistic	Optimistic	Optimistic	Pessimistic	Optimistic	Pessimistic
Nuclear energy	Optimistic	Optimistic	Optimistic	Pessimistic	Optimistic	Pessimistic	Optimistic	Pessimistic
Wind & Solar	Optimistic	Optimistic	Optimistic	Optimistic	Pessimistic	Optimistic	Pessimistic	Pessimistic
Bioenergy	Optimistic	Optimistic	Optimistic	Optimistic	Pessimistic	Optimistic	Pessimistic	Pessimistic
Policy dimension								
Baseline								
50% Emissions Target								
Complementary Policies + 50% Emissions Target								
0% Emissions Target								
10% Emissions Target								
20% Emissions Target								
30% Emissions Target								
40% Emissions Target								
60% Emissions Target								
70% Emissions Target								
80% Emissions Target								
Transportation (CAFE)								
Electricity (RPS)								
Electricity (CES)								
Complementary Policies (RPS & CAFE)								

Note: Emissions targets represent percentage reduction in 2050 relative to 2005 levels.

LEGEND

Baseline Scenarios
50% Emissions Target Scenarios
Alternate Emissions Targets
EMF scenarios not in this paper

This chapter focuses on two sets of scenarios. First, the 50 percent emissions target scenarios across all technology assumptions plus the 50 percent emissions target scenarios combined with regulatory policies. These scenarios present a common emissions target, linearly reducing GHGs from baseline 2013 levels to 50 percent below 2005 levels by 2050, across a wide range of technology assumptions and complementary policies, and allow an examination of the range of carbon prices and technology futures compatible with this target. Second, the chapter explores the range of carbon prices and energy systems that result from some of the alternative emissions targets.

Table 4.2 lists the modeling teams involved in EMF 24. Not all of the participating models submitted the full set of runs, and some models are limited to the electricity sector. Here we summarize results from eight of the models (ADAGE, CIMS, EC-IAM, FARM, GCAM, NewERA, US-REGEN, and USREP), without explicitly identifying results by model.[4]

Note that the scenario design and model baselines for EMF 24 were locked down in early 2012, so the baselines do not reflect policies that were later adopted (e.g., the light-duty vehicle and corporate average fuel economy standards that were published in October 2012). Additionally, developments in energy markets such as the shale gas boom have altered baseline emissions projections since the EMF 24 scenarios were developed (e.g., the Energy Information Administration's Annual Energy Outlook (AEO) for 2013 projects 2020 CO_2 emissions to be 6 percent lower than the then current AEO 2011 projections).

TABLE 4.2 Modeling teams

Model	*Full name*	*Institution*
ADAGE	Analysis of the Global Economy Model	Research Triangle Institute
CIMS	Canadian Integrated Modeling System	Simon Fraser University
EC-IAM	Environment Canada Integrated Assessment Model	Environment Canada
FARM	Future Agricultural Resources Model	U.S. Department of Agriculture
GCAM	Global Change Assessment Model	Pacific Northwest National Laboratory/Joint Global Change Research Institute
NEMS	National Energy Modeling System	Energy Information Administration
NewERA	NewERA	NERA Economic Consulting
ReEDS	Regional Energy Deployment System	National Renewable Energy Laboratory
US-REGEN	U.S. Regional Economy, GHG, and Energy Model	Electric Power Research Institute
USREP	U.S. Regional Energy Policy Model	MIT Joint Program on the Science and Policy of Global Change

3. Exploration of the 50 percent reduction target

The 50 percent emissions target scenarios in EMF 24 impose a GHG emissions cap from 2013 through 2050 that linearly reduces emissions from reference levels in 2013 to 50 percent below 2005 levels in 2050. Banking is allowed, but borrowing is not. And scenarios do not allow for international or domestic emissions offsets. The results indicate the emissions, or emissions reductions, associated with different emissions price trajectories (assuming all reductions are made in covered sectors).

With the exception of CO_2 emissions from land use and land use change, the cap covers all 'Kyoto' gases in all sectors of the economy that the particular model represents.[5] For comparison, the American Clean Energy Security Act of 2010 covered a smaller percentage of emissions and allows extensive offsets, so even though that bill required an 83 percent reduction in covered emissions below 2005 levels by 2050, actual U.S. emissions under the bill are more similar to the 50 percent emissions target scenario considered here.[6]

3.1 Emissions

Figure 4.1 shows covered CO_2 emissions with baseline scenarios on the left and 50 percent emissions target scenarios on the right. The left-hand panel displays 58 separate model runs, 8 baseline scenarios, and 8 participating models (three of the scenarios were not run by two of the models). The right-hand panel displays 74 separate model runs, 10 policy models and 8 participating models (again with three scenarios not run by two of the models). The dark lines represent mean, and plus and minus one standard deviation, covered CO_2 emissions levels in each year.

Taking the left panel first, in 2020 the baseline covered CO_2 emissions range from approximately 5,400 to 6,800 $MtCO_2$, with a mean value of 6,006 $MtCO_2$. This mean is 2 percent below 2005 U.S. CO_2 emissions levels.[7] The range in 2050 is from approximately 5,200 to 8,600 $MtCO_2$, with a mean value of 7,085 $MtCO_2$, 16 percent above 2005 levels.

Generally the models show that emissions are growing more slowly than the economy – although population and per capita income are growing over this period, the resulting upward pressure on emissions is partly offset by declining energy intensity of GDP (due to improvements in energy efficiency) and declining carbon intensity of energy (due in part to a progressive shift away from carbon-intensive coal to other fuels). The spread in baseline emissions largely reflects different scenarios for the latter two factors (all of the models use similar population and GDP growth rates). In the context of a carbon tax, this uncertainty in baseline emissions leads to uncertainty about the level of carbon tax required to achieve a particular target, or alternatively, uncertainty about the emissions levels that would be achieved by a specific carbon tax.

Turning to the right panel, across all of the 50 percent emissions target scenarios, covered CO_2 emissions range from approximately 4,100 to 5,600 $MtCO_2$ in 2020, with a mean value of 5,083 $MtCO_2$, 17 percent below 2005 U.S. CO_2 emissions

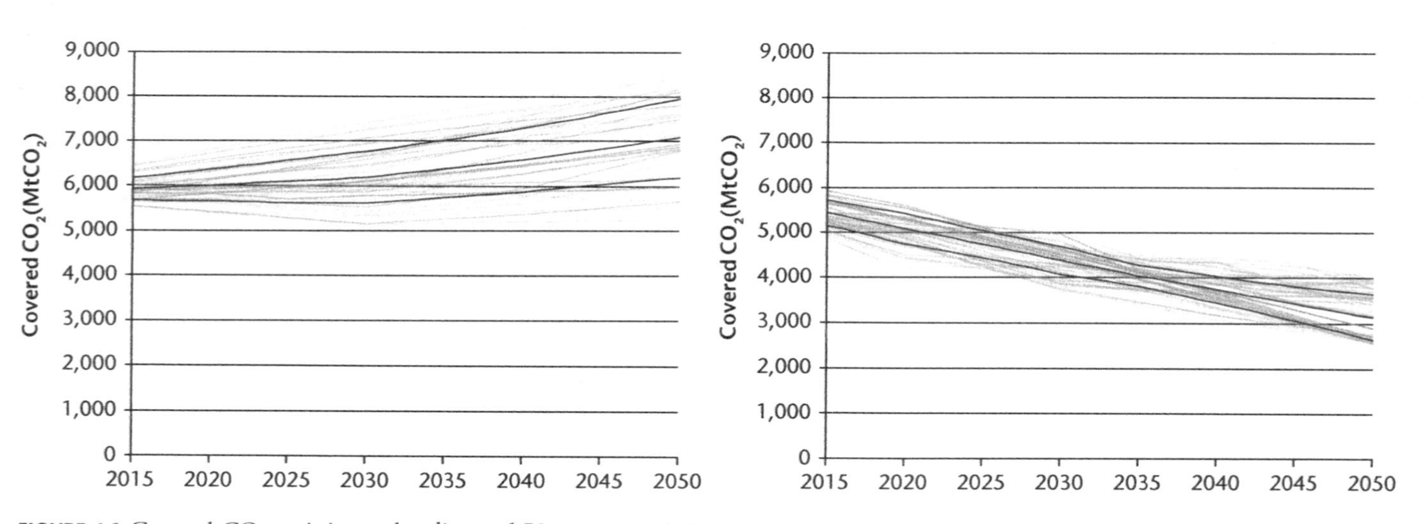

FIGURE 4.1 Covered CO_2 emissions – baseline and 50 percent emissions target scenarios

levels. The 2050 range is from approximately 2,500 to 4,000 $MtCO_2$, with a mean value that is 49 percent below 2005 U.S. CO_2 levels at 3,138 $MtCO_2$.

There are two reasons for the spread in emissions across model runs in these policy scenarios – even though the same emissions targets are applied across all model runs. First, the policy allows for permit banking (though not borrowing), allowing covered emissions in a particular year to deviate from the annual emissions target (so long as cumulative emissions are equal to the cumulative emissions allowed under the cap). Second, the policy is specified as a cap on all Kyoto gases represented by a particular model; however, not all participating models incorporate non-CO_2 emissions. Because abatement opportunities differ between CO_2 and non-CO_2 emissions, in models capturing both emissions sources, proportionate reductions in CO_2 emissions may differ from proportionate reductions in GHGs required under the cap.

3.2 Carbon prices

For the 74 model runs covering all of the 50 percent emissions target scenarios we now examine the resulting carbon prices to see what carbon tax levels could be consistent with this emissions target.

Figure 4.2 displays the carbon price paths for all of the 50 percent emissions target runs across all models. The dark lines represent mean, and plus and minus one standard deviation.

The average carbon price in 2020 is \$35/tCO_2 with standard deviation of \$30/tCO_2 (prices in year 2005). And for 2050, the average price is \$163/tCO_2 with standard deviation of \$96/tCO_2.

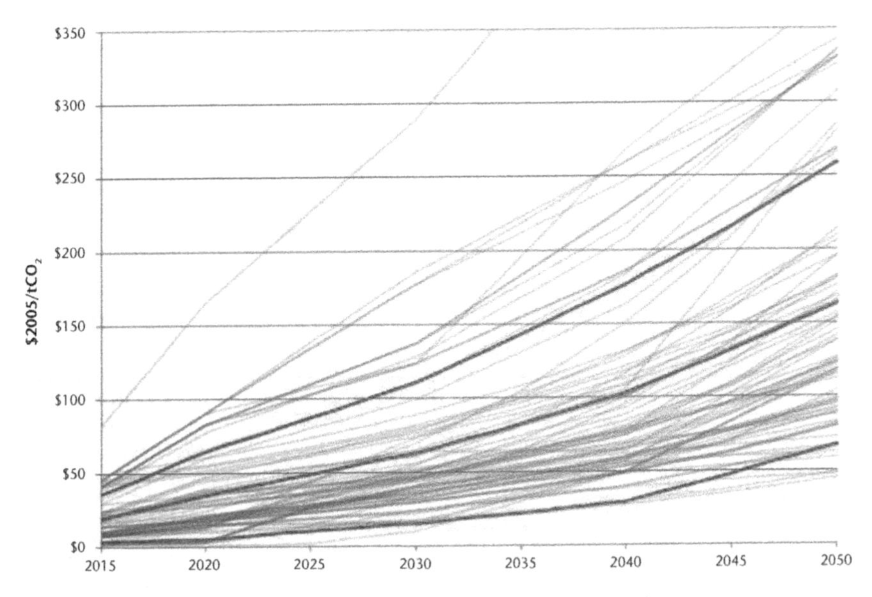

FIGURE 4.2 Carbon prices – all 50 percent emissions target scenarios

This figure therefore underscores the large uncertainties involved in choosing a carbon tax to meet specific emissions goals. Correspondingly, once a carbon tax is specified, there will be large uncertainties in the resulting emissions levels. One policy implication is that, just as many mechanisms have been considered for addressing price uncertainty under a quantity-based policy, it may be important to consider mechanisms for dealing with quantity uncertainty under a price-based policy if meeting specific emissions targets is of concern. One possibility would be to combine a carbon tax with a 'carbon budget' specifying the cumulative carbon emissions allowed to date along a path to meet an emissions goal. If at any point actual emissions exceed the carbon budget to that date, then a clock would start and if cumulative emissions in 5 or 10 years are not below the carbon budget allowed to that date, then a mechanism would kick in to adjust the tax or tax growth rate upward. This could provide a degree of certainty about meeting the emissions target, while avoiding any unanticipated price shocks.

Table 4.3 lists the mean and standard deviation across models for the carbon price in 2020 and 2050 across the eight models for six specific scenarios with different assumptions about technologies and other policies (see Table 4.A.1 in the Appendix for detailed descriptions of the technology assumptions in each scenario).

As expected, generally lower carbon prices are required to reach the emissions targets under the optimistic technology assumption, and higher carbon prices are required under the pessimistic technology assumption. The standard deviations

TABLE 4.3 Carbon prices – mean and standard deviation by scenario

Policy assumption	*Technology assumption*	*2020*		*2050*	
		Mean ($2005/t$CO_2$)	*Std. dev. ($2005/t$CO_2$)*	*Mean ($2005/t$CO_2$)*	*Std. dev. ($2005/t$CO_2$)*
All	All	$35	$30	$163	$96
50% Target	Optimistic	$30	$26	$133	$66
50% Target	Pessimistic	$60	$48	$283	$114
50% Target	CCS/Nuke	$33	$25	$145	$85
50% Target	Renewable	$36	$27	$173	$77
50% Target + Comp.	CCS/Nuke	$22	$26	$113	$92
50% Target + Comp.	Renewable	$25	$25	$143	$90

Note: The top row gives the mean and standard deviation of the carbon price across all scenarios that model the 50 percent emissions reductions target, including all scenarios that vary technology assumptions, and scenarios that include complementary policies along with the 50 percent target. Breakouts of specific policy and technology assumptions are given for all combinations that were completed by the full set of models included in this chapter. The 'CCS/Nuke' technology assumption represents the scenarios with optimistic assumptions for CCS, nuclear, and end-use technology, and pessimistic assumption for other technologies; the 'Renewable' technology assumption represents the scenarios with optimistic assumptions for wind, solar, biomass, and end-use technology, and pessimistic assumptions for other technologies.

show that even within a specific scenario, considerable variation in the carbon price across the models remains. Carbon prices are fairly similar under the two mixed technology assumptions, the optimistic CCS/nuclear assumptions (which also limit renewable technologies) and the optimistic renewable assumptions (which do not allow CCS or new nuclear). This is an important result, the comparison between the optimistic technology and pessimistic technology cases clearly demonstrate the important role technology plays in enabling cost-effective emissions reductions, and the mixed technology assumptions show that there is no single silver bullet technology or technology portfolio, but instead there are multiple different technology pathways that can lead to cost-effective emissions abatement.

Under both the optimistic CCS/nuclear and the optimistic renewable technology assumptions in Table 4.3, the carbon prices fall dramatically when the emissions target policy is combined with complementary policies including a CAFE standard, RPS, and a CCS requirement for new coal-fired power plants. Under the optimistic CCS/nuclear technology assumptions the addition of the regulatory policies lowers the mean carbon price by 33 percent in 2020 and 22 percent in 2050. Under the optimistic renewable assumptions, the reductions are 31 percent and 17 percent in 2020 and 2050, respectively.

Although the carbon prices fall with the addition of the complementary policies, the overall economic cost of achieving these emissions reductions increases.[8] In the presence of an emissions target policy, the complementary policies have little impact on emissions levels, but instead simply favor a particular set of technologies. From the perspective of a carbon tax policy, the presence of complementary policies can be thought of in two ways: they can lower the carbon tax necessary to achieve a desired emissions target, or they can increase the emissions reductions achieved under a particular carbon tax.

3.3 Primary energy

Next we turn to examining how the U.S. energy system is transformed in all of the 50 percent emissions target scenarios in order to demonstrate the diverse ways in which the emissions target can be achieved. Figure 4.3 compares the

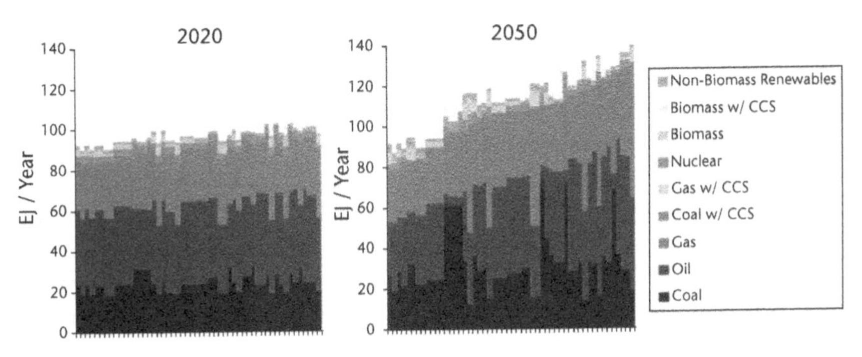

FIGURE 4.3 Primary energy – all baseline scenarios – total fossil in 2050 sort

primary energy in 2020 and 2050 for the 58 baseline model runs. In this figure the model runs are sorted by quantity of fossil primary energy without carbon capture and storage (CCS) in 2050. The variation across model runs is due both the technology assumptions in each scenario and general differences between the eight models and their core driving assumptions such as the evolution of energy intensity. Taken together these runs represent a range of plausible business-as-usual futures for the U.S. energy system. In 2020, fossil fuels represent between 84 and 99 exajoules (EJ) of primary energy, and by 2050 the range expands to 78 to 130 EJ. The biggest differences are seen in the projections for the coal primary energy, which ranges from 19 to 33 EJ in 2020 and from 13 to 74 EJ in 2050. The amount of oil primary energy has a similar spread of results, between 30 and 46 EJ in 2020 and between 5 and 57 EJ in 2050.[9] Natural gas primary energy varies between 23 and 33 EJ in 2020 and between 25 and 66 EJ in 2050.

For the non-fossil fuels, the technology assumptions in different baseline scenarios explicitly drive some of the variation in primary energy. For example, the pessimistic nuclear scenario requires the phase-out of nuclear power with no new construction and no lifetime extensions beyond 60 years. This assumption drives the lower end of the range of nuclear energy primary energy, which falls between 2 and 4 EJ in 2020, and between 0 and 12 EJ in 2050.[10] Finally, the primary energy from all renewables ranges from 1 to 7 EJ in 2020, and from 1 to 13 EJ in 2050.

Turning to primary energy in the policy scenarios, Figure 4.4 shows primary energy across all 74 separate model runs for 50 percent emissions target scenarios. The expansion and contraction of different primary energy carriers are indicative of how the required emissions reductions are achieved. The first difference to notice is that the top white light gray bars represent the energy efficiency and demand reduction, which is between 3 and 21 EJ in 2020 and between 7 and 70 EJ in 2050. Primary energy from fossil without CCS falls across all runs to between 71 and 85 EJ in 2020 and between 45 and 75 EJ in 2050. The most dramatic and consistent shift across all runs is the reduction in primary energy from coal without CCS, which falls to between 2 and 25 EJ in 2020 and 0 to 15 EJ in 2050. Primary energy from oil and natural gas fall less dramatically, with 29 to 43 EJ of oil primary energy in 2020 and 7 to 50 EJ in 2050, and 21 to 32 EJ of natural gas primary energy in 2020 and 15 to 47 EJ in 2050.

In the policy scenarios we break out primary energy from sources with and without carbon capture and storage. CCS technology is not allowed by assumption in some scenarios, so the lower bound of primary energy with CCS is always zero. In cases where it is allowed, the maximum penetration is 5 EJ of primary energy with CCS in 2020, and 46 EJ in 2050. Nuclear energy provides between 1 and 4 EJ in 2020, and between 0 and 17 EJ in 2050, with the lower end being driven by scenarios that assume nuclear phase-out. Biomass renewables provide between 0 and 7 EJ of primary energy in 2020, and between 0 and 35 EJ in 2050. Finally,

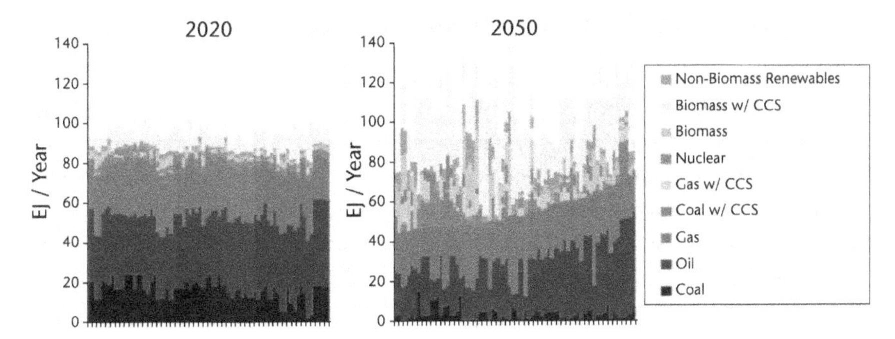

FIGURE 4.4 Primary energy – all 50 percent emissions target scenarios – total fossil in 2050 sort

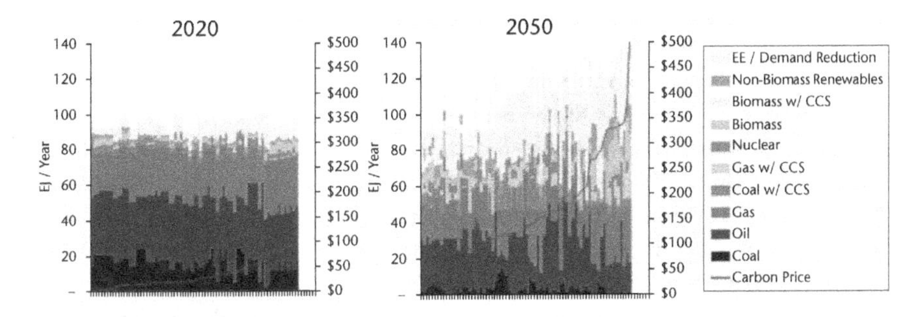

FIGURE 4.5 Primary energy – all 50 percent emissions target scenarios – carbon price sort

non-biomass renewables account for between 1 and 5 EJ of primary energy in 2020, and between 1 and 19 EJ of primary energy in 2050.

In order to tie the primary energy data back to the carbon price trajectories from Figure 4.2, Figure 4.5 displays primary energy across all 50 percent emissions target scenarios sorted by the carbon price in the corresponding year. One pattern that emerges is that many of the runs with lower carbon prices tend to have lower reference case primary energy in 2050, and the reverse is true for the higher carbon price runs, as can be seen by the total height of the bars including the energy efficiency and demand reduction category. Total reference case primary energy varies more across models than across scenarios within any particular model. Looking at the wide range of carbon prices and technology futures in the EMF 24 database that are consistent with the 50 percent emissions target emissions goals, we see that even with a broad set of sensitivities on technology assumptions and complementary policies, the range of possible outcomes projected by any single model is far narrower than the range presented in a multi-model comparison.

4. Exploration of alternative emissions targets

In this section we move away from the 50 percent emissions target scenarios examined in the previous section, and look at the carbon prices and energy systems that are consistent with a range of alternative emissions goals. In addition to the 50 percent emissions target scenarios that cover all technology assumptions in EMF 24, the scenario matrix includes scenarios with policies that linearly reduce GHG emissions to 0, 10, 20, 30, 40, 60, 70, and 80 percent below 2005 levels by 2050. Aside from the emissions target levels all other aspects of these policies are identical to the 50 percent emissions target scenario (see the Appendix Table 4.A.2 for full descriptions of the policy assumptions). Unlike the 50 percent emissions target scenarios, these alternative targets are only analyzed under two sets of technology assumptions: first, a CCS and nuclear world, with optimistic assumptions on CCS, nuclear, and energy efficiency, and pessimistic assumptions for biomass, wind and solar; second, a renewables world, with optimistic assumptions for energy efficiency, biomass, wind and solar, and pessimistic assumptions for CCS and nuclear. In this section we examine two of these alternative emissions target scenarios, the 0 percent emissions target scenario and the 80 percent emissions target scenario.

Figure 4.6 depicts results for the 0 percent emissions target scenarios, which hold emissions constant at 2005 levels through 2050. Seven models submitted results for these scenarios. For three of the models, this emissions target is non-binding, so the carbon price is zero. Across all of the runs, the mean carbon price is \$4/tCO_2 in 2020 and \$15/tCO_2 in 2050 (in year 2005).

The U.S. energy system remains unchanged in the three models where this policy is non-binding, while for the remaining models we find a variety of responses. For primary energy from coal without CCS the largest change from the reference case is a 6 EJ reduction in primary energy in 2020 and a 15 EJ reduction in 2050, and across all the models the mean reduction is 2 EJ in 2020 and 5 EJ in 2050. The largest reduction in oil without CCS primary energy is 1 EJ in 2020 and 4 EJ in 2050, and the mean change is 0 EJ in 2020 and a 1 EJ reduction in 2050. Natural gas without CCS primary energy shows a mean reduction of 0 EJ in 2020 and 1 EJ in 2050, and the largest reduction is 1 EJ in 2020 and 3 EJ in 2050. Only one model projects CCS to penetrate, finding 1 EJ of CCS in 2020 and 21 EJ of CCS in 2050.

The non-fossil sources of primary energy show much smaller changes in the 0 percent emissions target scenario. In 2020 and in 2050 the mean change is zero for nuclear, biomass, and non-biomass renewables. The largest changes are an addition of 1 EJ of nuclear and 1 EJ of biomass primary energy in both 2020 and 2050. Finally there is a mean 2 EJ of energy efficiency and demand reduction in 2020 and 5 EJ in 2050, with a maximum of 7 EJ in 2020 and 20 EJ in 2050.

The 80 percent emissions target scenarios, which reduce emissions linearly to 80 percent below 2005 levels by 2050, are shown in Figure 4.7. Seven models submitted results for these scenarios. The mean carbon price across all of the 80 percent emissions target runs is \$65/tCO_2 in 2020 and \$439/tCO_2 in 2050, with a standard deviation of \$40/tCO_2 in 2020 and \$228/tCO_2 in 2050. Note that the mean is

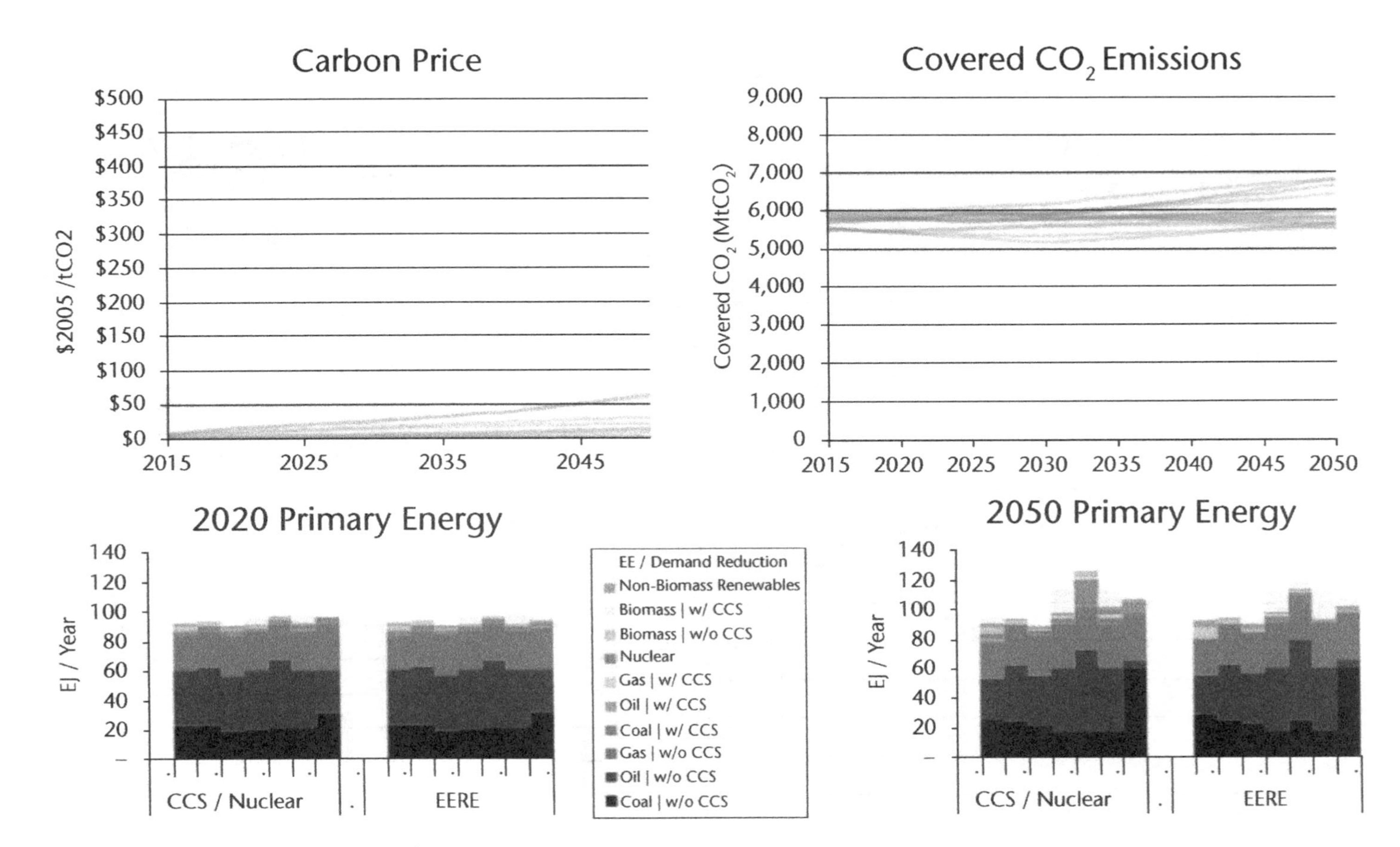

FIGURE 4.6 Emissions, carbon price, and primary energy – 0 percent emissions target scenarios

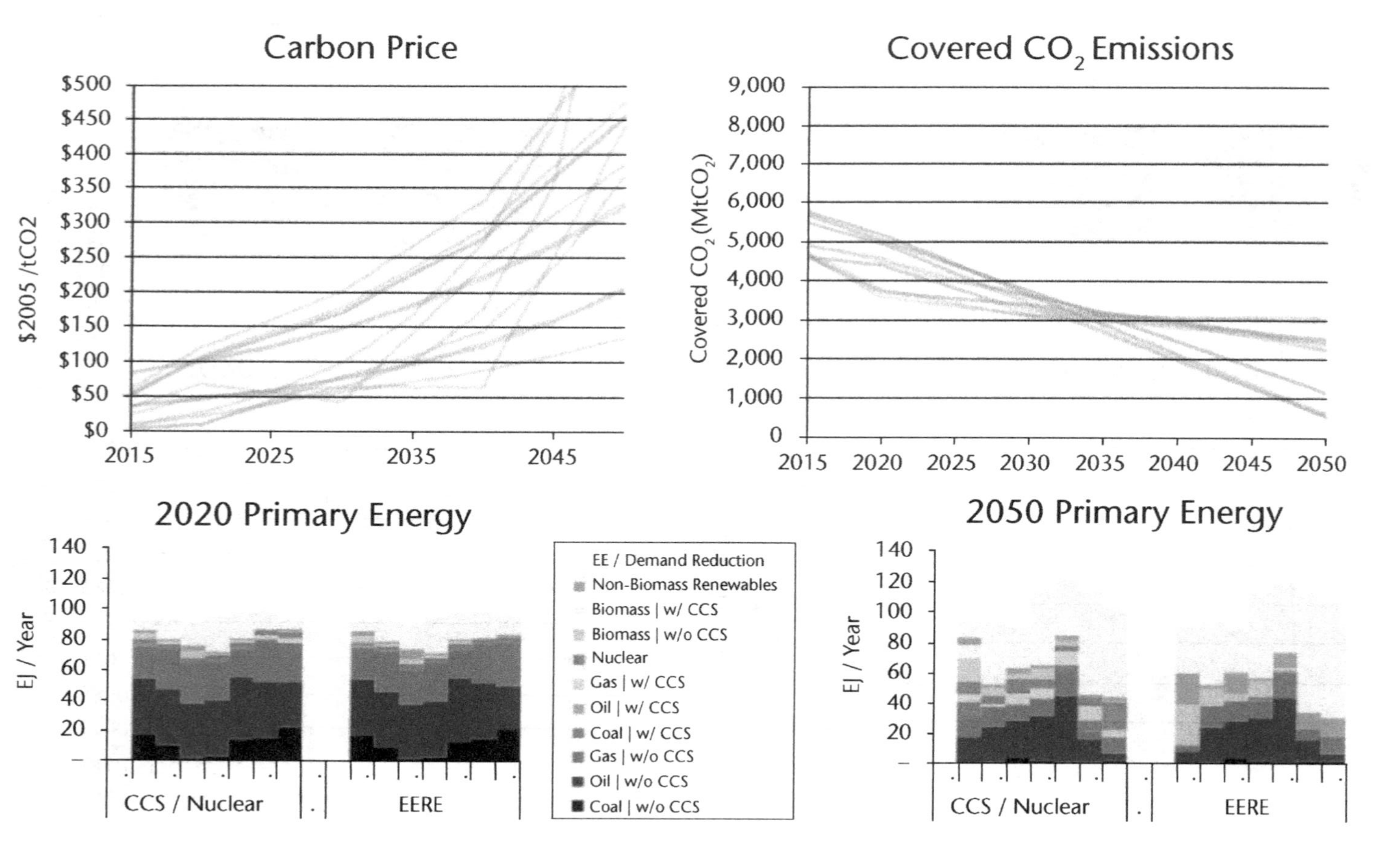

FIGURE 4.7 Emissions, carbon price, and primary energy – 80 percent emissions target scenarios

driven up by one model that exhibits a large price spike in 2050 to near \$900/tCO_2, the remaining models find prices between \$135/tCO_2 and \$593/tCO_2. In 2045 the carbon prices are more closely grouped with a mean of \$296/tCO_2, and a \$108/tCO_2 standard deviation.

The U.S. energy system is completely transformed in the 80 percent emissions target scenarios. In 2020 there is a mean of 12 EJ of primary energy from coal without CCS remaining, with a minimum of 2 EJ and a maximum of 22 EJ, and by 2050 the mean is only 2 EJ, with a minimum of 0 EJ and a maximum of 4 EJ. Oil sees less of a change than coal in this scenario, with between 29 and 41 EJ remaining in 2020, with a mean of 36 EJ, and between 5 and 44 EJ in 2050, with a mean of 22 EJ. Note that the low end of oil primary energy in 2050 is from a model that also sees very little oil remaining in the baseline. In 2020 primary energy from natural gas without CCS falls between 19 and 29 EJ, with a mean of 25 EJ, and in 2050 the range is from 4 to 16 EJ, with a mean of 11 EJ. The pattern of large reductions in primary energy from coal and smaller reductions in oil and gas primary energy occurs because the power sector is able to largely decarbonize by shifting to low or zero carbon technologies, whereas the transportation sector has fewer low-cost options in these models for switching away from oil.

In the 80 percent emissions target scenarios where it is available, CCS plays a large role. Three models project CCS penetrating in 2020 with between 1 and 5 EJ of primary energy from fossil with CCS. In 2050 all models have primary energy from fossil with CCS, with a minimum of 2 EJ and a maximum of 16 EJ. One model also includes an additional 8 EJ of biomass with CCS, for a total of 24 EJ of primary energy with CCS.

Primary energy from non-fossil sources grows in the 80 percent emissions target scenarios. Nuclear primary energy only increases in the scenarios where it is not assumed to be phased out, with little change in 2020, but an increase of between 2 and 9 EJ in 2050. Biomass without CCS increases between 0 and 3 EJ in 2020 and between 0 and 12 EJ in 2050. Non-biomass renewables increase between 0 and 5 EJ in 2020 and between 0 and 15 EJ in 2050. Finally, the biggest impact on primary energy in the 80 percent emissions target scenario is the reduction in the overall level of primary energy due to energy efficiency and demand reduction, which accounts for between 7 and 24 EJ in 2020 with a mean of 15 EJ, and between 9 and 78 EJ in 2050 with a mean of 46 EJ.

5. Conclusion

An important distinction between carbon taxes compared to pure cap-and-trade policies is that carbon taxes provide a degree of cost certainty at the expense of certainty about emissions outcomes, whereas cap-and-trade policies provide the reverse, emissions certainty and cost uncertainty. If a carbon tax is implemented with a goal of achieving a specific emissions goal (e.g., an international agreement that specifies required emissions reductions, or simply a domestically stated emissions goal), the runs contained in the EMF 24 database show that a very wide range of carbon taxes is potentially consistent with any given emissions goal. Factors influencing this range include assumptions about the business-as-usual emissions trajectory, assumptions about how the cost and performance of technologies evolve, and general differences across models. If the emissions goal must be met, and the carbon tax produces less abatement than expected, then the policy must be adjusted to meet the goal. Cap-and-trade policies have generally included features designed to address price uncertainty (e.g., safety valve, price collar, offsets), and carbon tax policies may need to consider equivalent mechanisms to address uncertainty in cumulative emissions reductions, while maintaining predictability in prices.

The EMF 24 database also sheds some light on the interaction between a policy that prices carbon and other complementary policies. The models find that policies such as a CAFE standard or RPS, when combined with a quantity-based emissions target, do not change the amount of emissions reductions, but instead change the way in which those reductions are achieved, which generally lowers allowance prices, but increases overall costs. When these complementary policies are combined with a carbon tax, there are two ways to consider the interaction. The complementary policies generate emissions reductions, and thus reduce the carbon tax that would be needed to reach any specific emissions goal, but may raise the overall cost of reaching that goal. Alternatively, the complementary policies increase the total amount of abatement achieved under any particular carbon tax.

The variety of technology assumptions and range of models in EMF 24 demonstrate the diversity of futures for the U.S. energy system that can be consistent with emissions reduction goals. The differences in energy futures are at least partially accounted for by how the energy system is assumed to evolve in a business-as-usual scenario, the specifics of the policy assumptions, the evolution of the cost and performance of different technologies, and a host of other factors that characterize the broad differences between models.

Notes

* The views and opinions of this author herein do not necessarily state or reflect those of the United States Government or the Environmental Protection Agency.

1. For some recent discussions in the literature see, for example, Ramseur et al. (2012), McKibbin et al. (2012), Parry and Williams (2011), and Rausch and Reilly (2012).
2. This chapter uses the penultimate data submissions to the EMF 24 exercise. The final submissions may be subject to some modest revisions as we finalize the study that are not expected to impact the core results presented in this study. The final EMF 24 data is now publicly available, following publication of various papers in a special issue of the *Energy Journal* (Weyant et al. 2014).
3. For example, pessimistic CCS assumptions allow no implementation of the technology and pessimistic nuclear assumptions allow no new construction of nuclear power plants. Conversely, optimistic assumptions for nuclear and CCS specify that the technologies are available, but the cost and performance characteristics are left to the modeler's choice.
4. The complete database of EMF 24 is publicly available, following publication of the EMF 24 special issue of the *Energy Journal* (Weyant et al. 2014).
5. This refers to the six main gases included in the Kyoto Protocol: carbon dioxide (CO_2), methane (CH_4), nitrous oxide (N_2O), hydrofluorocarbons (HFCs), perfluorocarbons (PFCs), and sulphur hexafluoride (SF_6).

6. Fawcett et al. (2009) discusses this further in the context of the similarly designed EMF 22 scenarios.
7. As noted above, the baseline scenarios for EMF 24 were locked down in early 2012. More recent projections would tend to forecast lower baseline emissions. This is in part due to changes in energy markets such as the natural gas boom, but also in part due to policies that have been adopted such as the light-duty vehicle corporate average fuel economy standard, which in this study is roughly represented in a policy scenario.
8. The policy overview paper for EMF 24 (Fawcett et al. 2014) includes an extensive discussion of the costs of alternative scenarios, and alternative possibilities for measuring economic cost.
9. Note that the extreme high amount of coal primary energy and corresponding extreme low amount of oil primary energy are results from a single model that tends to project a move in the transportation sector from petroleum fuels to coal-to-liquids. The model differences that drive these results are further explored in Clarke et al. (2014).
10. Primary energy values for non-fossil technologies are calculated using the direct equivalent method.

References

Clarke, L. C., Fawcett, A. A., McFarland, J., and Weyant, J. P. (2014). Overview of the EMF 24 Technology Scenarios. *Energy Journal* 35.

Fawcett, A. A., Calvin, K. V., de la Chesnaye, F. C., Reilly, J. M., and Weyant, J. P. (2009). Overview of EMF 22 U.S. Transition Scenarios. *Energy Economics* 31:S198–S211.

Fawcett, A. A., Clarke, L. C. and Weyant, J. P. (eds.) (Forthcoming). Energy Modeling Forum 24. Energy Journal Special Issue.

Fawcett, A. A., Clarke, L. C., Rausch, S., and Weyant, J. P. (2014). Overview of the EMF 24 Policy Scenarios. *Energy Journal* 35.

McKibbin, W., Morris, A., and Wilcoxen, P. (2012). The Potential Role of a Carbon Tax in U.S. Fiscal Reform. *Brookings Climate and Energy Economics Discussion Paper.*

Parry, I., and Williams, R. (2011). Moving US Climate Policy Forward: Are Carbon Taxes the Only Good Alternative? *Resources for the Future Discussion Paper* 11–02.

Pizer, W. (2005). Climate Policy Design under Uncertainty. *Resources for the Future Discussion Paper* 05–44.

Ramseur, J., Legget, J., and Sherlock, M. (2012). Carbon Tax: Deficit Reduction and Other Considerations. *Congressional Research Service Report* R42731.

Rausch, S., and Reilly, J. (2012). Carbon Tax Revenue and the Budget Deficit: A Win-Win-Win Solution? *MIT Joint Program on the Science and Policy of Global Change Report* No. 228.

Appendix

TABLE 4.A.1 EMF 24 technology assumptions

Policy	*Description*
50% Emissions Target	This represents the assumption of a national policy that allows for cumulative greenhouse gas emissions from 2012 through 2050 associated with a linear reduction from 2012 levels to 50 percent below 2005 levels in 2050. The cumulative emissions are based on the period starting from, and including, 2013 and through 2050. With the exception

(Continued)

TABLE 4.A.1 (Continued)

Policy	*Description*
	of CO_2 emissions from land use and land use change, the cap covers all Kyoto gases in all sectors of the economy that the particular model represents. This includes non-CO_2 land use and land use change emissions and emissions of GHGs not covered under many U.S. climate bills. To be explicit, CO_2 emissions from land use and land use change are not included in the cap. For models that do not operate on annual time steps, the first year with a positive price on carbon should be the first model time step after 2012 (e.g., 2015 in a model with 5-year time steps), but the cumulative emissions are still to be based on an assessment of the emissions associated with a linear path starting from, and including, 2013 and through 2050. Banking of allowances is allowed, but borrowing of allowances is not permitted. Note that the emissions target scenarios with alternative targets use identical assumptions to allow for cumulative greenhouse gas emissions associated with a linear reduction from 2012 levels to *X* percent below 2005 levels in 2050, where *X* is the percentage reduction target associated with the scenario.
Renewable Portfolio Standard (RPS)	The RPS applies only to the electricity sector. In this case, renewable energy includes all hydroelectric power and bioenergy. The RPS is defined as 20 percent by 2020, 30 percent by 2030, 40 percent by 2040, and 50 percent by 2050. Banking and borrowing are not allowed. If modelers are unable to meet these requirements within their model, they should create a scenario that includes a less aggressive RPS, but one that can be met by the model.
New Coal	This policy requires that all new coal power plants capture and store 90 percent or more of their CO_2 emissions.
Transportation Regulatory Policy	The transportation policy is a CAFE standard for light-duty vehicles (LDV) that specifies a linear increase in fuel economy of new vehicles, starting in 2012, to 3 times 2005 levels in 2050. If modelers do not have the ability to represent a CAFE policy, they can alternatively represent the policy as a cap that covers all LDV in the transportation sector, as defined in the particular model. This alternative policy is defined as a linear reduction in LDV emissions from 2012 levels to 55 percent below 2010 levels in 2050. Banking and borrowing are not allowed. It is understood that with rebound effects and differences in reference scenario, this LDV emissions cap policy structure will not be identical to the CAFE policy; however, we expect them to be similar (the 55 percent reduction in LDV emissions

(*Continued*)

TABLE 4.A.1 (Continued)

Policy	*Description*
	under the cap is consistent with the emissions reductions achieved in a test run of GCAM), and there are benefits to explicit analysis of CAFE standards. Note that biofuels, electricity, and hydrogen are assumed to be zero-emissions fuels for calculating the emissions cap.
Emissions target + Regulations	Combines all policies listed above.

TABLE 4.A.2 EMF 24 policy assumptions

	Pessimistic	*Optimistic*
End Use	The pessimistic technology scenario should represent evolutionary assumptions about the availability, cost, and performance of technologies that would reduce energy consumption at the end use or enhance opportunities for fuel switching. The precise assumptions are left to modeler's choice.	The optimistic technology scenario should represent plausibly optimistic assumptions about the availability, cost, and performance of technologies that would reduce energy consumption at the end use or enhance opportunities for fuel switching. For consistency between scenarios, modelers are encouraged to develop assumptions that would lead to roughly a 20 percent decrease in final energy in 2050 relative to the low-tech, no-policy case. The precise assumptions are left to modeler's choice.
CCS	No implementation of carbon capture and storage technology.	CCS is available. The cost and performance characteristics are left to modeler's choice.
Nuclear	Phase out of nuclear energy after 2010. Phase out is defined as no construction of new nuclear power plants beyond those already under construction or planned (excluding proposed plants). This reflects the concept of the 'off' case being triggered by public skepticism about nuclear technology. In addition, modelers are encouraged to assume no lifetime extensions beyond 60 years as representing an environment that generally discourages the development and deployment of nuclear energy.	New nuclear energy is fully available. The cost and performance characteristics are left to modeler's choice.
Wind & Solar	Pessimistic techno-economic assumptions for solar and wind energy are left to modeler's choice. However, modelers should choose their assumptions carefully, with the goal of developing a scenario that represents evolutionary technology development.	Optimistic techno-economic assumptions for solar and wind energy technologies are left to modeler's choice. However, modelers should choose their assumptions carefully, with the goal of developing a scenario that represents plausibly optimistic technology development.
Bioenergy	The pessimistic scenario should represent a scenario of bioenergy supply on the lower end of what is deemed sustainable. The precise assumptions are left to modeler's choice. However, modelers should choose their assumptions carefully, with the goal of developing a scenario that represents evolutionary technology development.	The optimistic scenario should represent a scenario of bioenergy supply on the higher end of what is deemed sustainable. The precise assumptions are left to modeler's choice. However, modelers should choose their assumptions carefully, with the goal of developing a scenario that represents plausibly optimistic technology development.

5

MACROECONOMIC EFFECTS OF CARBON TAXES

*Roberton C. Williams III and Casey J. Wichman**

KEY MESSAGES FOR POLICYMAKERS

- A carbon tax (taken by itself) will likely impose a small but significant long-term drag on the economy.
- Using the carbon tax revenue in ways that promote long-run economic growth will offset most of that negative effect (or perhaps even lead to a net economic gain).
- Potential pro-growth uses of carbon tax revenue include cuts in the rates of other taxes, reductions in the budget deficit, and forms of government spending that can boost long-term economic growth (e.g., research, education, and infrastructure).
- The reduction in greenhouse gas emissions caused by a carbon tax can also promote long-term growth, by limiting economic damage from climate change. But this effect depends on global emissions, not merely on US emissions, so other nations' policy responses to a US carbon tax are important.
- Short-run effects of carbon taxes on unemployment and the business cycle are less well understood. But in general, reducing the budget deficit during an economic downturn tends to worsen unemployment and exacerbate the downturn. This suggests that during a downturn, it may be advantageous to delay implementation of a carbon tax or to pair it with short-term cuts in other taxes or increases in government spending.

1. Introduction

A critical concern about a carbon tax (particularly given the sluggish recovery from the Great Recession) is how the resulting increases in energy prices will affect the overall level of economic activity and its rate of growth over time – and to what extent potentially harmful effects can be ameliorated through use of carbon tax revenues.

This chapter discusses the evidence on these issues, looking first at effects on economic growth and employment over the longer term and then turning to shorter-term impacts. The focus is on aggregate effects on the whole economy: effects on individual industries or regions, while important, are beyond the scope of this chapter (for some discussion see Chapter 9).

2. Longer-term effects on growth and employment

In this section, we look at the long-term effects – with "long term" defined based on an economic (not climate) perspective. More specifically, we focus on effects that show up over a sufficiently long timeframe that business-cycle fluctuations – economic expansions and recessions – become relatively unimportant. In more normal economic times, the longer term would be any time more than a few years into the future, though under current conditions – with an unusually persistent global economic slump – the longer term may be further off, but it seems safe to say that any time more than a decade into the future could be considered longer term.

Over this timeframe, the overall economic effects of a carbon tax are driven primarily by its supply-side impacts: policy-induced changes in the longer-run productive capacity of the economy resulting from advances in technology or changes in the supply of inputs into production (capital, labor, raw materials, etc.). This point is not unique to a carbon tax. During an economic downturn, productive capacity may be underused (workers may be unemployed, machines may sit idle, etc.), and thus demand-side effects can be highly important in the short run. But over the longer term, the ups and downs of the business cycle even out, and the level of economic activity is driven primarily by the productive capacity of the economy.

There is a wide variety of metrics used to assess the long-run level of economic activity. Here, we focus on the two most widely used of those metrics: gross domestic product (GDP) and economic welfare. GDP measures the total value of goods and services produced within a country. It is directly measured, widely reported, and relatively easy to understand. However, GDP fails to measure the value of most non-market goods and services (that is, goods and services that are not bought and sold), even though they have substantial value.[1] This means that GDP can sometimes be misleading.[2]

Economic welfare is a much broader measure of how well-off households are, which includes (at least in theory) everything that individuals value – including both the market goods and services measured by GDP and all of the non-market items that GDP omits. This makes it a more complete and more accurate measure of whether a policy truly makes households better or worse off. But it cannot be directly measured (there is no obvious price for non-market goods and services), and it is somewhat more complex, and thus more difficult to explain and to understand. In this chapter, we will consider both of these measures, though we focus more on GDP, primarily because it is more familiar for a policy-oriented audience. But one should bear in mind that GDP can be misleading – and in this context, looking at GDP will tend to slightly overstate the cost of a carbon tax.[3]

In reviewing these longer-term economic effects, we look first at the effects of the carbon tax itself, then at the effects of potential uses of revenue from the carbon tax, and finally at how the environmental effects of the tax – primarily mitigating potential damage from climate change – could affect the economy.

2.A. *Macroeconomic effects of the carbon tax (prior to use of revenues)*

Fossil fuels, and electricity produced from them, are used pervasively throughout the economy, and thus a carbon tax would have widespread effects. Even industries that directly emit little or no carbon dioxide (CO_2), such as auto manufacturing, are still affected, because they use as inputs goods produced by other industries that do emit CO_2 (e.g., steel). Industries that are formally subject to a carbon tax (e.g., fuel suppliers) can be expected to pass along at least part (and perhaps all) of that tax to consumers of their products. This is part of why a carbon tax is an economically efficient way to reduce carbon emissions: not only do the direct users of fossil fuels have an incentive to reduce emissions, but the pass-through of the tax means that industries and consumers who buy goods that are carbon-intensive in production also have an incentive to shift to less carbon-intensive alternatives (e.g., buying more energy-efficient appliances in order to use less electricity).

However, this also means that a carbon tax implicitly acts as a tax on factors of production (primarily labor and capital). Some portion of the tax is "passed backward," lowering wages for labor, returns on capital, and the prices of other inputs in production. Another portion is "passed forward," raising the prices of both consumer and capital goods. Either way, the effect is to lower the real return to those factors of production, thus reducing the incentive to work, save, and invest.[4] This leads to somewhat lower levels of GDP, employment, and other measures of economic activity.

For example, Goulder and Hafstead (2013) suggest that imposing a carbon tax with an initial rate of $10/ton[5] and rising at 5 percent/year would cause the level of GDP 20 years later to be roughly 0.6 percent lower than it would have been in the absence of the tax.[6] That may sound bigger than it really is: keep in mind that it is a difference in GDP levels, whereas we are used to hearing about GDP growth rates. A 0.6 percent difference in GDP levels over 20 years translates to less than a 0.03 percent difference in average annual GDP growth rates over that time – an effect small enough that it would be impossible to notice (though still large enough to significantly influence the overall cost of the policy). Put differently, real GDP in

20 years will be roughly 55 percent higher than today without a carbon tax, versus 54 percent higher than today with a carbon tax.[7] Estimates from other studies (see the end of section II.B.1) suggest a broadly similar result: a small (though not insignificant) drag on economic growth.

It is not surprising that the macroeconomic effects are relatively small, because the energy price impacts of the level of carbon tax considered here are also relatively small: smaller than (sometimes much smaller than) changes in annual energy prices from market volatility over the last 10 years. For example, a $20 per ton carbon tax would increase gasoline prices by 18 cents per gallon, but pump prices have varied between $1.60 and $3.60 per gallon; the tax would add about 1.0 cent per kWh to the average price of electricity, but prices have varied between 7.4 and 9.9 cents per kWh; and it would add $0.9 per thousand cubic feet to the price of natural gas, though wellhead prices varied from $2.70 to $8.00 per thousand cubic feet.[8]

Moreover, this economic effect could be largely (if not completely) offset if the revenue from the carbon tax is used in a way that boosts economic activity, as we discuss next.

2.B. Potential pro-growth uses of carbon tax revenue

Figure 5.1 shows estimates from seven different studies of the amount of revenue that a carbon tax would raise. The amount of revenue varies, depending on the initial carbon tax rate and how rapidly it rises, but in every case the revenue is substantial and rising over time. And except for the very lowest rate on the graph, the revenue starts at more than $100 billion/year.[9] There are many potential uses for this revenue, including some that could significantly boost economic growth.

The idea that revenue from an environmental tax could be used to boost economic activity or economic efficiency has been a key focus of the "double dividend" literature in environmental economics. The name comes from the idea that imposing an environmental tax could produce two "dividends": first, a reduction in pollution emissions; and second, a boost in GDP and/or economic efficiency from the use of the environmental tax revenue.

We consider three general categories of uses of carbon tax revenue that could potentially boost economic activity: cuts in other taxes (such as payroll taxes or corporate or personal income taxes), reducing the government budget deficit, or financing valuable public spending.

2.B.1. Cuts in other taxes

We first look at the effects of using carbon tax revenue to finance cuts in marginal tax rates on capital or labor, with the most obvious candidates being the largest taxes: payroll taxes and corporate and personal income taxes. This option has been studied the most, because it is the easiest pro-growth option to model.

The argument for why using carbon tax revenue to pay for cuts in other taxes boosts the economy is straightforward. Cutting corporate income taxes and personal

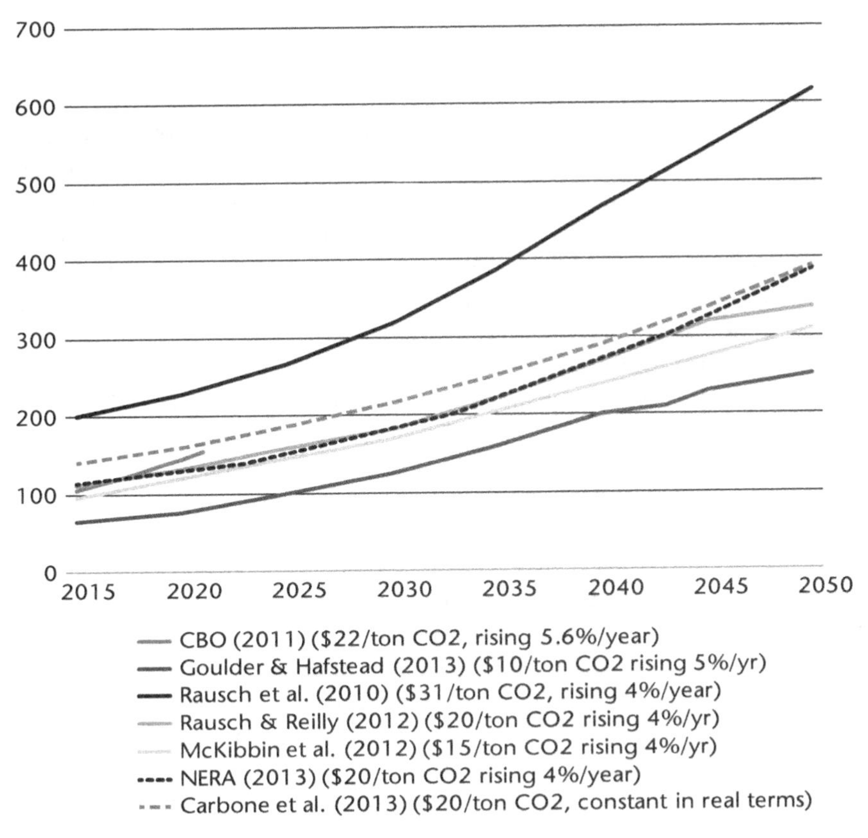

FIGURE 5.1 Projected gross revenue from a carbon tax (billions of $2012)

Sources: See citations in the reference list.

income taxes on capital gains, dividends, and interest increases the incentive to save and invest, thus promoting capital accumulation. Cutting taxes on labor income has a similar effect by boosting incentives to work and to invest in human capital, thus increasing labor supply and labor force productivity. In either case, the result is an increase in the productive capacity of the economy, and thus a long-run boost in the level of economic activity. This also corresponds to a boost in economic efficiency, as those tax cuts reduce the tax distortions in labor and capital markets.

The general finding in recent studies is that the economic boost from cutting other taxes is enough to offset most – but not quite all – of the economic drag from the carbon tax.[10] The reason is that shifting taxes off labor and capital and onto carbon reduces the overall efficiency of the tax system a bit (if one ignores environmental benefits), and thus slightly reduces economic activity. From a pure revenue-raising standpoint, broad-based taxes such as income or payroll taxes are more efficient than taxes with narrower bases: the broader the tax base, the harder it is for individuals and firms to alter their behavior in ways that lower their tax payments, and hence the less the tax distorts behavior. A carbon tax is easier to avoid (by shifting to cleaner fuels, improving energy efficiency, driving less, and so on) and thus is less efficient at raising revenue.[11]

Economic models suggest, therefore, that the net effect of a carbon tax (with revenues used to cut labor/capital taxes) on the overall economy would be slightly negative, though much smaller than the effect of the carbon tax by itself. For example, Carbone et al. (2013) finds that imposing a carbon tax and using the revenue to fund cuts in taxes on labor still leads to a net reduction in GDP, but that the economic boost from the labor tax cuts offsets more than 80 percent of the effect of the carbon tax itself – so the net reduction in GDP is tiny.

Going one step further, there are reasons for believing the tax cut could more than offset the effect of the carbon tax, and thus actually have a slight positive overall effect on GDP. For example, Parry and Williams (2010) take into account the effects of tax preferences – exemptions and deductions (like those for employer medical insurance and owner-occupied housing) that are large and pervasive across the US tax system. These deductions and exclusions narrow the base of income and other taxes, making them less efficient – and thus boosting the economic gain from cutting them. As a result, the net effect of the carbon tax shift can be to increase GDP.[12]

The prospects for a net gain also depend on what other tax is cut. Most studies find that the biggest economic gains come from cutting taxes on capital, particularly from cutting the corporate income tax rate.[13] The less efficient the tax that gets cut, the bigger the economic gain from cutting it. And the corporate income tax is a particularly inefficient tax.

As a result, some models find that imposing a carbon tax and using the revenues to cut taxes on capital can yield a small net gain for the economy. For example, Carbone et al. (2013) finds that the net effect of a $30/ton carbon tax with the revenues used to cut taxes on capital is to increase GDP by about 1 percent in 20 years.[14]

Figure 5.2 displays the estimated effects on the level of GDP resulting from the imposition of a carbon tax, for a total of seven different model runs taken from four different studies. For the most part, these studies show quite similar results, even though they use different models, different modeling assumptions, and somewhat different carbon tax policies. In five cases shown in the graph, the long-term effect on GDP in 2050 is a drop of roughly 0.7 percent relative to a case without a carbon tax (though not accounting for the role of tax preferences which could lower the costs). One outlier is Carbone et al.'s (2013) estimate, just discussed, for the effect of a carbon tax with the revenue recycled to finance cuts in taxes on capital.

The other outlier, in the opposite direction, is the NERA (2013) study (performed for the National Association of Manufacturers), which estimates that a carbon tax sufficient to reduce carbon emissions by 80 percent would cause GDP to be 3.4 percent lower in 2050 than it would have been without the carbon tax – a quite substantial drop. The reason this case is so different is that it represents a much higher carbon tax rate than any of the other studies: the carbon tax rate in that case is approximately $1,000 per ton by 2050 (which might be impractically high), whereas none of the other studies have a rate over $60/ton by 2050.[15] The higher

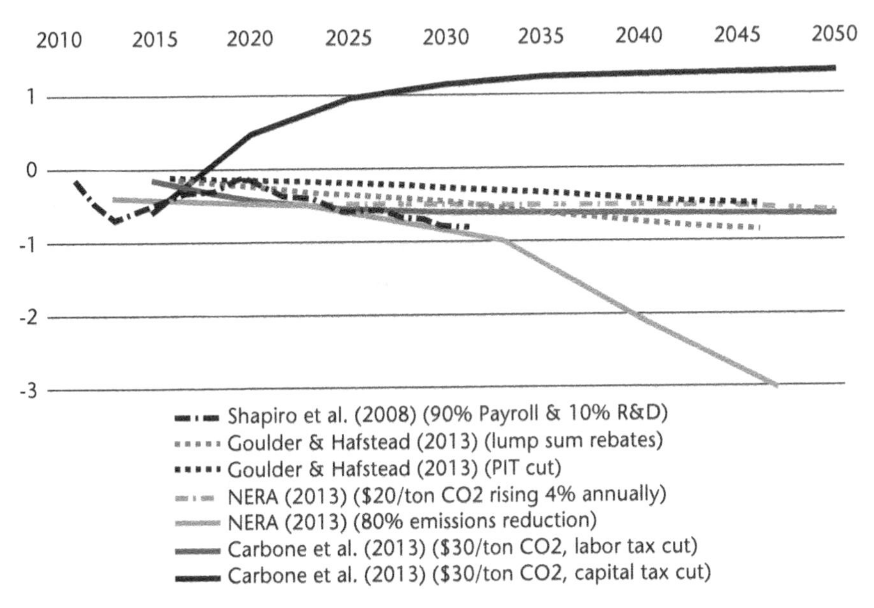

FIGURE 5.2 Projected impacts of carbon tax shifts on GDP (% change), selected results

the carbon tax rate, the more substantial the effect on the economy, so it is not surprising that such an extremely high rate would have major effects on the economy.

2.B.2. Reduce the federal budget deficit

The United States faces substantial fiscal deficits in short-, medium-, and long-term scenarios. While short-run deficits may be desirable and/or necessary because the economy still has a long way to go in recovering from the Great Recession, the outlook for medium- and long-run deficits are cause for concern (Gale and Harris, 2011). While the most recent projections from the Congressional Budget Office indicate that the deficit will shrink slightly over the next few years, those projections also indicate that the deficit and the debt-to-GDP ratio are set to rise substantially over the longer term (see Chapter 1).

Large government budget deficits can retard economic growth in a variety of ways.[16] Government borrowing creates additional demand in capital markets, thus potentially driving up interest rates and crowding out private investment. The need to eventually pay off the debt – or just to pay the interest on it – means that tax rates will need to rise in the future, cutting economic growth then (and perhaps also affecting the economy today, as workers and investors anticipate future tax increases). And a larger debt-to-GDP ratio increases the risk of a debt crisis – a risk that seems tiny for the United States, though nonetheless worth considering because the potential consequences would be dramatic.

Of course, if the government is optimally choosing tax and deficit policy, then the gain from using a dollar to cut the budget deficit will be the same as the gain from using that dollar to cut taxes. But very few observers seem to think that the longer-run path that the deficit is currently on is optimal. If the deficit is larger than optimal, then the benefits of using carbon tax revenue to cut the deficit will exceed those of using that revenue to cut other taxes.

Evidence on the effects of using carbon tax revenue to cut the deficit is far sparser than on the effects of using it to cut other taxes, though there are a few recent studies on this.[17] For example, results in Carbone et al. (2013) suggest that over the long term, the economic gain from using carbon tax revenues to reduce the deficit is generally substantially larger than the gain from using those revenues to cut taxes now.[18] This result is driven primarily by the effects of higher future taxes: because the deficit is currently higher than a long-run sustainable level, future taxes will need to be higher than taxes today. The higher the tax rate, the more harmful a tax increase will be, so it is more efficient to raise additional revenue now (at today's lower rates) than it will be later. The study doesn't fully capture the other effects of deficits (mentioned above), and thus likely understates the gains from deficit reduction.[19]

Carbone et al. (2013) also indicate one reason why addressing the budget deficit is so difficult: even though using carbon tax revenues to fund cuts in the deficit leads to substantially larger gains over the long term than using those revenues to fund tax cuts today, today's voters tend to be better off with tax cuts today, whereas those who benefit more from deficit reductions are too young to vote (or not yet born).

2.B.3. Funding public spending

A third potential pro-growth use for carbon tax revenue would be to fund particularly valuable government spending. If the government has set the mix of taxes and spending efficiently, then the gains from funding an additional dollar of spending will equal the gains from a dollar of tax cuts. But again, it is unlikely that the right balance is struck due to lack of information on the benefits of extra spending programs, let alone the influence of constituencies in the determination of spending levels.

Relatively few studies model the effects of using environmental tax revenues to fund additional public spending, and we are not aware of any that go beyond purely theoretical models.[20] Moreover, the broader literature on the effects of public spending is also relatively thin. Nonetheless, that literature does give some indications of what types of spending could give a particular boost to longer-term economic activity.

Areas of public spending that seem particularly promising are research, education, and infrastructure (see Chapter 12 for discussion of the apparently high return from transportation infrastructure). The common thread among these is that they all represent investments that can boost the future productive capacity of the economy. In effect, increasing spending in these areas represents a boost to the stock of capital – but in the form of knowledge, human capital, and public infrastructure, rather than private physical capital.[21]

However, because we are going beyond the existing literature on environmental taxes – and because the benefits of public spending are inherently difficult to measure – this section is necessarily quite speculative. Further research in this area could be quite valuable (though also quite difficult).

2.C Environmental effects on economic growth

A third channel through which carbon taxes may affect the economy is via their environmental effects (primarily in mitigating potential damage from climate change, but also reducing emissions of other pollutants). Climate change is predicted to have a wide range of effects that might influence the economy. Climate can directly affect economic productivity, particularly for agriculture. Sea level rise can damage coastal property and/or require defenses to protect it. To the extent that imposing a carbon tax can mitigate those and other effects, it can influence the economy.[22]

While the impact of environmental quality on economic growth is difficult to model, there is a small body of research addressing this issue. Williams (2002) outlines the theory behind the relationship between environmental improvements and economic growth. Pollution can reduce productivity, and hence economic activity, for two reasons. The first and more obvious reason is that the lower productivity leads directly to lower economic output. In addition, that lower productivity implies lower returns to labor and capital, and thus reduces incentives to work and to save. In effect, reduced productivity caused by pollution acts like a tax (but with the "revenue" simply being lost, rather than going to the government). This change in incentives magnifies the effect of pollution on economic activity. Effects on human health can also change incentives to save and invest.

Thus, to the extent that a carbon tax reduces future damages from climate change, that could affect the economy. Practical magnitudes of the macroeconomic effects of pollution are difficult to estimate. The leading work in this area is Barrage (2013), which uses a simple model of the world economy linked together with a simple model of the climate. The paper considers four different ways in which climate change potentially affects economic growth – via effects on mortality, morbidity, labor productivity, and health care expenditures. Its results suggest that climate change could have a significant negative effect on the economy – and thus that mitigating climate change could be beneficial for the economy. Again, future research in this area could be very valuable.

3. Short-term effects on growth and unemployment

Given the current state of the U.S. economy – no longer in recession, but only slowly recovering – short-term macroeconomic effects are a particular area of interest. However, there is relatively little research on how a carbon tax affects unemployment or short-term economic growth. Most models of environmental taxation are full-employment models: models that assume anyone who wants to work (at prevailing wages) can find a job. Such models work well for analyzing the effects of taxes over the longer run, as business cycles tend to even out over the longer term. But they are incapable of looking at unemployment, or of realistically modeling effects on short-term growth.

Moreover, within the small set of environmental tax models that do model unemployment, nearly all of those models focus on unemployment caused by rigid labor market rules, overly generous union contracts, and similar factors – a type of unemployment that is important in Europe, but far less relevant in the United States.[23] In the United States, the relevant issue is cyclical unemployment: unemployment caused by downturns in the business cycle. And we are unaware of any completed research studies on how carbon taxes (or any other environmental taxes) affect cyclical unemployment.

However, we can draw some tentative conclusions from the broader macroeconomic literature. Research suggests that tax increases (or cuts in government spending or transfers) during an economic slump tend to increase short-run unemployment and reduce the rate of economic growth. And recent research suggests that those effects are larger than previously thought.[24] This suggests some need for caution in the near term, particularly if the carbon tax revenue is used for deficit reduction (rather than funding increased government spending or cuts in other taxes, either of which would tend to stimulate the economy and thus offset the contractionary effect of the carbon tax itself).

Given the longer-run worries about the budget deficit, however, one potentially promising approach might be to put a policy in place today that would gradually phase in deficit reduction.[25] One way to do this would be to gradually phase in a carbon tax. But this would do little or nothing to limit carbon emissions in the short run. An alternative would be a policy that imposes a carbon tax immediately, but pairs it with a temporary cut in income or payroll taxes. Such a policy could be revenue-neutral in the short run – thus limiting or avoiding any contractionary short-run effect – but would reduce the deficit in the longer run.

Once again, however, we remind the reader that in this section on short-run effects, we are going beyond the existing research on the effects of carbon taxes. Research on the effects of a carbon tax in a model with cyclical unemployment would be highly valuable.

4. Conclusions

The literature on the economic effects of carbon taxes suggests that a carbon tax by itself will tend to have a modest negative effect on aggregate economic activity (as would be the case for other taxes raising the same revenue). However, the revenue from the carbon tax can be used in ways that would promote economic activity, such as cutting taxes on labor and capital, thereby offsetting much or all of (or perhaps even more than offsetting) the drag on the economy from the carbon tax. Moreover, a carbon tax could help to mitigate future economic damage from climate change.

However, wise use of carbon tax revenue is critical. If the revenue is used in ways that do not promote growth, the economic effects will be more negative, and the overall cost of the policy substantially higher. This does not necessarily mean that the revenue should be used only in ways that boost the economy: there may be good reasons for other uses (for example, distributional considerations may favor devoting some revenue for increased transfers to low-income households to offset the burden of higher energy prices, or it may be politically necessary to compensate firms in carbon-intensive industries). But policymakers should be aware of the tradeoffs such decisions entail.

Notes

* We acknowledge helpful comments and suggestions from participants in the AEI-Brookings-IMF-RFF Conference on the Economics of Carbon Taxes, and especially from Ian Parry.

1. For example, the value of leisure time is ignored in measuring GDP. A meal bought at a restaurant counts in GDP, but one cooked at home does not (except for the value of the ingredients). Childcare provided by daycare centers counts in GDP, but a parent caring for his or her own child does not.
2. For example, if a drop in the stock market forces a retired person to go out and get a job (when he would prefer to stay retired), that shows up as a boost to GDP, even though he is worse off.
3. One reason for this is that GDP ignores the value of leisure and non-market time. Many studies find that imposing a carbon tax will induce households to work slightly less and consume slightly more leisure. Ignoring the value of that increased leisure will cause GDP to overestimate the cost of such a shift.
4. Higher consumer good prices mean that any given nominal wage buys less (i.e., a lower real wage) and higher prices for capital goods (e.g., new factory equipment) make investment more expensive, thus lowering the real rate of return to capital. The lower real wage reduces the incentive to work, and the lower real return to capital reduces incentives to save and invest.
5. All carbon tax rates in this chapter are given in dollars per ton of carbon dioxide.
6. This estimate assumes that the carbon tax revenue is returned via transfer payments that do not affect incentives for saving or investment.
7. This assumes that in the absence of a carbon tax, real GDP will grow at a 2.2 percent annual rate (the rate used by the CBO in its most recent long-term budget estimates).
8. These estimates come from Krupnick et al. (2010) and from www.eia.gov.
9. Note that this amount (and all of the estimates in Figure 5.1) represents gross revenue: the revenue from the carbon tax, without any adjustment for how that tax will affect the revenues from other taxes or the cost of goods and services purchased by the government. In general, net revenue (the revenue left after adjusting for those two effects) will be somewhat less. The CBO, for example, assumes that net revenue from a carbon tax will be 25 percent less than gross revenue.
10. Some of the first papers to reach this conclusion were Bovenberg and de Mooij (1994a and b), Bovenberg and Goulder (1996), Bovenberg and van der Ploeg (1994), Goulder (1995) and Parry (1995).
11. From an environmental standpoint, of course, these shifts are the goal of the tax. But from a pure revenue-raising standpoint, they make it less efficient.
12. If carbon tax revenues are used to expand tax preferences, rather than cut tax rates, the economic benefits would be reduced considerably. Moreover, a carbon tax that emerges

from a protracted legislative process may itself contain significant exemptions raising the cost of the policy for a given amount of revenue (though the scope for exemptions seems more limited if the tax is applied to fuel supply – see Chapter 3).

13. See, for example, Bovenberg and Goulder (1997), Carbone et al. (2013), Goulder and Hafstead (2013), and McKibbin et al. (2012).
14. However there are tradeoffs. The corporate income tax is also a relatively progressive tax (i.e., as a proportion of income it imposes a larger burden on wealthier households), so imposing a carbon tax and cutting only the corporate income tax could run counter to distributional objectives (see Chapter 8). This could be offset by combining the cut in corporate income taxes in a broader fiscal package with other measures (such as scaling back tax preferences for the wealthy) to make the overall impact more distributionally neutral.
15. Moreover, a rate of $1,000/ton seems far higher than what could be politically possible.
16. See Elmendorf and Mankiw (1999) for an extensive discussion of the effects of deficits. This issue is also addressed in Chapter 1 of this volume.
17. Modeling the effects of budget deficits is much more difficult than modeling the effects of tax cuts, which is the main reason that far fewer studies look at using environmental tax revenue for deficit reduction.
18. However, much depends on what tax is being cut now, and on which taxes will increase in the future if the deficit remains high. If tax cuts now would cut a particularly inefficient tax (such as the corporate income tax), and continued high budget deficits would eventually be addressed with a particularly efficient tax (such as a value-added tax), this would lessen the benefit from reducing the deficit over reducing taxes.
19. Carbone et al.'s model does not include any risk of a debt crisis, and thus cannot examine that potential effect. And it assumes that the supply of foreign capital is perfectly elastic, thus greatly minimizing any potential crowding out of private investment.
20. Note that we are implicitly drawing a distinction here between government spending and government transfers. Many papers model the effects of using environmental tax revenue to fund increases in government transfers. In general, these find that using the revenue to increase transfers does little or nothing to encourage economic activity, because increased transfers do not increase the incentives to work or save. We are also drawing a distinction between modeling a change in government spending and studies that model wasting the revenue from a carbon tax (i.e., spending it in a way that provides no benefits for anyone).
21. Jorgenson and Fraumeni (1992) find that investment in human and non-human capital accounted for the overwhelming proportion of economic growth in the United States after World War II and that educational investments dominate the investment requirements for increased future growth rates. Munnell (1992) suggests that in addition to providing immediate economic stimulus, investment in public infrastructure has a strong longer-run impact on economic output and long-run growth.
22. The direct effect of US emissions reductions on climate damages in the United States is likely to be relatively small. Climate damages depend on global greenhouse gas emissions. US CO_2 emissions are currently less than 20 percent of global emissions, and that share is falling. However, if US action to reduce emissions induces other countries to take similar actions, then the effect could be substantially magnified.
23. Such studies tend to find that imposing an environmental tax and recycling the revenue to cut other taxes will reduce unemployment (for examples, see Bovenberg and van der Ploeg (1994 and 1996) and Bye 2002). But it is not at all clear that cyclical unemployment would be affected in the same way.
24. See Blanchard and Leigh (2013). Their estimates suggest that during the recent economic downturn, fiscal multipliers exceeded 1 (that is, a reduction in the budget deficit equal to 1 percent of GDP would reduce near-term GDP by more than 1 percent). However, they also suggest that fiscal multipliers are likely to be substantially lower in more normal economic conditions.

25. Jones and Keen (2009), for example, argue that during a recession, carbon pricing should be pre-announced and phased in gradually, so as to provide the economy more time to adjust to the new policy.

References

Barrage, L., "Optimal dynamic carbon taxes in a climate-economy model with distortionary fiscal policy," Working Paper, Yale University, Department of Economics (2013).

Blanchard, O. and D. Leigh, "Growth forecast errors and fiscal multipliers," IMF Working Paper, WP/13/1, January (2013).

Bovenberg, A. L. and R. A. de Mooij, "Environmental levies and distortionary taxation," *The American Economic Review* 84(4):1085–1089 (1994a).

Bovenberg, A. L. and R. A. de Mooij, "Environmental taxes and labor-market distortions," *European Journal of Political Economy* 10:655–683 (1994b).

Bovenberg, A. L. and L. H. Goulder, "Optimal environmental taxation in the presence of other taxes: General equilibrium analyses," *The American Economic Review* 86(4):985–1000 (1996).

Bovenberg, A. L. and L. H. Goulder, "Costs of environmentally motivated taxes in the presences of other taxes: General equilibrium analyses," *National Tax Journal* L(1):59–87 (1997).

Bovenberg, A. L. and F. van der Ploeg, "Consequences of environmental tax reform for unemployment and welfare," *Environmental and Resource Economics* 12:137–150 (1994).

Bovenberg, A. L. and F. van der Ploeg, "Optimal taxation, public goods and environmental policy with involuntary unemployment," *Journal of Public Economics* 62:59–83 (1996).

Bye, B., "Taxation, unemployment, and growth: Dynamic welfare effects of 'green' policies," *Journal of Environmental Economics and Management* 43:1–19 (2002).

Carbone, J. C., R. D. Morgenstern, R. C. Williams III, and D. Burtraw, "Deficit reduction and carbon taxes: Budgetary, economic, distributional and economic impacts," Resources for the Future Report, August (2013).

Elmendorf, Douglas W. and N. Gregory Mankiw, 1999. "Government debt," in J. B. Taylor & M. Woodford (eds.), *Handbook of Macroeconomics*, vol. 1. Amsterdam: Elsevier, 1615–1669.

Gale, W. G. and B. H. Harris, "Reforming taxes and raising revenue: Part of the fiscal solution," *Oxford Review of Economic Policy* 27(4):563–588 (2011).

Goulder, L. H., "Environmental taxation and the double dividend: A reader's guide," *International Tax and Public Finance* 2:157–183 (1995).

Goulder, Lawrence H. and Marc A. C. Hafstead, "Tax reform and environmental policy: Options for recycling revenue from a tax on carbon dioxide," Working Paper, Department of Economics, Stanford University (2013).

Jones, Benjamin and Michael Keen, "Climate policy and the recovery," IMF Staff Position Note SPN/09/28 (2009).

Jorgenson, D. W. and B. M. Fraumeni, "Investment in education and U.S. economic growth," *The Scandinavian Journal of Economics* 94:51–70 (1992).

Krupnick, Alan J., Margaret Walls, Ian Parry, Tony Knowles, and Kristin Hayes, *Toward a New National Energy Policy: Assessing the Options*. Washington, DC: Resources for the Future and National Energy Policy Institute (2010).

McKibbin, W., A. Morris, P. Wilcoxen, and Y. Cai, "The potential role of a carbon tax in U.S. fiscal reform," Climate and Energy Economic Discussion Paper, Brookings Institution, July (2012).

Munnell, A. H., "Infrastructure investment and economic growth," *The Journal of Economic Perspectives* 6(4):189–198 (1992).

NERA Economic Consulting, *Economic Outcomes of a U.S. Carbon Tax* (2013).

Parry, Ian W., "Pollution taxes and revenue recycling." *Journal of Environmental Economics and Management* 29:64–77 (1995).

Parry, Ian W. H. and Roberton C. Williams III, "What are the costs of meeting distributional objectives for climate policy?" *The B.E. Journal of Economic Analysis & Policy* 10(2): Article 9 (2010).

Rausch, Sebastian and John Reilly, "Carbon tax revenue and the budget deficit: A win-win-win solution?" MIT Joint Program on the science and policy of global change, Report No. 228, August (2012).

Rausch, Sebastian; Metcalf, Gilbert E.; Reilly, John M. and Paltsev, Sergey. "Distributional Implications of Alternative U.S. Greenhouse Gas Control Measures." *B.E. Journal of Economic Analysis & Policy*: Vol. 10: Iss. 2 (Symposium), Article 1 (2010).

Shapiro, Robert, Nam Pham, and Arun Malik, "Addressing climate change without impairing the U.S. economy: The economics and environmental science of combining a carbon-based tax and tax relief," U.S. Climate Task Force, June (2008).

United States Congressional Budget Office. *Reducing the Deficit: Spending and Revenue Options.* Washington, D.C., Pub. No. 4212, March (2011).

Williams III, R. C., "Environmental tax interactions when pollution affects health or productivity," *Journal of Environmental Economics and Management* 44:261–270 (2002).

6

THE DISTRIBUTIONAL BURDEN OF A CARBON TAX

Evidence and implications for policy

*Adele Morris and Aparna Mathur**

KEY MESSAGES FOR POLICYMAKERS

- The distributional effect, or economic incidence, of a carbon tax is the combination of the ways a carbon tax affects the things people buy and their sources of income.
- Estimating the retail price increases that would arise from a carbon tax is fairly straightforward. Economists use input-output tables of the goods and services in the economy and data on households' spending on energy and other products whose prices rise in response to carbon pricing to assess the burdens on different households.
- Estimates of consumer price effects suggest low-income households could be hit relatively harder than high-income households; the estimated burden as a share of income on the poorest quintile is several times that on the top quintile. These results indicate that the bottom income quintile (as a group) could be held harmless by targeting about 11 percent of the carbon tax revenues directly to those households.
- A carbon tax looks much less regressive when its burden is expressed relative to household consumption rather than income. It is also less regressive if one takes into account the effects of the tax on the sources of household income, such as from returns on capital assets, wages, and transfer payments. However, assessing the effect of a carbon tax on the sources of income is difficult and remains a subject of active research.
- Using carbon tax revenues to fund proportional reductions in other federal taxes would produce a less progressive revenue system because most taxes are borne disproportionately by higher-income groups. However,

policymakers can use any number of adjustments to the tax system and social safety net programs to construct a more progressive package of reforms. This highlights the case for including a carbon tax as part of a more comprehensive (and badly needed) overhaul of the fiscal system.

- The economic benefits from using carbon tax revenues to cut other taxes are large, implying potentially high economic costs to diverting revenues for compensation or earmarked spending on other government activities.
- The burden of a federal carbon tax would vary somewhat regionally, but it is not clear what, if anything, policymakers should do to account for this. For example, states that currently use a lot of coal tend to have relatively low prices for electricity, so the tax would tend to even out electricity prices across the country. Federal transfers to states may not necessarily reach the most burdened households.

Introduction

A carbon tax policy that is perceived as unfair is less likely to pass and less likely to stay passed. The mere threat of repeal would undermine the price expectations that are critical to inducing large long-term investments in emissions abatement. This chapter reviews the evidence on the potential distribution of burdens of a carbon tax across different households with an eye to informing policy approaches that could enhance both the fairness and net social benefits of a carbon tax.

One way or another, a carbon tax is ultimately borne by people (and not purely legal entities like corporations or small businesses). Most of a carbon tax would likely be "passed forward" to consumers in the form of higher prices for energy and other products. Economists can estimate these consumer price increases straightforwardly using input/output tables of goods and services in the economy, and they can estimate the ultimate effects on households by matching the price increases with survey data about how much different household income groups spend on fossil energy (electricity, gasoline, natural gas, etc.) and on other products whose prices would increase.

However, in practice the distributional outcomes are likely to be more complicated than those simple calculations of higher retail prices would suggest. For one thing, the carbon tax also affects households' sources of income, not just how they spend their income. For example, a carbon tax could lower capital income and wages, and higher overall prices could increase transfers, such as Social Security payments, to poor households. Moreover, the final economic incidence of a carbon tax hinges critically on how the government uses the revenue, what other tax and spending reforms accompany the carbon tax, and what other changes are made to environmental policies.

Even if we know the effects of a particular proposal across different household groups, it is not obvious how to judge its fairness. Many would argue that the very poorest should not be made worse off, but what about slightly less poor or middle-class households? Moreover, the distribution of burdens across different household groups depends heavily on whether the burdens are measured relative to household income or consumption, and it is not obvious which measure of socioeconomic status (income or consumption) best characterizes well-being. Finally, the average effects of a policy on a particular group of households could mask important variations within the group.

Despite these complexities, we can offer a way to think about and, to a limited extent, quantify the potential distributional impacts of U.S. carbon pricing both before and after the government uses the revenue, thereby providing a framework for more details as the research matures. The focus here is primarily on burdens borne by different household income groups, though we also briefly review impacts on different regions of the United States. Chapter 9 considers impacts on different industries.[1]

We should emphasize that a concern about the distributional impacts is not a good reason to delay putting a meaningful price on greenhouse gases. Keeping fossil energy prices below their full social cost is economically inefficient, and high-income households benefit more from inefficiently low prices than low-income households do because they consume more at those inefficiently low prices.

The discussion proceeds as follows. Section 2 discusses some basic concepts of tax incidence. Section 3 discusses the evidence on the incidence of the carbon tax itself, before accounting for how the government uses the revenues. Section 4 discusses the different ways carbon tax revenue might be used and how the use of revenue affects both the fairness and economic efficiency of the overall policy package. Section 5 discusses how the carbon tax incidence could vary across the country, and Section 6 briefly summarizes.

1. Basic concepts of tax incidence

The economic burden or incidence of a tax refers to whose economic welfare is reduced by this tax, and by how much. This is quite different from the formal or legal incidence – fuel suppliers, for example, may be responsible for remitting tax payments to the Internal Revenue Service, but they may bear little economic incidence if they can charge higher prices. In fact, the point in the fossil fuel supply chain where the tax is levied (e.g., on fuel suppliers or emissions from industrial smokestacks) matters for the practical implementation of the charge, but makes little or no difference to its ultimate economic incidence.

The total burden of a carbon tax refers here to the overall cost to households of the carbon tax, excluding environmental and other benefits of protecting the climate. Those benefits could be quite large relative to the costs, but they may principally accrue to future generations;[2] one should not infer anything about the net benefits of the carbon tax from the focus here on its economic incidence.

If a carbon tax burdens lower-income households as a share of their income (or some other measure of socioeconomic status) relatively more than it burdens higher-income households, then the policy is "regressive," even if wealthier people bear a greater burden in absolute terms. If the policy imposes the same burden relative to income across all household income groups, then the tax is termed "distributionally neutral." And if higher-income households are hit relatively harder (as a share of income) than poorer households, then the policy is "progressive."

Although here we focus (as does most of the economic literature) on comparisons between scenarios of having a carbon tax and not having a carbon tax, a fuller discussion would compare a carbon tax to alternative ways to achieve the same environmental and fiscal objectives. Box 6.1 briefly remarks on this.

BOX 6.1 THE INCIDENCE OF CARBON TAXES VS. ALTERNATIVE POLICIES

The incidence of a carbon tax and that of its main market-based alternative, an emissions trading system (ETS) or cap-and-trade system, are potentially quite similar. For example, policymakers could auction ETS allowances to generate about the same revenue stream as under a (similarly scaled) carbon tax. Thus in both cases, the incidence depends on how the revenue is used. If policymakers give free allowances to energy companies (as in the U.S. sulfur trading program and early phases of the EU ETS), this creates windfall profits for these firms since they will pass the price on carbon through to their customers anyway. These profits will raise stock prices and thereby compound the regressive effects of the emissions price.[1] The Waxman-Markey bill included an ETS that would have distributed some of the carbon allowances free to local electricity and natural gas distribution companies with the proviso that the value of allowances would hold down residential energy prices to limit burdens on households (thereby channeling these policy rents to households rather than firms).[2] Policymakers could achieve the same effect with a carbon tax by using some of the revenue to subsidize residential energy bills. This is inefficient, however, because it blunts the incentive to conserve electricity, one of the more cost-effective ways to lower emissions.

In the absence of a price on GHGs, policymakers could cut emissions with a number of policy tools, such as Clean Air Act regulations, clean energy mandates, and energy efficiency standards. These other policies also tend to increase energy prices less than carbon taxes would because they don't price emissions below the regulated constraints. By raising revenues, however, carbon taxes offer a means to protect poor households, reduce the federal budget deficit, and lower more distortionary taxes.

A carbon tax could be more or less regressive than other ways to cut budget deficits. It is likely to be less regressive than cutting discretionary or entitlement spending that disproportionately benefits the poor, but it might be more regressive than raising progressive income taxes.

[1]Dinan and Lim Rogers (2002), Parry (2004).

[2]The bill, H.R. 2454, passed the House of Representatives in 2008. A companion bill failed in the Senate the next year.

2. Carbon tax incidence prior to use of revenues

This section focuses on the distributional pattern of a carbon tax on households by income class, not accounting for how the government uses the revenues. First, we discuss the absolute burden of an illustrative carbon tax and the burden relative to income across different income groups. Next we discuss burdens relative to consumption (an alternative measure of socioeconomic status) and the potential effect of the tax on the sources of household income (such as capital income and transfer payments).

The absolute incidence of a carbon tax

A number of studies have estimated the likely incidence of a carbon price by using input/output tables to trace through the effects of the tax through the economy: higher after-tax fuel prices, higher prices for intermediate inputs (e.g., steel), and ultimately higher retail prices for all consumer products. The studies assume the carbon tax raises fuel prices in direct proportion to the amount of carbon in them, and they typically assume all the price increases are passed forward into the consumer prices faced by households. These studies estimate the burden on different household income groups by mapping the estimated price changes to data on how different groups of households spend their income. Usually these calculations assume consumers don't change what they consume in response to the tax, which is reasonable if the carbon price is modest and the focus is on short-run distributional effects. They also assume that the carbon tax doesn't have any broader economic effects, for example, that would lower other tax revenues.

Table 6.1 below shows just this sort of calculation for an illustrative $15 per ton CO_2 tax, assuming it was imposed in 2010. These estimates are from Mathur and Morris (2014), a study that is typical in this literature.[3] The total burden of the tax in 2010 (i.e., the revenues) is $102.3 billion. Not surprisingly (given that people with more income tend to spend more on energy), the burden increases steadily up the income scale from $5.0 billion (5 percent of the total) for the bottom income decile to $19.1 billion (19 percent of the total) for the top income decile.

The table suggests that if policymakers directed about 11 percent of the tax towards the poorest two deciles, for example, through social safety net programs, then those households would on average be no worse off after the carbon tax than they were before. Of course, individual households within those groups might be better or worse off than average depending on their particulars.

Incidence relative to income

Economists usually express the distribution of burdens against a measure of households' socioeconomic status, such as annual income. Lower-income households in the United States generally devote a higher share of their household budget to energy and other goods whose prices would rise upon imposition of a carbon tax than higher-income households do. There are two reasons for this. First, poorer

TABLE 6.1 Burden of a $15 per ton CO_2 tax by income decile, 2010

Decile	*Burden ($ billions)*	*Cumulative burden ($ billions)*	*Percent of total burden (%)*	*Cumulative % of burden*
Bottom	5.0	5.0	5	5
Second	6.5	11.5	6	11
Third	7.0	18.5	7	18
Fourth	8.2	26.7	8	26
Fifth	9.3	36.0	9	35
Sixth	10.0	46.0	10	45
Seventh	11.2	57.2	11	56
Eighth	12.1	69.3	12	68
Ninth	13.9	83.2	14	81
Top	19.1	102.3	19	100

Source: Mathur and Morris (2014).

Notes: The table reports the total burden of the carbon tax to each household income decile. The calculations in Table 6.1 are based on a static analysis of how the burden of the tax would be distributed across households if there were no behavioral response to the tax in terms of consumers shifting away from products rich in carbon to less carbon-intensive products. This may overstate the long-run burden of the tax but by a modest amount. More important, the calculations do not account for revenue recycling. Figures are in 2010 dollars.

households spend a greater share of their current income than higher-income families, who save relatively more. That means that poorer households' budgets are generally more exposed to general increases in prices than richer households'. Second, poorer households spend proportionately more of their income directly on electricity and other fuels than higher-income households do.

As Table 7.1 of Chapter 7 indicates, households in the bottom income quintile spend far more of their budget (21.4 percent) on energy than the top income quintile (4.1 percent). The two big energy items for all households are motor fuels and electricity. Other energy expenditures include natural gas, fuel oil, and other fuels. Of all the fuels, the electricity budget share falls most with higher income. The bottom line is that a carbon tax that raises the price of fossil fuels is going to look strongly regressive when burdens are expressed relative to income and the analysis focuses only on how households use their income.

Figure 6.1 illustrates this point. It shows the burden to income ratio for a $15 carbon tax by income decile, reported in Mathur and Morris (2014). The carbon tax looks quite regressive, with the bottom income decile suffering a burden equal to 3.5 percent of their income, or about six times the burden (relative to income) for the top income decile (0.6 percent). Figure 6.1 also shows the breakdown of the burden across increases in energy prices (the darker bars) and increased prices for everything else (the lighter bars). The direct effect is more important, particularly for lower-income households. Accounting for the indirect effect makes the carbon tax look a little less regressive overall.

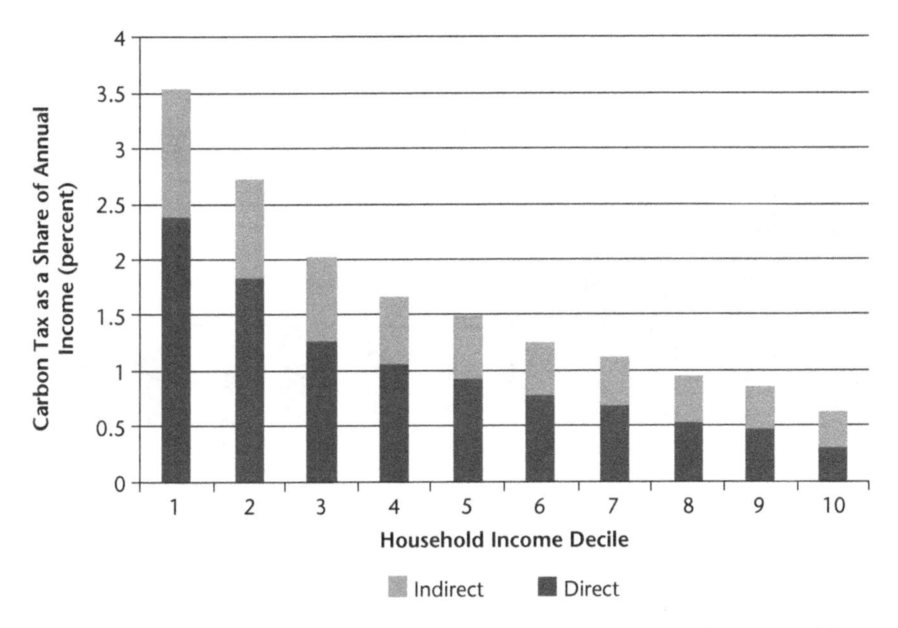

FIGURE 6.1 Estimated burden (relative to income) of a $15 carbon tax by household income decile, 2010

Source: Mathur and Morris (2014).

Incidence under other measures of socioeconomic status

Annual income may not be the best way to characterize a household's well-being. For example, some people (e.g., graduate students, retirees) may be in a stage of life where their annual income is relatively low, despite a lifetime of reasonably good living standards. An ordinarily well-off person may also have low income in a particular year due to transitory factors, like temporary unemployment, illness, maternity leave, and so on.

To account for this, analysts sometimes measure tax burdens against other measures of households' socioeconomic status, such as their annual consumption, that vary less over a lifetime than income does.[4] The consumption approach considerably reduces the measured regressivity of a carbon tax because low-income households have a high consumption to income ratio relative to high-income households.

Figure 6.2, again taken from Mathur and Morris (2014), illustrates this point. It is the same scenario as Figure 6.1, but the burden of the tax is expressed as a share of annual household consumption, rather than annual income. Under this measure, the same $15 per ton CO_2 tax would have burdened the poorest 10 percent of households on average by 2.1 percent of their annual consumption and the wealthiest decile by 1.3 percent of their annual consumption. Thus, the relative burden for the bottom decile in this approach is 60 percent higher than for the top decile (rather than 6 times as high).

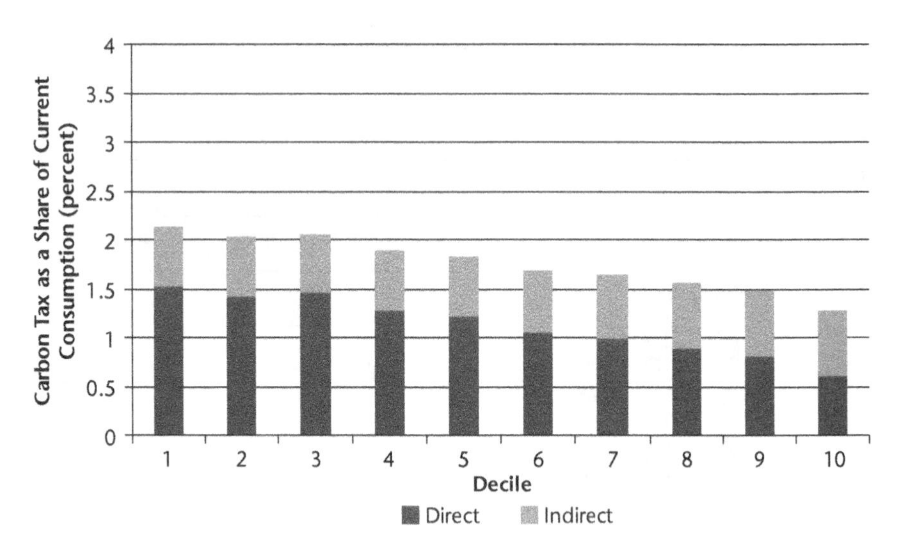

FIGURE 6.2 Burden of a $15 carbon tax by consumption decile, 2010

Source: Mathur and Morris (2014).

Incidence on the sources of income

So far, we've discussed results that assume the entire burden of a carbon tax falls on consumers via higher retail prices. However, a carbon tax may have complex effects on the sources of household income, with implications for how fair a carbon tax is likely to be in practice.

First, rather than being fully passed forward into higher consumer prices, some of the burden of the tax may be "passed back" to producers in the form of lower producer prices. Just how much is passed back depends on the relative slopes of demand and supply curves in energy markets – see Box 6.2. To the extent that producer prices do fall, this may lower the returns to capital in energy industries – implying lower stockholder wealth – and lower wages or other costs for workers.

Further, some sources of income, such as Social Security, Supplemental Security Income, and food stamps, are indexed to overall price levels. Thus, if higher fossil energy prices translate into higher overall price levels, some households may automatically receive more income, which will at least partially compensate them for the tax. After-tax income can also be affected by higher prices because income tax schedules are indexed to consumer price levels.

To parse all this out, a good place to start is the rough sketch of income sources across the socioeconomic ladder in Figure 6.4. The chart shows the (pre-tax) income to different household groups by source according to the Congressional Budget Office. All income classes on average receive the majority of their income from wages (and businesses they operate themselves, like sole proprietorships). Most capital income (such as dividends and capital gains) goes to the richest households. Income from private retirement accounts is included in the "other" category. In

BOX 6.2 IMPACTS OF A CARBON TAX ON CONSUMER AND PRODUCER PRICES

The extent to which a carbon tax is passed forward in higher prices to fuel users, as opposed to being passed backward in lower prices to fuel suppliers, depends on the relative slopes of demand and supply curves. Consider Figure 6.3, which shows the market for an energy product, perhaps coal. Without the tax, the price would be P_1. After the tax, the fuel price paid by consumers is P_2 and the price received by fuel suppliers is P_3, with the difference between these prices reflecting the tax on the fuel, equal to carbon dioxide emissions per unit of fuel use times the tax rate. The tax causes fuel output to fall from Q_1 to Q_2.

If the fuel supply curve is flat, all of the tax is passed forward into higher consumer prices. But if the supply curve is upward sloping then some of the tax is passed backward in lower producer prices, the more so the steeper the slope of the supply curve relative to that of the demand curve. Fuel supply curves are generally thought to be flatter than demand curves – so the majority of the tax is passed forward – but just how much flatter is uncertain (much depends on the mobility of capital across sectors which is high in the longer term but is more limited in the near term).

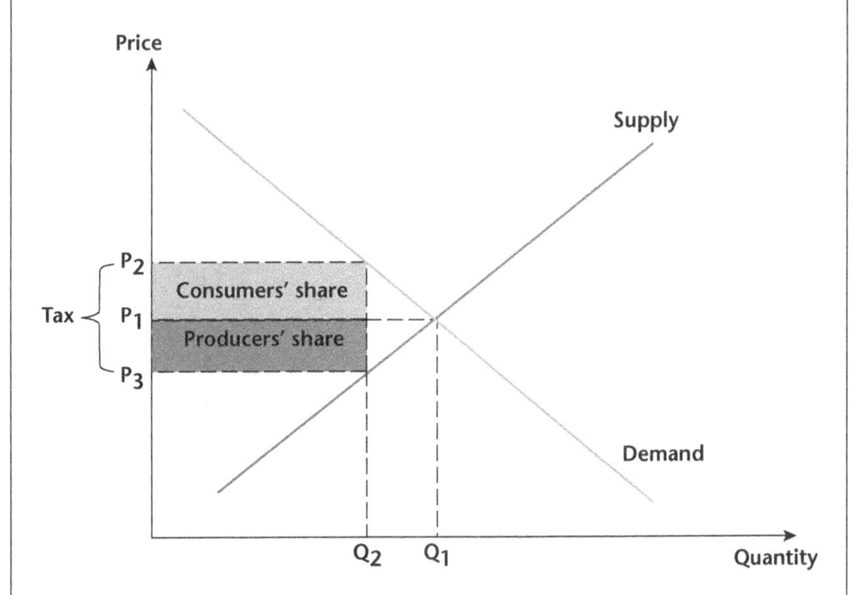

FIGURE 6.3 The effect of a carbon tax on consumer and producer prices

Tax revenue is given by the areas of the shaded rectangles combined. The amount of tax effectively borne by consumers is the lighter shaded rectangle (the increase in consumer price times fuel consumption) while that borne by producers (which ends up as lower wages for workers, lower prices to suppliers of inputs, and lower returns to shareholders and investors through lower corporate profits, dividends, and capital gains) is the darker shaded rectangle (the reduction in producer price times fuel output).

contrast, the poorest households receive proportionately more income in the form of transfer payments (including Social Security).

Figure 6.4 suggests (not surprisingly) that to the extent the carbon tax reduces income to capital, this will largely affect higher-income households. This means that carbon taxes are less regressive in practice, when producers bear at least some of the cost, than if producers could pass everything to consumers, as assumed in the earlier discussion. At the same time, if higher price levels trigger increases in transfer payments (the green bars), that will augment a larger share of income for poorer households than richer households, again a progressive effect.

The distributional pattern from a carbon tax's effects on wages is a little complicated. Households that receive relatively more of their income from labor and less from transfers will face relatively higher burdens, all else equal. According Figure 6.4, labor income (wages and benefits) comprises 55 percent of the lowest

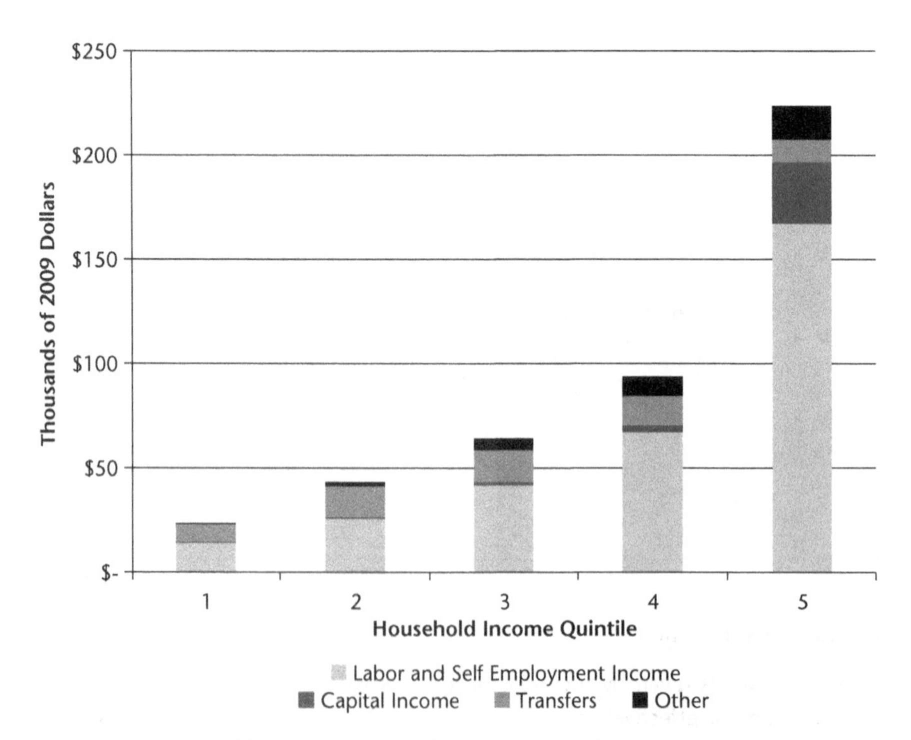

FIGURE 6.4 Source of household income by income quintile, 2009

Source: Congressional Budget Office (2012).

quintile's total income, and 69 and 64 percent, respectively, of the fourth and fifth quintiles' income, suggesting any reduction in wages is mildly progressive. However, we haven't accounted for the extent to which lower-income households tend to work disproportionately more or less in energy-intensive industries (e.g., coal mining) than in other sectors of the economy, and that could affect the results.

Despite the broad insights from Figure 6.4, the effect of a carbon tax on the sources of income is hard to analyze more concretely. The portion of the carbon tax that might be passed back in lower producer prices is still a matter of ongoing research, along with how this portion may decline as investors reallocate capital over the long run. And to the extent that returns on capital are determined internationally, this will tend to cushion the fall in the domestic returns. The effects on capital income also depend on the degree to which capital and labor can be substitutes in production, which is also complicated. Moreover, the data on the sources of income for different groups of people are not consistent across different government surveys. And it is not clear what the effect of a new tax could be on all sorts of important but complex sources of income, like payments employers make to their employees' retirement accounts.

Some studies find that effects of the tax on the sources of income can offset some or all of the regressivity on the uses of income, at least across part of the income distribution.[5] For example, Fullerton et al. (2011) find the burden to income ratio across the income distribution is U-shaped – low-income household still suffer a disproportionately larger burden, but so do high-income households (due to the drop in capital income). Categorizing households by annual consumption produces somewhat different results. In this case, Fullerton et al. (2011) find that the overall burden of carbon pricing is progressive across the bottom half of the distribution and regressive across the top half. (In both cases, the indexing of transfers adds significant progressivity to carbon tax incidence.)

Some recent studies use computational models of the economy to estimate how the burden could fall on workers and shareholders. They also take into account the indexing of transfer payments and how households could change their consumption patterns over time in response to new relative prices, substituting away from higher-cost energy-intensive items. Again, these studies find substantially different conclusions about the regressivity of a carbon tax than studies that just consider the uses of income. For example, Rausch et al. (2011) find that progressivity on the sources of income is sufficiently strong as to offset regressivity on the uses side, leaving a carbon tax distributionally neutral overall.

3. Distributional tradeoffs in the disposition of the revenue

This section explores the tradeoffs between using carbon tax revenues in the most economically efficient or pro-growth way and using it to achieve other (especially distributional) objectives. The discussion compares the carbon tax to what we know about the incidence of other taxes. It explains the tradeoffs inherent in protecting low-income households, the strong economic case for using much of the revenue for cutting other taxes or reducing debt ratios, and then considers some other possibilities for revenue use.

Who pays federal taxes?

Higher-income households pay a disproportionately large share of most federal taxes. Table 6.2 illustrates how different a carbon tax would be in this regard than most other federal taxes. It shows the share of an illustrative $15 carbon tax borne by different household income groups (the same estimates from Morris and Mathur (2012) as in Table 6.1, summed into income quintiles) with analogous information from the Congressional Budget Office about other federal taxes in 2008 and 2009. The top income group, assuming all of the revenue is passed forward to consumers, would pay about 32 percent of the carbon tax and the poorest 20 percent of households would pay about 11 percent. In contrast, the top income group pays the overwhelming majority of individual income taxes and (through their role as shareholders) corporate income taxes, 94.1 and 77.2 percent, respectively. The bottom two income quintiles each pay under 4 percent of the corporate income tax and receive refundable income tax credits, such as the earned income and child tax credits, sufficient to make their net individual income tax burdens negative.

Payroll taxes (including employer and employee contributions to Social Security and Medicare taxes) are more evenly distributed, with the top income quintile contributing 45.3 percent of the revenues and the bottom quintile 5.3 percent. Federal excise taxes (primarily on tobacco, alcohol, and motor fuels) are more like the carbon tax, another kind of excise tax, with the bottom quintile paying 11.0 percent of revenues and the top quintile 34.3 percent.

Even if one accounts for the potential effect of a carbon tax on the sources of income (which the estimates in Table 6.2 do not), imposing a carbon tax and using

TABLE 6.2 Distribution of carbon tax relative to other federal taxes

	Income quintile				
	bottom	*second*	*third*	*fourth*	*top*
Percent of revenue from carbon tax					
Carbon tax (estimate for 2010 assuming 100% falls on consumers)	11.2	14.9	18.9	22.8	32.3
Percent of revenue from other federal taxes (2008 and 2009)					
Individual income tax	–6.6	–3.5	2.7	13.4	94.1
Payroll taxes	5.3	9.7	15.4	24.0	45.3
Corporate income tax	1.8	3.2	5.8	10.2	77.2
Excise taxes	12.2	15.1	18.8	21.3	32.1

Source: Table 6.1 above with data from Mathur and Morris (2014) and CBO (2012), Table 2.

the revenue for proportional reductions in other federal taxes combines a potentially regressive new tax with an assuredly regressive tax cut. This underscores the desirability of thinking carefully about how to embed a carbon tax in a fiscal package that protects the poor but also uses as much revenue as possible in a pro-growth way.

Protecting low-income households vs. cutting other taxes

One way to ensure a carbon tax is progressive is to return all of the carbon tax revenue to households in equal rebates, also called lump-sum transfers.[6] But rebates don't lower any existing taxes, so they don't increase the after-tax returns from working or for investing and saving and thereby encourage more productive economic activity. This would be a large forgone opportunity. Research shows that using carbon tax revenue to reduce other taxes can greatly improve the economics of a price on carbon and climate policy more generally.[7] This is called "revenue recycling" with a "tax swap."

The most efficient form of revenue recycling would offset the most distortionary taxes, meaning the ones that create the greatest economic drag for the last dollar they bring in.[8] Unfortunately, the most efficient tax swap may be the most regressive.[9] Most experts believe the most distortionary taxes are likely those on capital income, like dividends, capital gains, and corporate earnings.[10] With a federal corporate tax rate of 35 percent and an average state rate of 6.3 percent, the combined U.S. corporate income rate is roughly 39.1 percent, the highest statutory corporate tax rate in the developed world today.[11] For comparison, Hassett and Mathur (2011) show that the U.S. corporate tax rate was only slightly higher than the OECD median in 1981. The United States is an outlier in its tax treatment of corporate income, and many argue that the tax system is likely harming U.S. economic competitiveness and driving multinational corporations to shift taxable profits abroad.[12]

So how much revenue would policymakers need to set aside to make poor households no worse off? Studies show that if policymakers target about 11 percent of the revenue each year to households whose income falls below 150 percent of the poverty level, however that level is defined in each year, it would ensure that roughly the poorest fifth of households (on average) remain no worse off. Mathur and Morris (2014) estimate that 11 percent of carbon tax revenue would be necessary to hold the bottom two deciles of households by income harmless (see above and Morris (2013) for further discussion), while Chapter 7 suggests 12 percent, leaving almost 90 percent of the revenues for other purposes.[13]

Alternatively, instead of preventing an absolute decline in the well-being of the bottom income group, the government's objective may be to ensure the burden of a carbon tax is more or less proportional to income for all groups. Grainger and Kolstad (2010) investigate how the potential regressivity of a $15 tax on CO_2 could be ameliorated by returning the revenue in a progressive way. Policymakers can target transfers, finance cuts in regressive payroll or excise taxes, target income tax cuts to lower-income groups, or spend more on government programs targeted to lower-income groups. The authors estimate that the carbon tax could be made distributionally neutral by directing transfers (or income tax credits) in the amounts of $119, $112, $105, and $76 to individuals in the first four income quintiles, respectively. The net result of the carbon tax and transfers would burden individuals proportionately at around 1 percent of their net annual income. It would also offset the regressive effects of the carbon tax while leaving $49.6 billion in net revenues for the government.

A number of papers investigate or advocate using a carbon tax to reduce capital income taxes in order to maximize the efficiency gain from the tax swap. Dinan and Lim Rogers (2002) found that using carbon revenues to reduce corporate income taxes could reduce the economic cost of limiting carbon emissions by about 60 percent. Analyzing a 15 percent cut in emissions from the carbon price associated with a cap-and-trade program, CBO estimates that the downward hit to GDP could be reduced by more than half if the government sold allowances and used the revenues to lower corporate income taxes rather than to provide lump-sum rebates to households.[14]

Metcalf (2008) considers how a carbon tax could be used to reduce capital income taxes through corporate tax integration, meaning a reform that would tax corporate earnings only when shareholders receive dividends and realize capital gains. He finds that the tax could improve the efficiency of the system and that price increases throughout the economy are likely to be modest. Using a general equilibrium model, McKibbin et al. (2012a) find that using the carbon tax revenue to buy down taxes on capital income could slightly boost GDP, employment, and wages through the first few decades of the tax, in part as a result of the tax swap's beneficial effect on U.S. investment.

Marron and Toder (2013) estimate that cutting the corporate tax rate from 35 percent to 28 percent would reduce U.S. tax revenues by about $800 billion over the next 10 years. Cutting the rate from 35 percent to 25 percent would reduce tax revenues by about $1.15 trillion. While some of that lost revenue could be made up by expanding the corporate income tax base through elimination of corporate tax preferences of various kinds, these preferences have organized supporters who will oppose their elimination. Even if some tax preferences can be phased out, tax reform, whether it be corporate tax reduction or other desirable tax reform, will likely require new revenue elsewhere in the budget, and a carbon tax is a natural fit.

Morris (2013) offers a specific proposal along these lines. She analyzes a tax that starts at $16 per ton of CO_2 and rises by 4 percent annually over inflation, along with a cut in $6 billion per year in clean energy subsidies. She finds it could finance a long-term reduction in corporate income tax rates from 35 to 28 percent and still reduce the deficit by over $800 billion over two decades, even after reserving 15 percent of revenue for the protection of the poor.

Other scholars have investigated using carbon tax revenue to reduce personal income or payroll taxes. Parry and Williams (2012) find very striking differences between the costs of a $33 per ton carbon tax with revenues used for personal income tax cuts (a net annual *gain* of $6 billion or $9 per ton of CO_2 reduced) and the same policy when all revenues are returned as lump-sum transfers (an annual cost of \$45 billion or \$90 per ton of CO_2 reduced). Their analysis accounts for some broader distortions created by income taxes (not only reductions in work effort but also incentives for excessive spending on goods that receive favorable tax treatment, like employer medical insurance and owner-occupied housing) that are not included in other studies.

Metcalf (2007) proposes using a carbon tax to cut the income tax tied to payroll taxes paid by workers. Specifically, he proposed an environmental tax credit equal to the employer and employee portions of the payroll taxes paid by the worker in the current year, up to a cap. Capping the rebate makes the tax cut more progressive, and the payroll tax cut is greatest for low-wage workers.

Rausch and Reilly (2012) compare payroll and personal income tax approaches to other carbon tax recycling options and find that of all the scenarios, the payroll tax approach has the most evenly distributed benefits across households as a share of income. The payroll tax swap also makes all household groups better off than baseline policies.[15] However, using all the revenue for payroll tax cuts generated less aggregate welfare gain through 2035 than reducing corporate income taxes.

Other ways to use the revenue

Here we briefly touch on use of revenues for deficit reduction, subsidies to household energy bills, and earmarks for other purposes.

Using carbon revenue to reduce the federal deficit may produce even larger benefits than using it to cut current taxes, as this reduces the need for future tax increases and may alleviate upward pressure on interest rates and reduce the risks of financial crises.[16] However, in the absence of a carbon tax, policymakers might increase other taxes or reduce spending, so the most realistic fiscal policy baseline against which a carbon tax should be compared isn't obvious. Gauging the distributional implications of using a carbon tax for deficit reduction is even more difficult given uncertainty over what future tax increases might be avoided due to lower accumulated federal debt.

In general policymakers should be cautious about earmarking some of the revenues from a carbon tax for certain specific purposes as this can significantly increase overall costs to the economy. One potential exception, as discussed in Chapter 10, could be ramping up government funding for basic research and early deployment of emissions-abating technologies, though the costs and benefits of these interventions would need to be carefully assessed, including relative to other potential investments in research and development.

One particular pitfall would be to use the carbon tax revenue to directly reduce households' energy costs. For example, provisions in the Waxman-Markey emissions trading bill would have used the value of emissions allowances to subsidize residential energy bills (see Box 6.1). But these subsidies reduce consumers' inclination to conserve energy, thereby undermining the environmental benefits of the carbon tax. Moreover, as emphasized in the introduction, lowering energy prices is a very inefficient way to target low-income households. In general, if we're worried about effects on certain households, it's better to channel the revenue to them through a tax reduction or rebates, or for that matter pretty much any way other than through their energy bills.

4. Regional incidence

One concern about a single national price on carbon is that areas of the United States that are heavily dependent on fossil fuels, especially coal, will be more burdened than other regions. And indeed the fossil energy intensity of economic activity does vary a lot across the country, as shown in Figure 6.5 below. The figure shows the emissions intensity of the economy of states, measured as tons of energy-related CO_2 emissions per million dollars of Gross Domestic Product (GDP), vary by an order of magnitude across states. In these figures, the emissions include those released where fossil fuels are combusted, including for generating electricity that may be used in another state; they do not include carbon in fuels that are burned in other states. Even so, the most CO_2-intensive states include West Virginia, North Dakota, and Wyoming, all large fossil fuel producers. The states with the least CO_2-intensive economies, including New York and California, have about one-tenth the emissions intensity of the most emissions-intensive states.

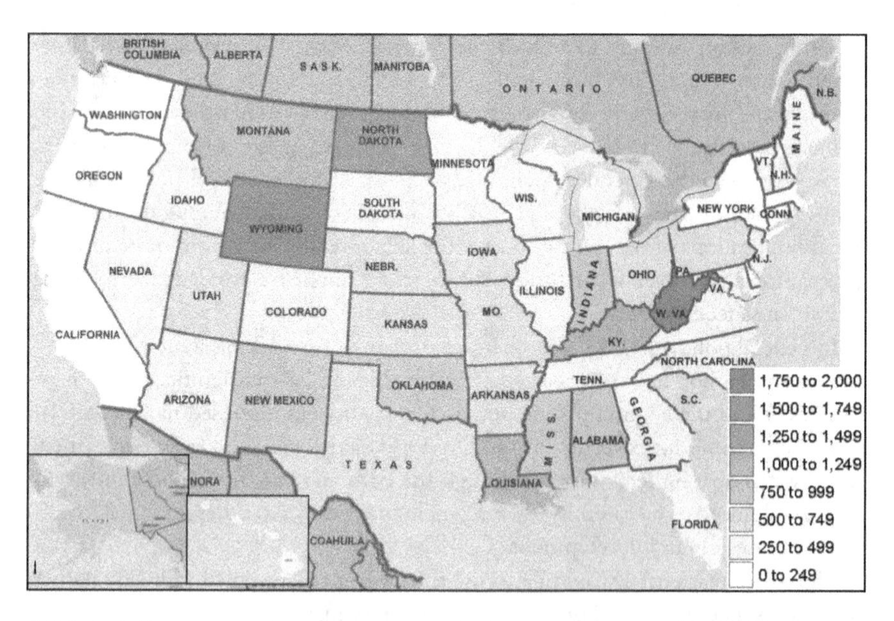

FIGURE 6.5 Energy intensity of state economies in tons of energy-related CO_2 emissions per million of dollars of GDP, 2010

Source: Emissions data come from the U.S. Environmental Protection Agency, www.eia.gov/environment/emissions/state/analysis.

Notes: The figures include emissions from combustion facilities like power plants and the carbon in fuels used in transportation and residential and commercial HVAC.

To be sure, emissions intensity isn't perfectly correlated with the burden of a carbon tax. We noted earlier that the majority of the tax burden will be passed forward to consumers, so states that emit carbon in the production of goods and services that are consumed elsewhere could pass most of that burden to consumers in other states. And just because a state's economic activity is carbon intensive (hence a high emissions per GDP ratio), it doesn't necessarily mean that the goods that its residents buy are carbon intensive. One particular exception, however, is for coal-fired electricity, and the research shows that factor affects the regional distribution of a carbon tax.

Some regional analyses show that the burdens of a carbon tax as a share of income won't vary nearly as much as one might guess from this map.[17] First, some of these regional analyses assume all of the costs of a carbon tax are passed along to consumers. People in different regions use different mixes of fuels to heat and cool homes, and they also vary in their gasoline consumption. In other words, areas where electricity prices may go up most may be the same places where they use relatively less gasoline. For instance, gas consumption is highest in the East North Central region, electricity used more in the West South Central region, and home heating oil consumption is highest in New England. In addition, households in most regions consume similar baskets of non-energy goods, resulting in similar patterns of indirect energy consumption. The biggest item that varies on the consumption side is electricity, which is highly coal intensive in some areas. Studies show that a carbon tax could fall slightly harder than average on households in East Central states (such as Wisconsin, Illinois, Michigan, Indiana, and Ohio) because of their higher overall energy consumption as a share of income.

Areas that currently use relatively more coal in their electricity sectors have relatively low electricity rates, so disproportionately raising electricity rates in those areas would serve to flatten the differences in electricity prices across the country. Palmer et al. (2012) find that a $25 per ton tax of CO_2 would raise national electricity prices by an average of 12 percent, but that the increase could be as little as 4 percent in southern California or as much as roughly 33 percent in Missouri, Kansas, and most of Appalachia. Rausch et al. (2011) also show the South Central regions would have the largest share (17 to 33 percent) of households that experience a net income loss of 1 percent or more. This is driven primarily by the large share of coal in the electricity mix in that area. They also find that wage rates fall the most in the Mountain states as well as the North and South Central states as the impact on industrial activity and hence labor demand in these regions is relatively large due to high energy and emissions intensities. Conversely, the Northeast and West regions with less carbon-intense economies experience relatively small reductions in wage rates.

In theory, policymakers could target carbon tax revenue regionally, for example, through state grants, to address disparities in the burden. However, as we have seen, the burdens are complex and depend on the extent to which one accounts for both sources and uses of income and other factors. In addition, transferring funds to states, even if factoring in regional burdens, doesn't necessarily target the households that bear the burden; it depends on how states use the revenue. One can also imagine that the Congressional negotiations around the formula that might govern this targeting would be contentious.

5. Summary

Estimates of the impacts of carbon taxes on consumer prices and the share of income spent on energy across different households suggest a carbon tax could be decidedly regressive, with the burden as a proportion of income several times higher for the bottom income quintile as for the top income quintile. But the estimated incidence is far less regressive if analysts measure burdens relative to household consumption and take into account (as best as possible) how the tax affects households' sources of income.

Policymakers could ensure that low-income households (as a group) are made no worse off by targeting a small share of the carbon tax revenues, perhaps as little as 11 percent. A robust literature shows that using the preponderance of carbon tax revenues for reductions in other taxes (or the budget deficit) could greatly lower the overall cost of imposing a carbon tax. Some studies indicate that a carefully designed tax swap might produce net economic benefits, not even counting the environmental benefits of the carbon tax. But a tax swap, including the broader fiscal package in which it might feature, should be designed carefully to protect low-income households, lower particularly distortionary taxes, and resist inefficient spending.

The burden of a federal carbon tax would vary somewhat regionally, but it is not clear what, if anything, policymakers should do to account for this. For example, states that currently use a lot of coal tend to have relatively low prices for electricity, so the tax would tend to even out electricity prices across the country. And federal transfers to states may not necessarily reach the most vulnerable households.

Notes

* The authors gratefully acknowledge the research assistance of Danny Cohen, Will Daniel, Nathan Joo, and Daniel Hanson.

1. This chapter draws from Morris (2009), Mathur and Morris (2014), and Morris and Munnings (2013).
2. Benefits include lower damages from extreme weather events, reduced sea level rise, and the like. Poor households, coastal communities, farmers, and other sub-groups may be relatively more vulnerable to these risks, and thus the benefits of GHG abatement may have important socioeconomic, regional, and industry-specific distributional effects.
3. For other studies like this, see Burtraw et al. (2009) and Hassett et al. (2009).
4. For more discussion see, for example, Hassett et al. (2009), Parry et al. (2007).
5. Rausch et al. (2010), Rausch and Reilly (2012), and Fullerton et al. (2011).
6. For example, CBO (2009) estimated that rebating the value of allowances (akin to the potential tax revenue) in the Waxman-Markey emissions trading bill, would result in a net gain of 0.7 percent of income to the bottom income quintile, with all other income quintiles showing less than 1 percent drop in their purchasing power.
7. Some estimates suggest that using carbon tax revenue to lower the deficit or other taxes can lower the overall costs of the program by 75 percent relative to a program that gives the revenue away. See Parry (1997).

8. See Goulder et al. (1999), Parry et al. (1999), Parry and Oates (2000), Parry and Bento (2000), and CBO 2007.
9. See Dinan and Lim Rogers (2002).
10. Feldstein (2006).
11. Of course, given special deductions, accelerated depreciation, and other tax provisions the *effective* tax rate may be much lower for many firms than the statutory tax rate (see Chapter 8).
12. See Chapter 8.
13. It would be hard to compensate all poor households perfectly, however. The burden may evolve over time as households adjust to new prices, and the impacts might vary systematically by region, household demographics and rural/urban location, and other factors that influence households' consumption patterns. That implies a wide distribution of burdens within each quintile, meaning that average compensation is still under-compensation for a significant number of households (Rausch et al. 2011).
14. Elmendorf (2009).
15. Rausch and Reilly (2012), Figure 7, p. 15 and Table 3, p. 7.
16. See Chapters 1 and 5 and Rubin et al. (2004).
17. Mathur and Morris (2014) demonstrate this.

References and suggestions for further reading

Aldy, Joseph E. and Robert N. Stavins. 2012. "The Promise and Problems of Pricing Carbon: Theory and Experience." *Journal of Environment and Development* 21(2): 152–180.

Bovenberg, A. Lans and Lawrence H. Goulder. 2001."Neutralizing the Adverse Industry Impacts of CO_2 Abatement Policies: What Does It Cost?" In Carlo Carraco and Gilbert E. Metcalf (eds.), *Behavioral and Distributional Effects of Environmental Policy.* Chicago: University of Chicago Press, 45–85.

Bovenberg, A. Lans and Lawrence H. Goulder. 2002. "Environmental Taxation and Regulation." In Alan J. Auerbach and Martin Feldstein (eds.), *Handbook of Public Economics.* Amsterdam: Elsevier, 1471.

Bull, Nicholas, Kevin A. Hassett, and Gilbert E. Metcalf. 1994. "Who Pays Broad-Based Energy Taxes? Computing Lifetime and Regional Incidence." *The Energy Journal* 15(3): 145–164.

Burtraw, Dallas. 2009. Testimony at *Climate Change Legislation: Allowance and Revenue Distribution: Hearing before the U.S. Senate Committee on Finance,* One Hundred Tenth Congress, August 4, 2009. http://finance.senate.gov/sitepages/hearing080409.html

Burtraw, D., K. Palmer, R. Bharvirkar, and A. Paul. 2001. "The Effect of Allowance Allocation on the Cost of Carbon Emissions Trading." Washington, DC: RFF Discussion Paper 01–30.

Burtraw, D., R. Sweeney, and M. Walls. 2009. "The Incidence of U.S. Climate Policy: Alternative Uses of Revenues from a Cap and-Trade Auction." *National Tax Journal* 62(3): 497–518.

Burtraw, D. and Sarah Jo Szambelan. 2009. "U.S. Emissions Trading Markets for SO2 and NO*x*." October 2009, RFF DP 09–40. http://rff.org/RFF/Documents/RFF-DP-09–40.pdf

Congressional Budget Office. 2007. "Trade-Offs in Allocating Allowances for CO_2 Emissions." CBO Economic and Budget Issue Brief, April 25, 2007. www.cbo.gov/ftpdocs/89xx/doc8946/04–25-Cap_Trade.pdf

Congressional Budget Office. 2009. "The Economic Effects of Legislation to Reduce Greenhouse-Gas Emissions." September 2009. www.cbo.gov/publication/41266

Congressional Budget Office. 2012. "The Distribution of Household Income and Federal Taxes, 2008 and 2009." Supplemental Data, Publication Number 4441, July 10,

2012. www.cbo.gov/sites/default/files/cbofiles/attachments/43373–06–11-Household IncomeandFedTaxes.pdf

Consumer Expenditure Survey. 2013. "Table 1101:Quintiles of Income before Taxes: Annual Expenditure Means, Shares, Standard Errors, and Coefficient of Variation, 3rd Quarter 2011 through 2nd Quarter 2012." Washington, DC: Bureau of Labor Statistics.

Dinan, Terry. 2012. "Offsetting a Carbon Tax's Costs on Low-Income Households." Working Paper 2012–16, Congressional Budget Office, Washington, DC, November 2012. www.cbo.gov/sites/default/files/cbofiles/attachments/11–13LowIncomeOptions.pdf

Dinan, Terry M. and Diane Lim Rogers. 2002. "Distributional Effects of Carbon Allowance Trading: How Government Decisions Determine Winners and Losers." *National Tax Journal* 55(2): 199–221.

Edenhover, Ottmar and Matthias Kalkul. 2011. "When Do Increasing Carbon Taxes Accelerate Global Warming? A Note on the Green Paradox." *Energy Policy* 39(4, April): 2208–2212. www.zef.de/module/register/media/b59d_Increasing-Resource-Taxes.pdf

Elmendorf, Douglas W. 2009. Director, Congressional Budget Office, *The Distribution of Revenues from a Cap-and-Trade Program for CO_2 Emissions*, testimony before the Committee on Finance United States Senate. May 7, 2009.

Energy Information Administration. 2006a. "Emissions of Greenhouse Gases in the United States 2005." Washington, DC: EIA.

Energy Information Administration. 2006b. "Energy Market Impacts of Alternative Greenhouse Gas Intensity Reduction Goals." Washington, DC: EIA.

Energy Information Administration (EIA). 2009. "Energy Market and Economic Impacts of ACESA." SR/OIAF/2008–01 (April).

Environmental Protection Agency (EPA). 2009. "Analysis of the American Clean Energy and Security Act of 2009 H.R. 2454" in the *One Hundred Eleventh Congress,* June 23, 2009. www.epa.gov/climatechange/Downloads/EPAactivities/HR2454_Analysis.pdf

Feldstein, Martin. 2006. "The Effect of Taxes on Efficiency and Growth." NBER Working Paper No. 12201, May 2006. www.nber.org/papers/w12201.pdf?new_window

Fullerton, D. and G. Heutel. 2007. "The General Equilibrium Incidence of Environmental Taxes." *Journal of Public Economics* 91: 571–591.

Fullerton, D., G. Heutel, and G. E. Metcalf. 2011. "Does the Indexing of Government Transfers Make Carbon Pricing Progressive?" *American Journal of Agricultural Economics* 94(2): 347–353.

Goulder, L. H., I. W.H. Parry, R. C. Williams III, and D. Burtraw. 1999. "The Cost-Effectiveness of Alternative Instruments for Environmental Protection in a Second-Best Setting." *Journal of Public Economics* 72(3): 329–360.

Grainger, Corbett A., and Charles D. Kolstad. 2010. "Who Pays a Price on Carbon?" *Environmental and Resource Economics* 46(3): 359–376. http://dx.doi.org/10.1007/s10640-010-9345-x

Harberger, A. C. 1962. "The Incidence of the Corporation Income Tax." *Journal of Political Economy* 96: 339–357.

Hassett, Kevin and Aparna Mathur. 2011. "Report Card on Effective Corporate Tax Rates: United States Gets an F." American Enterprise Institute, February 9, 2011. www.aei.org/article/economics/fiscal-policy/taxes/report-card-on-effective-corporate-tax-rates/

Hassett, Kevin A., Aparna Mathur, and Gilbert E. Metcalf. 2008. "The Consumer Burden of a Cap-and-Trade System with Freely Allocated Permits." American Enterprise Institute Working Paper No. 144, December 23, 2008, p. 5.

Hassett, Kevin, Aparna Mathur, and Gilbert Metcalf. 2009. "The Incidence of a U.S. Carbon Tax: A Lifetime and Regional Analysis." *The Energy Journal* 30(2, April): 155–178.

Ho, Mun S., Richard D. Morgenstern, and Jhih-Shyang Shih. 2008. "Impact of Carbon Price Policies on U.S. Industry." Discussion Paper 08–37. Washington, DC: Resources for the Future.

Krueger, Dirk and Fabrizio Perri. 2002. "Does Income Inequality Lead to Consumption Inequality? Evidence and Theory." NBER, Working Paper #9202.

Krupnick, Alan J., Ian W.H. Parry, Margaret Walls, Tony Knowles and Kristin Hayes, 2010. "Toward a New National Energy Policy: Assessing the Options." Washington, DC: Resources for the Future and National Energy Policy Institute.

Marron, Donald and Eric Toder. 2013. "Carbon Taxes and Corporate Tax Reform." February 11. Washington, DC: Urban–Brookings Tax Policy Center. www.taxpolicycenter.org/UploadedPDF/412744-Carbon-Taxesand-Corporate-Tax-Reform.pdf

Mathur, Aparna, and Adele C. Morris. 2014. "Distributional Effects of a Carbon Tax in Broader U.S. Fiscal Reform," *Energy Policy* 66: 326–334.

McKibbin, W., A. Morris, and P. Wilcoxen. 2012a. "The Potential Role of a Carbon Tax in U.S. Fiscal Reform." The Brookings Institution, July 24, 2012. www.brookings.edu/research/papers/2012/07/carbon-tax-mckibbin-morris-wilcoxen

McKibbin, W., A. Morris, and P. Wilcoxen. 2012b. "Pricing Carbon in the United States: A Model-Based Analysis of Power Sector Only Approaches." The Brookings Institution, October 5, 2012. www.brookings.edu/research/papers/2012/10/05-pricing-carbon-morris

McKibbin, W. and P. Wilcoxen. 2002. "The Role of Economics in Climate Change Policy." *Journal of Economic Perspectives* 16(2):107–129.

Metcalf, Gilbert E. 2007. "A Proposal for a U.S. Carbon Tax Swap: An Equitable Tax Reform to Address Global Climate Change." The Hamilton Project, Brookings Institution. October. www.hamiltonproject.org/files/downloads_and_links/An_Equitable_Tax_Reform_to_Address_Global_Climate_Change.pdf

Metcalf, Gilbert E. 2008. "Designing a Carbon Tax to Reduce U.S. Greenhouse Gas Emissions." *Review of Environmental Economics and Policy* 3(1, September): 63–83.

Metcalf, Gilbert. 2010. "Submission on the Use of Carbon Fees to Achieve Fiscal Sustainability in the Federal Budget." July 2010. http://works.bepress.com/gilbert_metcalf/86

Metcalf, Gilbert E., Sergey Paltsev, John M. Reilly, Henry D. Jacoby, and Jennifer Holak. 2008. "Analysis of U.S. Greenhouse Gas Tax Proposals." Report No. 160 (2008), MIT Joint Program on the Science and Policy of Global Change. http://globalchange.mit.edu/files/document/MITJPSPGC_Rpt160.pdf

Metcalf, Gilbert and David Weisbach. 2009. "The Design of a Carbon Tax." *Harvard Environmental Law Review* 33(2): 499–556.

Morris, Adele. 2009. "Equity and Efficiency in Cap-and-Trade: Effectively Managing the Emissions Allowance Supply." Brookings Institution Policy Brief 09–05, October 2009. www.brookings.edu/research/papers/2009/10/cap-and-trade-emissions-allowance-morris

Morris, Adele C. 2013. "Proposal 11: The Many Benefits of a Carbon Tax." In Michael Greenstone, Max Harris, Karen Li, Adam Looney, and Jeremy Patashnik (eds.), *15 Ways to Rethink the Federal Budget.* Washington, DC: The Hamilton Project, 63–69. <www.brookings.edu/research/papers/2013/02/benefits-of-carbon-tax>

Morris, Daniel and Clayton Munnings. 2013. "Progressing to a Fair Carbon Tax: Policy Design Options and Impacts to Households." Resources for the Future Issue Brief 13–03, April 2013. www.rff.org/RFF/Documents/RFF-IB-13–03.pdf

Murray, B. C., W. N. Thurman, and A. Keeler. 2000. "Adjusting for Tax Interaction Effects in the Economic Analysis of Environmental Regulation: Some Practical Considerations." U.S. Environmental Protection Agency. White Paper. www.epa.gov/ttnecas1/workingpapers/tie.pdf

National Research Council. 2012. *Climate Change Evidence, Impacts, and Choices.* National Academies of Sciences, 2012. http://nas-sites.org/americasclimatechoices/files/2012/06/19014_cvtx_R1.pdf

Palmer, Karen L., Anthony Paul, and Matthew Woerman. 2012. "The Variability of Potential Revenue from a Tax on Carbon." Resources for the Future Issue Brief 12–03, May 2012.

Paltsev, S., J. Reilly, H. Jacoby, A. Gurgel, G. Metcalf, A. Sokolov, and J. Holak. 2007. "Assessment of U.S. Cap-and-Trade Proposals," April 2007, *MIT Joint Program on the Science and Policy of Global Change,* Report No. 146. http://globalchange.mit.edu/pubs/abstract.php?publication_id=718.

Parry, Ian W. H. 1995. "Pollution Taxes and Revenue Recycling." *Journal of Environmental Economics and Management* 29: S64–S77.

Parry, Ian. 1997. "Reducing Carbon Emissions: Interactions with the Tax System Raise the Cost," *Resources,* Resources for the Future, Summer 1997.

Parry, Ian W. H. 2004. "Are Emissions Permits Regressive?" *Journal of Environmental Economics and Management* 47: 364–387.

Parry, Ian W. H. and Antonio M. Bento. 2000. "Tax Deductions, Environmental Policy, and the 'Double Dividend' Hypothesis." *Journal of Environmental Economics and Management* 39: 67–96.

Parry, I. and W. E. Oates. 2000. "Policy Analysis in the Presence of Distorting Taxes." *Journal of Policy Analysis and Management* 19: 603–614.

Parry, I., H. Sigman, M. Walls, and R. Williams. 2007. "The Incidence of Pollution Control Policies." In T. Tietenberg and H. Folmer (eds.), *International Yearbook of Environmental and Resource Economics 2006/2007.* Northampton, MA: Edward Elgar, 1–42.

Parry, Ian and Roberton Williams III. 2011. "Moving US Climate Policy Forward: Are Carbon Taxes the Only Good Alternative?" www.rff.org/RFF/Documents/RFF-DP-11–02.pdf

Parry, Ian W. H. and Roberton C. Williams. 2012. "Moving US Climate Policy Forward: Are Carbon Tax Shifts the Only Good Alternative?" In Robert Hahn and Alistair Ulph (eds.), *Climate Change and Common Sense: Essays in Honor of Tom Schelling.* Oxford: Oxford University Press, 173–202.

Parry, I. W. H., R. C. Williams III, and L. H. Goulder. 1999. "When Can Carbon Abatement Policies Increase Welfare? The Fundamental Role of Distorted Factor Markets." *Journal of Environmental Economics and Management* 37(1): 51–84.

Pigou, Arthur C. 1952. *The Economics of Welfare.* Transactions Publishers.

Ramseur, Jonathan L., Jane A. Leggett, and Molly F. Sherlock. 2012. "Carbon Tax: Deficit Reduction and Other Considerations." Congressional Research Service, Report R42731, September 17, 2012.

Rausch, S., G. E. Metcalf, and J. M. Reilly. 2011. "Distributional Impacts of Carbon Pricing: A General Equilibrium Approach with Micro-Data for Households." *Energy Economics* 33: S20–S33.

Rausch, S., G. Metcalf, J. M. Reilly, and S. Paltsev. 2009. "Distributional Impacts of a U.S. Greenhouse Gas Policy: A General Equilibrium Analysis of Carbon Pricing." MIT Joint Program Report Series, Report 182, November 2009.

Rausch, Sebastian, Gilbert E. Metcalf, John M. Reilly, and Sergey Paltsev. 2010. "Distributional Implications of Alternative U.S. Greenhouse Gas Control Measures." *The B.E. Journal of Economic Analysis and Policy* 10(2): Symposium, Article 1.

Rausch, Sebastian and John Reilly. 2012. "Carbon Tax Revenue and the Budget Deficit: A Win-Win-Win Solution?" MIT Joint Program on the Science and Policy of Global Change, Report No. 228, August.

Rubin, Robert, Peter R. Orszag, and Allen Sinai. 2004. "Sustained Budget Deficits: Longer-Run U.S. Economic Performance and the Risk of Financial and Fiscal Disarray." Paper Presented at the AEA-NAEFA Joint Session, Allied Social Science Associations Annual Meetings, The Andrew Brimmer Policy Forum, "National Economic and Financial Policies for Growth and Stability," Sunday, January 4, San Diego, CA. www.brookings.edu/views/papers/orszag/20040105.pdf

Shapiro, Robert, Nam Pham, and Arun Malik. 2008. "Addressing Climate Change without Impairing the US Economy: The Economics and Environmental Science of Combining a Carbon-Based Tax and Tax Relief." www.sonecon.com/docs/studies/Carbon TaxReport-RobertShapiro-2008.pdf

Toder, Eric. 2012. "International Competitiveness: Who Competes against Whom and for What?" *Tax Law Review* 65: 505–534.

Viard, Alan. 2009. Testimony at *Climate Change Legislation: Allowance and Revenue Distribution: Hearing before the U.S. Senate Committee on Finance,* One Hundred Tenth Congress, August 4, http://finance.senate.gov/sitepages/hearing080409.html

Wallach, Philip A. 2012. "U.S. Regulation of Greenhouse Gas Emissions." The Brookings Institution, October 2012. www.brookings.edu/research/papers/2012/10/26-climate-change-wallach

7

OFFSETTING A CARBON TAX'S BURDEN ON LOW-INCOME HOUSEHOLDS*

Terry Dinan

KEY MESSAGES FOR POLICYMAKERS

- A carbon tax would raise significant revenues and effectively reduce U.S. carbon dioxide (CO_2) emissions but may run counter to policymakers' distributional objectives as it would create price increases that impose a larger burden, relative to income, on low-income households than on high-income households.
- Only a small fraction (roughly 10 percent) of the higher energy costs that low-income households would incur would be offset by automatic increases in payments that are pegged to the price index (such as Social Security and Supplemental Security Income).
- Policymakers could consider a wide range of options for assisting low-income households, including broad measures affecting the majority of households (e.g., reducing income tax rates or providing payroll tax rebates) and targeted measures (e.g., providing additional payments to those currently receiving electronic transfer benefits or increasing Earned Income Tax Credits).
- This chapter evaluates seven options according to multiple criteria, including: the percent of low-income households affected; whether it would provide comparatively larger benefits for lower-income households; administrative costs; and implications for economic efficiency – specifically, whether it would preserve emission reduction incentives and broader economic benefits from revenue recycling.
- No one option performs best according to all the criteria.

- Several of the broad options could lower the economy-wide cost of a carbon tax by providing increased incentives for individuals to work, save, and invest. Those efficiency-increasing options, however, would disproportionately benefit higher-income households.
- The targeted options could be most effective in reaching households that do not have earnings or that have too little earnings to file income taxes, but typically would not improve incentives to work or invest.
- All but one of the options would preserve the incentive to reduce CO_2 emissions.
- Policymakers could combine options. For example, they could use some of the carbon tax revenue to provide an additional payment to households currently receiving Supplemental Nutrition Action Benefits (commonly known as food stamps) and use the remaining revenue to lower marginal income tax rates, which would increases individuals' incentive to work and invest. Households in the lowest one-fifth of the income distribution could be made whole with 12 percent of the carbon tax revenue. Offsetting the cost for the next lowest fifth would take an additional 15 percent of gross revenue.

Introduction

Taxing the carbon content of fossil fuels offers an efficient method of reducing emissions of carbon dioxide (CO_2), the most prevalent greenhouse gas (see Chapter 3). The effectiveness of the tax stems from the fact that it would increase the prices of goods and services based on the amount of CO_2 emissions associated with the production and use of them: Goods that lead to relatively high emissions, such as coal-fired electricity, would see larger price increases than goods that have relatively low emissions, such as services. Those changes in relative prices are essential to the success of the program because they provide incentives for businesses to produce goods in a manner that result in lower emissions and for households to reduce consumption of energy-intensive goods that cause high emissions.

One significant concern about carbon taxes (or similar carbon pricing schemes), however, is that those price increases would hurt low-income households. Indeed, by some measures, a carbon tax would impose a larger burden (relative to income) on lower-income households than on their higher-income counterparts. Holding energy prices below their true costs (including production costs and environmental costs) would obviate that concern, but is inefficient for at least two reasons. First, such a strategy encourages excessive use of energy, and thus excessive environmental damage. Second, only a small fraction of those implicit energy subsidies (the gap between energy's true cost and its price) are received by low-income households because their better-off counterparts account for the large majority of energy consumption.

Policymakers could consider a variety of methods of compensating low-income households for the higher energy costs that they would face under a carbon tax. This chapter considers seven options (discussed in detail below), which entail using the carbon tax revenue to:

- reduce income tax rates,
- provide income tax rebates,
- provide payroll tax rebates,
- increase incentives for energy-saving investments,
- increase Earned Income Tax Credit (EITC) payments,
- provide an additional fixed payment to households that are eligible for Supplemental Nutrition Action Program (SNAP) payments, and
- increase payments made to households through the existing Low Income Home Energy Assistance Program (LIHEAP).

These options are not comprehensive but include both broad-based policies, which would benefit households throughout the income spectrum, and policies that would explicitly target low-income households.[1]

This chapter first examines the potential effects of a carbon tax on households in various income categories, absent any compensation. It then describes the limited extent to which low-income households might be protected from the effects of the carbon tax by automatic compensation they would receive from indexing of transfer payments and the income tax system. Next it evaluates each of the seven compensation schemes using a consistent set of criteria, which include what share of low-income households would benefit from the measure as well as the measure's effect on environmental outcomes and economy-wide costs. The chapter concludes with a discussion of the portion of the revenue generated by a carbon tax that might be needed to offset the costs incurred by low-income households.

Effects of a carbon tax on low-income households absent compensation

Several studies have examined how setting a price on carbon – either by taxing it or, equivalently, by implementing a cap-and-trade program – would affect households at different points in the income distribution. Most studies have found that such policies would be regressive, imposing larger burdens (relative to their income) on low-income households than on their higher-income counterparts. However, the degree of regressivity varies among studies based on:

- Assumptions about whether the full cost of the tax would be borne by households in the form of higher prices for the goods and services that they buy or whether some of the cost would be borne by households in the form of lower wages and returns to capital. To the extent that the carbon tax lowers wages and returns to capital it would be less regressive.[2]

- The measure used to reflect households' ability to bear the cost of the tax. Annual income is one commonly used measure of this; however, some analysts observe that annual income could understate (or overstate) households' ability to bear higher costs. For example, households with retired members may have relatively low current incomes even though they have accumulated substantial savings. Some studies have addressed that concern by comparing households' costs under a carbon tax to their annual consumption rather than to their annual income (under the assumption that consumption is a better indicator of households' income over an extended period of time). Such studies find that a carbon tax would still be regressive, but much less so than if costs were compared to annual income.[3]

In general, assuming that the full costs of the tax would be passed forward to consumers in the form of higher prices and comparing the corresponding increase in expenditures to households' annual incomes provides an upper bound on how regressive a carbon tax might be. Using that methodology, the Congressional Budget Office (CBO) found that pricing CO_2 emissions at $28 per metric ton (roughly $103 per metric ton of carbon) would impose a cost of $425 dollars per year on the average household in the lowest income quintile – that is, those households would need to pay $425 more to purchase the same goods and services that they would have bought in the absence of the tax – and a cost of $1,380 per year on the average household in the highest income quintile (see Figure 7.1; note that those annual costs were measured based on the size of the economy in 2010 and see Chapter 6 of this volume for a more in-depth discussion of incidence effects).

The increased cost due to the carbon tax would account for 2.5 percent of annual after-tax income for the average household in the lowest income quintile, compared with less than 1 percent of annual after-tax income for the average household in the highest quintile (see Figure 7.2).[4] The price increases caused by a carbon tax

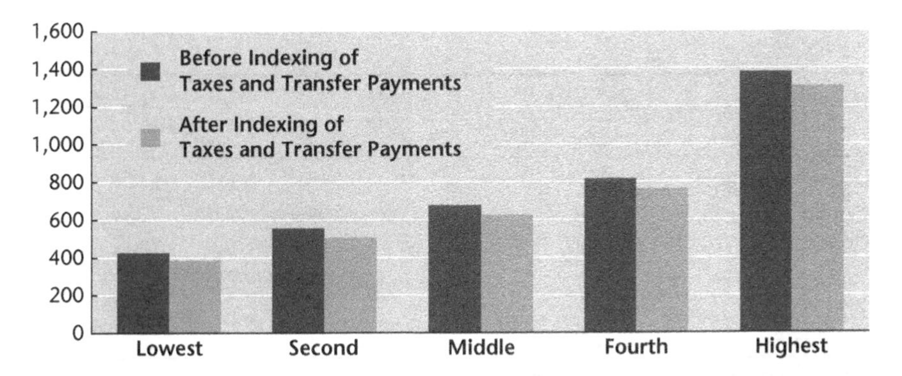

FIGURE 7.1 Estimated annual cost of a $28 tax per ton of CO_2, by income quintile, based on the 2010 economy (2009 dollars)
Source: Dinan (2012).

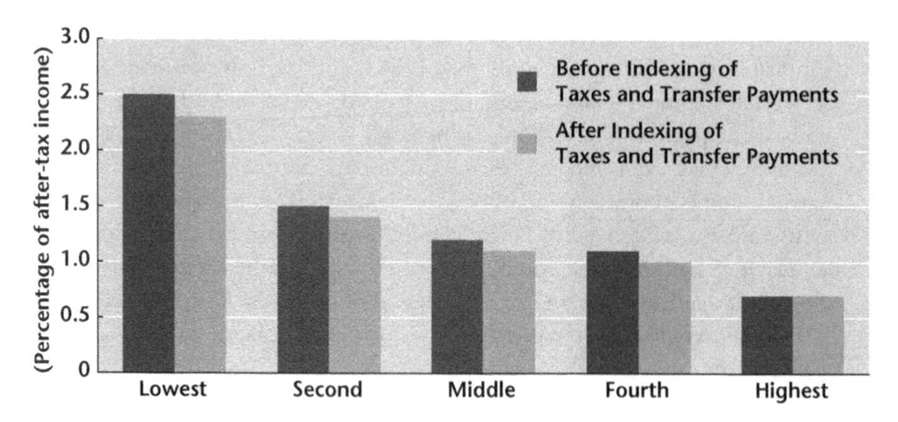

FIGURE 7.2 Estimated cost of a \$28 ton CO_2 tax, by income quintile (percentage of after-tax income)
Source: Dinan (2012).

account for a greater share of annual income for lower-income households than for higher-income households for two reasons: Lower-income households spend a larger fraction of their income, and energy-intensive goods (e.g., electricity used in the home, motor fuels) are necessities and, accordingly, make up a larger share of lower-income households' expenditures (see Table 7.1).

In contrast, the federal tax system (taking into account all federal tax payments by households in various income brackets), is progressive – that is, average tax rates generally rise with income. In 2009, households in the bottom fifth of the before-tax income distribution paid 1.0 percent of their before-tax income in federal taxes, households in the middle quintile paid 11.1 percent, and households in the highest quintile paid 23.2 percent. Average rates were higher for higher-income groups within the top quintile, and households in the top 1 percent of the before-tax income distribution faced an average rate of 28.9 percent.[5]

TABLE 7.1 Average annual household expenditures on energy-intensive items, by income quintile, 2007

	Quintile					*All households*
	Lowest	*Second*	*Middle*	*Fourth*	*Highest*	
Utility expenditures	1,203	1,596	1,840	2,181	2,847	1,934
Gasoline expenditures	1,046	1,768	2,418	2,988	3,696	2,384
Total spending on energy-intensive items	2,249	3,364	4,258	5,169	6,543	4,318
Total as a percentage of income	21.4	12.2	9.2	7.1	4.1	6.8

Source: Congressional Budget Office based on data from Bureau of Labor Statistics, Consumer Expenditure Survey, 2007 (www.bls.gov/cex/2007/Standard/sage.pdf).

Note: Energy-intensive items include natural gas, electricity, fuel oil, other heating fuels, gasoline and motor oil.

Compensation associated with indexing of government transfer payments and federal income taxes

Some households would receive automatic increases in government payments if a carbon tax caused the overall price level to rise.[6]

- Social Security payments and Supplemental Security Income (SSI) payments (stipends provided to low-income people who are 65 or older, blind, or disabled) are pegged to the consumer price index (CPI), so they would automatically increase as the price level rose.
- Supplemental Nutrition Action Payments (SNAP, formerly named Food Stamps), help low-income households purchase food. Those payments are pegged to food prices, so they would increase to the extent that a carbon tax increased grocery bills.

Based on 2006 data, cost-of-living increases for Social Security would offer some compensation for more than 40 percent of households in the lowest 20 percent of the income distribution and for roughly 30 percent of households in the second-lowest income quintile (see Figure 7.3). In addition, 9 percent of households in the lowest quintile and 3 percent of the households in the second-lowest quintile would receive automatic compensation in the form of higher SSI payments.[7] Finally, roughly 20 percent and 5 percent of households in the lowest and second-lowest quintiles, respectively, would receive compensation for higher food prices due to indexing of SNAP payments.

Cost-of-living increases for Social Security and SSI would only partially protect households receiving those benefits – to the extent that income from those

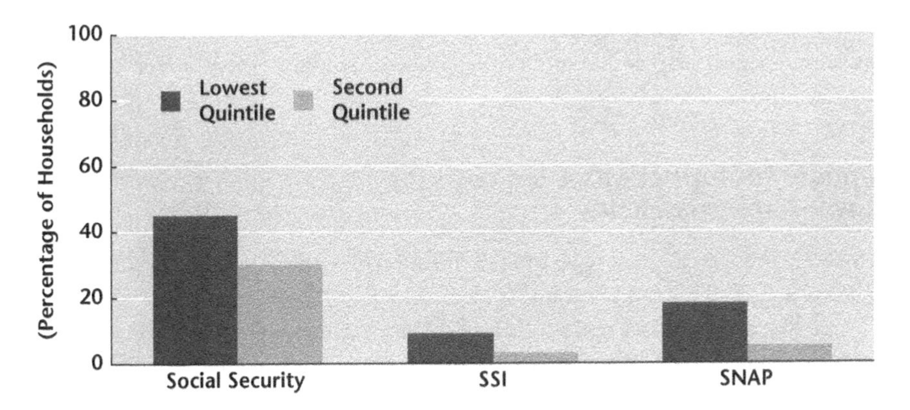

FIGURE 7.3 Low-income households that would benefit from automatic indexing of transfer programs (percentage of households)

Sources: Congressional Budget Office tabulations based on data from the March 2005 Current Population Survey; 2006 Statistics of Income Public Use File.

Notes: SSI = Supplemental Security Income; SNAP = Supplemental Nutrition Assistance Program.

TABLE 7.2 Average annual household expenditures on energy-intensive items, by age, 2007

	Under age 65	*Age 64 and over*	*All households*
Utility expenditures	1,947	1,880	1,934
Gasoline expenditures	2,607	1,461	2,384
Total spending on energy-intensive items	4,554	3,341	4,318
Total as a percentage of income	6.6	8.3	6.8

Source: Congressional Budget Office based on data from Bureau of Labor Statistics, Consumer Expenditure Survey, 2007 (www.bls.gov/cex/2007/Standard/sage.pdf).

Note: Energy-intensive items include natural gas, electricity, fuel oil, other heating fuels, gasoline and motor oil.

sources covers only part of their total expenditures. That effect would be exacerbated because expenditures on energy-intensive items are a higher share of total expenditures both for low-income households (see Table 7.1) and for the elderly (see Table 7.2).

In addition, the rise in the price level would reduce households' tax burden. Because the federal income tax system is largely indexed to the consumer price index (for example, an increase in the price index would reduce the income level at which a 15 percent tax bracket applies), an increase in consumer prices with no increase in nominal incomes would reduce federal income taxes.

On average, the automatic indexing of transfer payments and the tax system would reduce the cost that the carbon tax would impose on households in the lowest income quintile from $425 to $385 (see Figure 7.1) or from 2.5 percent to 2.3 percent of their after-tax income (see Figure 7.2). The average cost borne by households in the second-lowest quintile would fall from $555 to $505 (from 1.5 percent to 1.4 percent of after-tax income). In short, automatic indexing of the tax and benefit system would offset only a modest amount of the burden on low-income households.

Options for further offsetting costs for low-income households

This chapter compares two types of options for further offsetting the costs that low-income households would bear under a carbon tax.

The first type includes options that would direct some carbon tax revenue back to households in a manner that would benefit households in all income brackets, not just those at the lower end of the income distribution. Such possibilities include using carbon tax revenue to:

- Reduce federal income tax rates: A proportionate reduction could be made in all income tax rates or rate reductions could be limited to the lowest income tax brackets.

- Provide income tax rebates: An income tax rebate would provide the same fixed rebate to each recipient. Making the rebate fully refundable would allow households to receive it even if their income was too low for them to owe income taxes.
- Provide payroll tax rebates: Such a rebate could be implemented by exempting a fixed amount of workers' earnings from Social Security and Medicare taxes.
- Increase incentives for energy-saving investments: Providing tax credits for homeowners that make energy efficiency improvements is an example of this type of tax incentive.

The second type includes options that would specifically target low-income households. Those options include using carbon tax revenue to:

- Increase Earned Income Tax Credit (EITC) payments: EITC payments are available to low-income wage earners. Those payments could be increased to account for the average increase in low-income households' costs under a carbon tax or the rate at which the ETIC phases out could be modified.
- Provide an additional fixed payment (reflecting the average increase in low-income households' costs under a carbon tax) to households that are eligible for Supplemental Nutrition Action Program payments, and
- Increase payments made to households through the existing Low Income Home Energy Assistance Program (LIHEAP). Those higher payments would reflect increases in households' energy bills.

Criteria used for comparing policy options

When choosing among policies that could offset the cost increase that low-income households would incur under a carbon tax, policymakers might consider multiple criteria, including:

- *Targeting:*
  - What fraction of low-income households would the policy cover?
  - Would it provide more compensation (relative to their income) for low-income households than for high-income households?
- *Administration:* Could the compensation strategy rely largely on an existing administrative structure and thereby avoid creating new institutional structures or adding new compliance costs?
- *Economic Efficiency:*
  - Would the policy preserve incentives for households to reduce consumption of energy-intensive goods and services? Would it increase incentives for households to work and to invest?

This chapter evaluates each option on the basis of those criteria. The question of whether the compensation policy would increase incentives for individuals to work – for example, for stay-at-home parents to enter the workforce on a full- or part-time basis, for individuals to work additional hours, or to improve their earning potential by seeking additional training or education – and to invest is relevant because the carbon tax itself would discourage work and investment through the impact of higher energy prices on reducing the overall level of economic activity. More technically, the carbon tax would tend to reduce after-tax real wages, thereby reducing the labor supply, and to lower investment and output, which can impose significant costs on the economy. Using the revenue from the carbon tax to counter this adverse effect can significantly reduce the overall cost of the carbon tax policy (see Chapters 2 and 5 for a more detailed explanation). This chapter considers changes in the tax system that could offset the adverse economic effects of the carbon tax itself, but other options are possible as well, for example, using the revenue from a carbon tax to reduce the deficit or to fund investments in basic R&D (see Chapter 10) or in education. Those uses of the revenue would have positive economic effects that would help counter the economy-wide cost resulting from the carbon tax itself.

Results

No single option performs best according to all of the criteria; trade-offs are inevitable. For example, the economy-wide cost of a carbon tax would be considerably lower if the revenue were used to reduce marginal tax rates on labor, capital, or personal income than if it were used to offset costs imposed on low-income households. Reductions in marginal tax rates would help offset the disincentives to work and to invest that a carbon tax would otherwise create; however, they would provide larger relief (measured as a share of income) to higher-income households than to lower-income households. In general, as described below, options that would most effectively provide direct compensation to low-income households tend to be less effective at increasing households' incentives to work and to invest. Policymakers would not, of course, be limited to one form of compensation but could combine different options to meet multiple criteria.

The discussion below does not indicate the net effect (of the carbon tax and the revenue-use option) on households in each quintile or for the economy as a whole. It does, however, discuss available evidence as to whether each option would help offset the regressive effects of the carbon tax and the economy-wide cost of the tax. Furthermore, it discusses the feasibility of implementing various options.

Broad-based compensation options

Policymakers could choose to supplement automatic compensation that households would receive by enacting measures that would distribute some of the carbon tax revenue to households across the full income distribution. Several such broad-based compensation measures were considered in the context of cap-and-trade programs for CO_2 emissions – programs that, like a carbon tax, would lead to higher prices for energy-intensive goods.

Reduction in income tax rates

A proportional reduction in all individual income tax rates would provide the largest percentage increase in after-tax income and the largest dollar amount of tax reductions for taxpayers in the highest income tax brackets of the policies considered in this chapter; taxpayers in the 10 percent or 15 percent tax brackets, who constitute roughly two-thirds of taxpayers with taxable income, would receive minimal benefits. Limiting the rate reductions to the two lowest income tax brackets would cause all taxpayers with income exceeding those brackets to receive the same dollar benefit. For taxpayers whose taxable income fully falls within those brackets, taxpayers whose income put them near the top of the 15 percent bracket ($44,050 for a single taxpayer and $88,100 for a couple in 2012) would benefit the most. Reductions in income tax rates would not help low-income households that do not have sufficient income to owe income taxes. Only 30 percent of households in the lowest income quintile and 64 percent of households in the second-lowest income quintile would benefit from a reduction in income taxes (see Figure 7.4).

A reduction in corporate income tax rates would benefit owners of corporate stock in the short run, with benefits disproportionately going to higher-income households. As capital markets adjusted over the longer term, however, the economic gain from reducing the tax would spread across all types of capital. Over time, at least some of the economic gains could also be shifted to wage earners (as worker productivity improves with greater capital investments), although the degree of such shifting is uncertain. Nevertheless, any gains by low- and moderate-income households from a reduction in corporate taxes would be modest – even over the longer term – and insufficient to offset their increased energy costs.

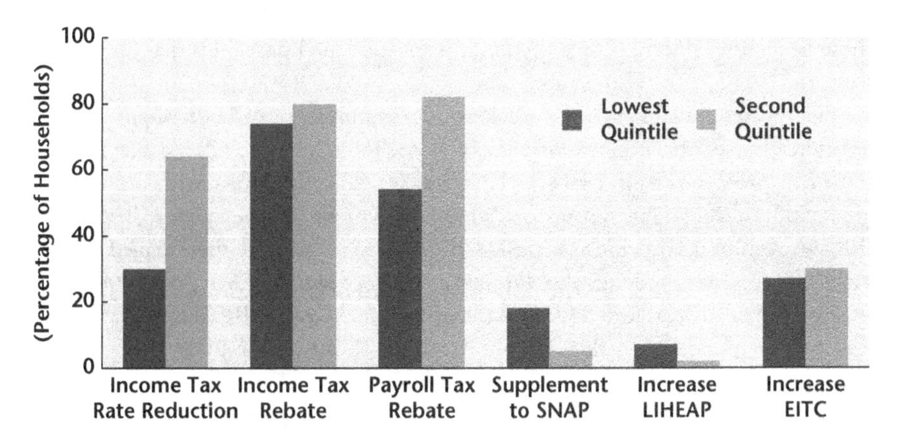

FIGURE 7.4 Low-income households that would benefit from various compensation options based on data on taxes, income, and benefits (percentage of households)

Sources: Congressional Budget Office tabulations based on data from the March 2005 Current Population Survey; 2006 Statistics of Income Public Use File.

Notes: SNAP = Supplemental Nutrition Assistance Program; LIHEAP = Low Income Household Energy Assistance Program; EITC = Earned Income Tax Credit.

Although a reduction in income tax rates would fail to provide compensation for many low-income households, it would be straightforward to administer (requiring no additional administrative costs) and would provide some efficiency advantages:

- Because lower rates would not be tied to households' consumption of energy-intensive goods, they would not offset the carbon tax-induced incentives to consume less of those goods.
- Further, a reduction in income tax rates would provide households with an added incentive to work and to save and invest because lower tax rates boost the after-tax return that households receive from working and investing.

Several researchers have found that using carbon tax revenue to reduce income tax rates could significantly lower the overall cost that a carbon tax would impose on the economy.[8] An income tax rate reduction is the only policy option considered here which would increase most households' incentives to work, save, and invest.

Income tax rebate

Unless an income tax rebate was refundable (that is, payable in excess of the amount of income tax owed), it would be of little or no value to taxpayers who filed income tax returns but owed no income tax – which was the case for approximately 30 million of the 106 million households that filed returns in 2006. Making a rebate refundable would expand the number of eligible households, but would raise administrative complications: Households that do not typically file income taxes would need to file in order to receive the rebate. For example, in 2006, an estimated 10 million households did not file a return. Households with very low income and those headed by elderly people account for most of the households that do not file a return.

The rebates available under the Economic Stimulus Act of 2008 might provide an indication of the number of eligible households likely to file an income tax return in order to claim a rebate. That Act provided taxpayers with a one-time rebate in mid-2008. The rebates, or "stimulus payments" as they were called by the IRS, were announced prior to the 2008-filing season, and were then mailed or sent via direct deposit to individuals who filed a 2007 tax return (during the 2008-filing season) and that had at least $3,000 in qualifying income. The IRS received approximately 156 million individual income tax returns during the 2008 filing season. That total represents an increase of 16 million returns (11.5 percent) over the number received in the previous year.[9] Much of that increase probably represents those filing solely to claim the rebate – the annual increases in returns received during the 2006 and 2007 filing seasons were just 1.6 percent and 3.0 percent, respectively. Although many households appear to have filed a return just to claim the rebate, the number that did so was a bit below expectation. When the Economic Stimulus Act of 2008 was enacted, the Joint Committee on Taxation estimated that $106.7 billion in stimulus payments would be paid in fiscal year 2008. A total of $94.1 billion was actually distributed in that year, although it is difficult to determine how much of the shortfall was attributable to eligible people failing to claim the rebate. The economic stimulus rebates were temporary. The percentage of eligible households that would file under a permanent program would probably be higher.

A refundable tax rebate of a fixed dollar amount would be progressive, providing greater relief as a percentage of income to low-income households. Rebates can be adjusted for differences in family size. They can also be targeted to lower-income taxpayers by reducing (phasing out) the amount of the rebate at higher incomes. For example, the individual income tax rebates that were part of the economic stimulus package enacted in 2008 were reduced by 5 percent of income in excess of $75,000 for individuals and $150,000 for couples. Phasing out a rebate reduces its budgetary cost but adds complexity to the calculation of tax liability and makes the true tax on additional income (the marginal tax rate) less transparent. Based on the $28 tax per ton of CO_2 used in Figure 7.1, a rebate of $385 would be sufficient (on average) to offset the cost that the tax would impose on households in the lowest income quintile (after accounting for indexing of federal income tax rates and transfer payments).

As seen in the experience with stimulus payments, the IRS would need to undertake substantial educational efforts to ensure that many wage earners and others who otherwise would not file income tax returns (because their incomes fall below the statutory requirements for filing) would file returns to obtain the rebate. The increase in administrative costs would be greatest if the rebates were paid to all households, rather than only to those that met certain income requirements. The recent economic stimulus rebates were payable to households without income tax liability if their combined income from earnings, Social Security, and veterans' disability payments was at least $3,000. Allowing households that made less than that amount to claim a refundable income tax rebate would increase administrative costs.

A fixed rebate (not dependent on earnings) would preserve households' incentives to reduce their consumption of energy-intensive goods, but would not provide households with any additional incentives to work or save and thus would not offset any of the overall economic costs associated with a carbon tax program. In contrast, the fixed rebate would boost households' incomes, thus creating a disincentive to work. The magnitude of that effect, however, is likely to be quite small.

Payroll tax rebate

One proposal would issue a payroll tax rebate for Social Security and Medicare taxes as means of offsetting the higher costs that households would face as a result of a carbon tax. Under that proposal, the rebate would apply to the tax on the first $3,660 of earnings. With a combined employee and employer tax rate of 15.3 percent, the maximum energy credit per worker would be $560.[10] Based on the $28 tax per ton of CO_2 used in Figure 7.1, that rebate would more than offset the average cost that the tax would impose on households in the two lowest income quintiles. However, many low-income households do not have earnings and, therefore, would not receive the rebate.

A payroll tax rebate would reach the approximately 159 million workers who are covered under the Social Security and Medicare programs, but would not benefit households without covered earnings.[11] Many of those non-covered households have low income or include retirees. Data from the 2008 Current Population Survey, produced by the U.S. Census Bureau, indicate that, although about 80 percent of all households would be eligible for a payroll tax rebate, only slightly more than half (54 percent) of the households in the lowest fifth of the income distribution would qualify. About three-quarters of the households in that quintile that would not qualify for a payroll tax rebate receive Social Security benefits and thus would be partially protected from higher energy costs by cost-of-living adjustments.

A payroll tax rebate (like any fixed-dollar rebate) would be progressive over most of the income distribution, providing benefits that were a larger percentage of income for lower-income households except for those with the very lowest income and little or no earnings. (However, because the rebate amount would depend upon the number of workers within each household, the rebate would not necessarily be equal for households with the same income.)

Administering a payroll tax rebate would be complicated by a number of issues. Adjusting payroll tax withholding would impose some administrative burden on employers, who also would lack the necessary information to adjust withholding for workers with more than one job. An alternative to adjusting payroll tax withholding would be to pay the rebate through the income tax system when workers file returns. Although that approach would be easier to administer, the timing of the rebate would not coincide with the timing of individuals' increased expenditures. Furthermore, because some workers who pay payroll taxes do not currently file income tax returns, some additional administrative costs would be incurred to process more returns.

A payroll tax rebate would preserve incentives to reduce CO_2 emissions and, by increasing workers' take-home pay, would provide modest incentives for greater participation in the labor force. It would not offer incentives for increased effort or hours worked on the job for people already in the labor force with earnings high enough to qualify for the maximum rebate.

Incentives for energy saving investments

Using revenue from a carbon tax to subsidize household investments that reduce CO_2 emissions would help mitigate households' costs of adapting to higher energy prices. For example, subsidizing weatherization improvements would enable households to use less energy for heating and cooling. Tax incentives for energy savings investments typically favor higher-income households over lower-income households. Data on the utilization of federal residential energy credits for 2006 indicate that both participation and average benefits increased with income (see Table 7.3). That pattern occurs for several reasons: Higher-income households are more likely to have the savings necessary to make the energy-saving investments and are more likely to own their own homes, thereby giving them an incentive to make such

TABLE 7.3 Utilization of federal residential energy credits, 2006

Adjusted gross income	*Number of tax returns (thousands)*	*Percentage of total*	*Percentage claiming residential energy expenses*	*Average usable residential energy credit (dollars)*
Under $15,000	37,614	27.2	0.1	111
$15,000 to $30,000	29,649	21.4	1	197
$30,000 to $50,000	24,907	18	3.2	208
$50,000 to $100,000	30,053	21.7	6.6	223
$100,000 to $200,000	12,110	8.7	8.9	252
$200,000 and above	4,088	3	6.9	305
Total	138,420	100	3.2	230

Source: Congressional Budget Office based on Internal Revenue Service, "Individual Income Tax Returns, Preliminary Data, 2006," *Statistics of Income Bulletin* (Spring 2008), Table 1, pp. 6–17.

investments. Further, higher-income households face higher tax rates, and therefore benefit more from incentives offered in the form of tax deductions (but not from tax credits which do not depend on marginal tax rates).

Moreover, such incentives could increase the total costs (both public and private) of reducing U.S. emissions of CO_2 because they would encourage households to choose certain alternatives over others when adjusting to higher energy prices. For example, a tax credit for solar heating would encourage the use of that technology even if it were not the most cost-efficient alternative in the absence of the credit. Creating a tax incentive system without distorting technology choices is difficult.

Options specifically targeting low-income households

Policymakers could choose to use some of the carbon tax revenue to fund programs that would specifically target low-income households. Those programs could be considered in lieu of, or in addition to, some of the broad-based options described above.

Increase EITC payments

An option based on the current tax system, and targeted specifically toward low-income households, would be to expand the Earned Income Tax Credit. The EITC is a refundable credit, payable to low-income families with earnings. Because the EITC is a refundable credit, wage earners can receive it regardless of whether or not they pay federal income taxes. In 2011, single parents with one child and income up to $36,052 ($41,132 for a married couple) were eligible for the credit. Single parents with two or more children could qualify with income up to $40,964 ($46,044 for a married couple). Childless workers between the ages of 25 and 65 were eligible for a much smaller credit but must have had income less than $13,660 ($18,740 for a married couple) to qualify.[12]

In 2009, taxpayers filed for the EITC on 27 million tax returns. The total amount of the credit was $59.2 billion, of which $54 billion (91 percent) was refundable. Forty-six percent of EITC recipients (accounting for 22 percent of total payments) were families whose income was less than $15,000.[13]

Increasing the EITC payments would be straightforward for the IRS to administer. If the increase was proportional to the existing credit, most of the benefits would go to low-income families with children and very few to childless workers. Increasing the EITC would not provide any benefits to households without earnings, however.

An expansion of the EITC could also yield economic benefits. For example, studies have found that increases in the EITC have had a positive effect on the participation of low-income single women in the labor force – because individuals need earnings to qualify for payments, the EITC encourages individuals to enter the workforce.[14] Although increasing the EITC would raise marginal tax rates for some workers, there appears to be little adverse effect on the number of hours worked by people who are already working.

Fixed payment for households eligible for SNAP

A fixed payment based on the same eligibility rules as those for the Supplemental Nutrition Assistance Program (SNAP, formerly known as the Food Stamp program) would be a way to target benefits to low-income households. To be eligible for SNAP in 2012, an applicant's monthly income must be at or below 130 percent of the poverty guideline ($2,422 for a family of four) and countable assets must be less than $2,000 ($3,325 for households with elderly or disabled members). In 2011 participation in SNAP was the highest it had ever been, with nearly 45 million people (or one in seven U.S. residents) receiving SNAP benefits each month.[15] In fiscal year 2010, the most recent year for which detailed demographic data are available, 85 percent of households receiving SNAP benefits had income (excluding SNAP benefits) below the federal poverty guideline and about three quarters of households receiving SNAP benefits included a child, a person age 60 or older, or a disabled person. Participation was exceptionally high in recent years due to the downturn in the economy. In 2006, SNAP recipients included 18 percent of households in the lowest income quintile and 5 percent of those in the second quintile (see Figure 7.4).

The fixed payment could be distributed to households through the same system as SNAP benefits, which are paid through an electronic benefit transfer system.

Those SNAP benefits are deposited electronically in individual accounts each month, and recipients use a card to debit their account when paying for groceries.

The fixed payment to SNAP recipients could be set equal to the increase in average cost that low-income households are expected to incur under the carbon tax. Such a payment would not undermine incentives for households to reduce their consumption of energy-intensive goods. However, it would not offset any of the economic efficiency costs of higher energy prices because it would not provide households with additional incentives to work or to save and invest.

Increased funding for LIHEAP

The Low Income Home Energy Assistance Program (LIHEAP) currently helps some low-income households pay their energy bills. Federal rules restrict LIHEAP assistance to households with income up to 150 percent of the federal poverty guideline (or 60 percent of state median income, if greater). States, however, can choose to set lower income limits, and as a result, eligibility requirements vary from state to state. In 2009, an estimated 7.4 million households received assistance through LIHEAP – about 16 percent of federally eligible households.

Providing assistance for carbon-tax-induced increases in energy bills to all low- and moderate-income households would require a major expansion of the program, a substantial increase in administrative costs, and possibly a major overhaul of the program. The current program is funded as a block grant from the federal government to the states and other entities, leaving wide latitude in the types of assistance provided.

Increasing LIHEAP subsidies could raise the overall cost of achieving any given level of emission reductions, because it would somewhat offset the price signals necessary to motivate households to undertake low-cost reductions.[16] Further, LIHEAP subsidies would not provide any additional incentives for households to work and invest.

Revenue needed to keep low-income households whole

Based on earlier CBO work, fully offsetting the additional cost that a carbon tax would impose on households in the lowest income quintile would take roughly 12 percent of the gross revenue collected by a carbon tax. Offsetting the cost for the second quintile would take an additional 15 percent of gross revenue.[17] Those calculations do not account for the extent to which a carbon tax might increase the federal government's own costs (for example, because of the increase in transfer payments and because of the federal government's own higher energy costs if the burden of the tax is passed forward to consumers in the form of higher prices) or might decrease other tax collections (for example, because of lower wages and profits if the burden of the tax is passed backward on the factors of production). Accounting for those effects would mean that the net increase in tax revenue available as a result of the carbon tax would be less than gross tax receipts; thus, the shares of net additional revenue required to protect households in the lowest two quintiles would be somewhat higher than the percentages described above.

Conclusions

Choosing among options for offsetting costs that low-income households might incur under a carbon tax would generally involve a trade-off between various criteria. For example, some options that specifically target low-income households, such as providing a supplement for all households eligible for SNAP payments, would not increase households' incentive to work and invest, but would direct relief at the most vulnerable households. In contrast, a proportional reduction in personal income tax rates would provide such an incentive but would direct larger benefits toward higher-income households (see Table 7.4).

Some policies could provide relief to low-income households, while also producing some degree of efficiency gains. For example, lowering payroll tax rates on a portion of earnings or reducing the rate at which the EITC phases out would target more relief to lower-income families than would a reduction in corporate tax rates, while offsetting some of the adverse economic effects of the program. In general, however, policymakers could best achieve multiple goals by combining different options for using the revenue. For example, using some of the revenue from a carbon tax to provide a supplemental payment to SNAP recipients could lessen the burden of a carbon tax on low-income households and using the remaining revenue to reduce personal income tax rates could mitigate the cost that the carbon tax might otherwise impose on the economy as a whole.

An important consideration in using revenues to provide assistance to households is the amount of new administrative or compliance costs. Using existing transfer programs or providing rebates through the income tax system would avoid creating new institutional structures for administering payments. Existing systems that already collect information on household income also are well suited to targeting assistance based on need. However, no single existing system would reach all households. Not everyone – especially members of low-income households and retirees – pays payroll taxes or files an income tax return. But people would need to file a return to participate in a rebate program based on the income tax system.

In theory, delivering rebates through a combination of the existing tax system and the existing transfer programs would better reach affected households than would relying on either approach by itself and would not require a new program. For example, data from the Current Population Survey indicate that about 95 percent of households would qualify for a payroll tax rebate (assuming that it was provided to individuals with earnings) or for an automatic cost-of-living increase in Social Security benefits, including 85 percent to 90 percent of households in the lowest income quintile. It is not easy, however, to coordinate rebates with existing

TABLE 7.4 Evaluation of policy options with respect to three criteria

	Targeting			*Administration*	*Economic efficiency*		
	Percent of households covered[a]		*Larger proportional benefits to lower-income households?*	*Requires significant increase in administrative costs?*	*Increase incentives to work?*	*Increase incentives to invest?*	*Preserve incentives to reduce emissions?*
	Lowest quintile	*Second quintile*					
Broad Based Measures							
Income tax rate reduction	30	64	No	No	Yes	Yes	Yes
Income tax rebate	74	80	Yes	Yes, if fully refundable	No	No	Yes
Payroll tax rebate	54[b]	82[b]	Yes	No[c]	Small increase	No	Yes
Energy saving investment tax credit	n.a.	n.a.	No	No	No	No	Yes
Targeted Measures							
Fixed payments for SNAP-eligible households	18	5	Yes	No	No	No	Yes
LIHEAP	7	2	Yes	No, if current recipients[d]	No	No	No
Expand EITC	27	30	Yes	No	Small increase	No	Yes

Source: Offsetting a Carbon Tax's Costs on Low Income Households.

Notes: n.a. = not applicable; SNAP = Supplemental Nutrition Assistance Program; LIHEAP = Low Income Household Energy Assistance Program; EITC = Earned Income Tax Credit.

[a] Indicates the percentage of households receiving income or benefits in 2006 from the program that the policy option would use for delivering compensation.

[b] Indicates share of households with earnings that would make them eligible for the payroll tax rebate.

[c] Administering a payroll tax credit could be straightforward, but preventing individuals with more than one job from receiving multiple rebates would raise complications.

[d] Providing additional payments to all households that are eligible for LIHEAP could entail a significant increase in administrative costs because only 16 percent of eligible households receive benefits.

programs to avoid overlap and to ensure that economically equivalent households receive roughly the same benefit.

Notes

* The views expressed in this chapter are those of the author and should not be interpreted as the views of the Congressional Budget Office. The author is grateful to Ed Harris, Kevin Perese, Joseph Kile, Frank Sammartino, Andrew Stocking, and David Weiner for assistance with data analysis and for helpful comments and suggestions.

1. There are many other ways in which the revenue from a carbon tax (or, equivalently, the value of allowances in a cap-and-trade program) could be used. For example, under a cap-and-trade program passed by the House of Representatives in 2009 (The American Clean Energy and Security Act of 2009, H.R. 2454), the federal government would sell some of the allowances and use the revenue to provide consumer refunds on a per-capita basis as well as to provide energy rebates to low-income households. Other allowances would be given to local distributers of electricity, who would have to use the value of the allowances for the benefit of their customers. Some businesses that were expected to face higher costs as a consequence of the cap-and-trade program would have received free allowances directly from the government. For a discussion of the distributional affects of H.R. 2454 see Congressional Budget Office (2009). Alternatively, some analysts and policymakers have advocated a "cap-and-dividend" approach, in which allowances would be sold and the proceeds returned to households on an equal per-capita basis. (See Barnes (2008.)
2. For example, see Rausch et al. (2011).
3. For example, see Hassett et al. (2009). Based on 2003 data they find that the burden that a carbon tax placed on households in the lowest income decile would be 4.6 times that of the burden placed on households in the highest decile when annual income is used as a measure of households' ability to pay. When annual consumption is used as a measure of ability to pay, the burden imposed on households in the lowest income quintile is 1.7 times higher than the burden imposed on households in the highest quintile. That ratio of burdens decreases even more, to 1.25, when a measure of consumption over the life cycle is used.
4. Congressional Budget Office (2009).
5. Congressional Budget Office (2012a).
6. For the purposes of this chapter, it is assumed that the overall price level will rise as a result of the tax. If the Federal Reserve were to take action to prevent that from happening, the overall price level would stay constant but the relative price of carbon-intensive goods and services would still rise. Either way, the change in relative prices would disproportionately affect low-income households to the extent that carbon-intensive goods and services, such as gasoline and electricity, constitute a relatively large share of their consumption.
7. For additional discussion of the effects of indexing and policy options for offsetting costs that households would incur under policies that set a price on CO_2 emissions, see Congressional Budget Office (2008), Dinan (2009), and Elmendorf (2009).
8. For example, see Parry and Williams (2012) and Rausch and Reilly (2012).
9. The significant gap between the number of returns filed in 2007 (140 million) and the number of households filing returns in 2006 (106 million) is explained by the fact that some households file multiple returns.
10. Metcalf (2007). For 2012, the employee tax rate for Social Security was temporarily reduced by two percentage points; therefore the combined employee and employer tax rate was 13.3 percent.
11. Such rebate would not have to affect the financial status of Social Security and Medicare or the future retirement benefits of workers. Workers would receive credit for their full

covered earnings, and the Social Security and Medicare trust funds could be credited for the full amount of the payroll tax. See www.ssa.gov/pressoffice/basicfact.htm.

12. See www.irs.gov/Individuals/EITC-Income-Limits,-Maximum-Credit-Amounts-and-Tax-Law-Updates.
13. See Department of Treasury, Internal Revenue Service (2011).
14. See Meyer (2007) and Eissa and Hoynes (2006).
15. Congressional Budget Office (2012b).
16. Under LIHEAP, states are directed to provide the highest level of assistance to households with the highest energy bills, relative to their income, thus, households with higher energy bills tend to receive higher payments; see 42 U.S.C. 8624(b)(5).
17. Those calculations are based on CBO's analysis of the cost that low-income households would incur under the cap-and-trade program that would have been established under H.R. 2454. The gross cost borne by all households – based on the $28 price on emissions that CBO estimated for 2020, but measured in the context of the 2010 economy – totaled $109.6 billion. The market value of allowances and the value of domestic and international offsets represent the gross revenue that would be collected under an equivalent tax and totaled $104.7 billion. Households in the lowest and second-lowest income quintiles incurred 11 percent and 14 percent of the total gross cost, respectively, accounting for 12 percent and 15 percent of the gross revenue. See Congressional Budget Office (2009).

References

Barnes, Peter, "Cap and Dividend, Not Trade: Making Polluters Pay," *Scientific American*, December 18, 2008.

Congressional Budget Office, "Options for Offsetting the Economic Impact on Low- and Moderate-Income Households of a Cap-and-Trade Program for Carbon Dioxide Emissions," Letter to the Honorable Jeff Bingaman, June 17, 2008; www.cbo.gov/publication/41704.

Congressional Budget Office, "The Estimated Costs to Households from the Cap-and-Trade Provisions of H.R. 2454," Letter to the Honorable Dave Camp, June 19, 2009; www.cbo.gov/publication/41194.

Congressional Budget Office, "The Distribution of Household Income and Federal Taxes, 2008 and 2009" (July 10, 2012a); www.cbo.gov/publication/43373.

Congressional Budget Office, "The Supplemental Nutrition Assistance Program" (April 2012b); www.cbo.gov/publication/43173.

Department of Treasury, Internal Revenue Service, Individual Income Tax Returns 2009 Publication 1304 (Rev. 07–2011); www.irs.gov/pub/irs-soi/09inalcr.pdf.

Dinan, Terry, "The Distributional Consequences of a Cap-and-Trade Program for CO_2 Emissions," Testimony before the Subcommittee on Income Security and Family Support, Committee on Ways and Means, U.S. House of Representatives, March 12, 2009; www.cbo.gov/publication/41168.

Dinan, Terry, "Offsetting a Carbon Tax's Costs on Low-Income Households: Working Paper 2012–16," Congressional Budget Office (November 13, 2012); www.cbo.gov/publication/43713.

Eissa, Nada and Hilary W. Hoynes, "Behavioral Responses to Taxes: Lessons from the EITC and Labor Supply," National Bureau of Economic Research Series: *Tax Policy and the Economy*, vol. 20, James M. Poterba, ed., Cambridge, Mass.: MIT Press, 2006, pp. 73–110.

Elmendorf, Douglas, "The Distribution of Revenues from a Cap-and-Trade Program for CO_2 Emissions," Testimony before the Committee on Finance, United States Senate, May 7, 2009; www.cbo.gov/publication/41183.

Hassett, Kevin A., Aparna Mathur, and Gilbert E. Metcalf, "The Incidence of a U.S. Carbon Tax: A Lifetime and Regional Analysis," *Energy Journal,* April 1, 2009, vol. 30 (2), pp. 155–77; www.highbeam.com/doc/1G1–197990922.html.

Metcalf, Gilbert E., *A Green Employment Tax Swap: Using a Carbon Tax to Finance Payroll Tax Relief,* The Brookings Institution, June 2007; http://pdf.wri.org/Brookings-WRI_GreenTaxSwap.pdf.

Meyer, Bruce D., "The U.S. Earned Income Tax Credit, Its Effects, and Possible Reforms," University of Chicago's Harris School of Public Policy Studies, and National Bureau of Economic Research, August 2007; http://harrisschool.uchicago.edu/research/working-papers-series/papers/us-earned-income-tax-credit-its-effects-and-possible-reforms.

Parry, Ian W.H. and Roberton C. Williams III, "Moving US Climate Policy Forward: Are Carbon Taxes the Only Good Alternative?" *Climate Change and Common Sense: Essays in Honour of Tom Schelling,* Robert W. Hahn and Alistair Ulph, eds., Oxford, England: Oxford University Press, 2012, pp. 173–202.

Rausch, Sebastian and John Reilly, "Carbon Tax Revenue and the Budget Deficit: A Win-Win-Win Solution?" MIT Joint Program on the Science and Policy of Global Change, Report No. 228, August 2012; http://globalchange.mit.edu/research/publications/2328.

Rausch, Sebastian, Gilbert E. Metcalf, John M. Reilly, and Sergey Paltsev, "Distributional Impacts of a U.S. Greenhouse Gas Policy: A General Equilibrium Analysis of Carbon Pricing," *U.S. Energy Tax Policy,* Gilbert E. Metcalf, ed., Cambridge, England: Cambridge University Press, 2011, pp. 52–107.

8

CARBON TAXES AND CORPORATE TAX REFORM

*Donald B. Marron and Eric Toder**

KEY MESSAGES FOR POLICYMAKERS

This chapter examines the pros and cons of using a carbon tax to help finance corporate tax reform. We find:

- Revenues from a plausible carbon tax would be large relative to corporate tax revenues and could thus help finance lower corporate tax rates, extension of business tax preferences, or other corporate tax reforms.
- Done well, such a tax swap could reduce the environmental risks of carbon emissions and improve the efficiency of America's corporate tax system.
- But a carbon-for-corporate tax swap poses a significant distributional challenge. A carbon tax would fall disproportionately on low-income families, while a reduction in corporate taxes would disproportionately benefit those with high incomes.
- Policymakers can offset some of those impacts through other policy measures, such as paying lump-sum tax rebates (for further analysis, see Chapter 7). But doing so would reduce the swap's efficiency benefits. Substituting other measures, such as a payroll tax cut, would mitigate the efficiency loss but provide relatively less relief to the lowest-income households than a lump sum rebate.
- Policymakers may also want to use some carbon revenues for deficit reduction. One option would be to aim for revenue neutrality over an initial period, after which a widening spread between growing carbon revenues and relatively stable corporate tax cuts would reduce the deficit.

Carbon taxes and corporate tax reform

Putting a price on carbon dioxide and other greenhouses gases would be the most cost-effective way to reduce future emissions and some of the potential risks of climate change. But pricing carbon has garnered little support from U.S. policymakers to date. At the same time, policymakers feel rising pressure to address America's inefficient and needlessly complex tax system and our long-term fiscal imbalances, but progress has been limited. One way forward might be to combine these policy challenges into a larger whole, with revenues from a new carbon tax being used to finance some combination of tax reform and deficit reduction.[1]

In this chapter, we consider the pros and cons of one potential pairing: combining a carbon tax with reform of the U.S. corporate tax system. Such a pairing would make good sense from the perspective of economic efficiency. By internalizing some of the external costs of carbon emissions, a carbon tax would reduce future carbon emissions and reduce or delay potential harm from climate change. In addition, corporate tax reform could boost the economy, offsetting some or all of the negative impact of new carbon taxes. Dinan and Lim Rogers (2002), for example, found that using carbon revenues to reduce corporate income taxes could offset about 60 percent of the economic cost of limiting carbon emissions, while Rausch and Reilly (2012) find that such a swap could actually increase economic output.

A carbon-for-corporate tax swap may also be timely for political reasons. Policymakers from across the political spectrum have expressed interest in corporate reform to help improve U.S. competitiveness and encourage domestic investment. During the recent presidential campaign, for example, both former Governor Mitt Romney and President Barack Obama proposed large cuts in the corporate income tax rate which, at 35 percent, has become very high by international standards (see below). Obama proposed lowering the rate to 28 percent (25 percent for manufacturing companies), while Romney proposed going down to 25 percent. Many members of Congress – most notably Representative Dave Camp, Chairman of the House Ways and Means Committee – have also proposed steep reductions in the corporate tax rate.

Most of these policymakers have proposed paying for these rate reductions by reducing "loopholes" and other business tax preferences, but details about which preferences they would remove have generally been lacking. Indeed, there is growing recognition that paying for large corporate rate reductions by rolling back tax preferences is harder than it sounds (see Box 8.1 and Joint Committee on Taxation 2011). That challenge is even larger when you consider the potential revenue cost of extending the research and experimentation credit and dozens of other popular business tax breaks, collectively known as the "extenders," that have recently been extended only through the end of 2013 (Marron 2012).

Thus, for both economic and political reasons, policymakers are in the market for a way to finance corporate tax reform. That alone makes new revenue sources, including a carbon tax, worth a close look.

This chapter thus proceeds as follows. Section 1 describes the rationale for corporate tax reform and notes that it involves more than just lowering the corporate tax

BOX 8.1 THE LIMITS OF TRIMMING TAX BREAKS TO CUT CORPORATE RATES

The difficulty of funding corporate rate reductions by eliminating business tax breaks can be illustrated by examining the size of revenue losses from corporate tax expenditures in relation to projected corporate receipts. On the surface, these revenue losses seem large; the Office of Management and Budget (OMB 2012) estimates revenue losses from corporate tax breaks total $880 billion between fiscal years 2013 and 2017, more than 40 percent of projected corporate tax receipts of $2.15 trillion. That figure includes the total revenue loss from all tax breaks for which corporate income revenue losses exceed personal income revenue losses. The $880 billion thus includes about $200 billion of revenue losses in the individual income tax from business tax expenditures that benefit owners of businesses taxed as "flow-through enterprises," including sole proprietorships, partnerships, limited liability companies, and subchapter S corporations.

That's a sizeable amount of money, even in Washington. But the practical potential appears much smaller when one considers the specific provisions. Over two-thirds of that revenue loss comes from just two provisions – accelerated depreciation and deferral of income in controlled foreign corporations. In both cases, the potential revenue gain from repeal is uncertain (the Joint Committee on Taxation scores them as costing much less than OMB), political leaders have proposed making them more instead of less generous, and there are plausible policy grounds to oppose repeal.

If policymakers repealed all corporate tax expenditures listed by OMB except these two and expensing of research costs, that would allow them to lower the corporate tax rate only to 31.5 percent if they want to maintain overall revenue. Enacting all the specific changes in the corporate tax base that President Obama has recommended would finance a cut of only 1.5 points, to 33.5 percent. In short, trimming tax breaks has limited potential for financing significant cuts in corporate tax rates.

rate. Section 2 considers how much revenue a carbon tax might generate and how that compares to the potential revenue loss in various corporate tax reforms. Section 3 documents a fundamental distributional mismatch in a carbon-for-corporate tax swap: introducing a carbon tax would particularly hit low-income households, while cutting corporate taxes would particularly benefit high-income ones. Policymakers may thus want to include additional policies, such as lump-sum rebates, to help low-income families, but doing so would reduce the economic benefits of the tax swap. Section 4 discusses how a carbon-for-corporate tax swap could leave room for deficit reduction, and section 5 concludes.

1. Corporate tax reform

There are three main concerns about the current corporate tax system: its harmful economic effects, uncertainty about important tax preferences, and the treatment of multinationals. We address each in turn.

There is accumulating evidence that the corporate income tax can be particularly harmful to the economy. Surveying the experience of developed countries, for example, researchers at the Organization for Economic Cooperation and Development (OECD) concluded that "corporate income taxes appear to have a particularly negative impact on GDP per capita," with much of that effect coming from negative impacts on "dynamic and innovative" firms (Johansson et al. 2008, p. 43).[2]

Such concerns have increasing traction in the United States, with many policymakers wondering whether our corporate tax system is contributing to a loss of international competitiveness (Toder 2012). Policy discussions about corporate tax reform usually begin with a simple and striking observation: the United States now has the highest statutory corporate tax rate in the developed world. With a federal rate of 35 percent and an average state rate of 6.3 percent, the combined U.S. rate is roughly 39.1 percent.[3] As Hassett and Mathur (2011) document, the U.S. corporate tax rate was only slightly above the OECD median in 1981, but has become an extreme outlier as other nations have cut their rates. That high rate raises legitimate concerns about the ability of the United States to compete for mobile investment capital and the incentive for multinational corporations to engage in complex strategies to shift taxable profits abroad (Toder 2012).

But statutory rates are not the only measure that matters. The incentive to invest in the United States instead of overseas depends more on relative marginal effective tax rates (METRs), where the METR measures the percentage reduction in the after-tax rate of return on a new investment due to corporate taxes. The METR depends on both the statutory rates and on tax preferences, such as accelerated cost recovery provisions or tax credits, which reduce taxable income below economic income or provide direct rebates.

Based on World Bank and OECD data, Hassett and Mathur (2011) report that the United States has a relatively high METR of 23.6 percent. In comparison, a simple average METR for OECD countries other than the United States is more than 6 percentage points lower (17.2 percent). But part of the reason for the large difference is the relatively lower METRs in small countries, which have less market power and are less competitive with the United States for real investments than larger economies. Using the data reported in Hassett and Mathur, we estimate a GDP-weighted METR in OECD countries excluding the United States of 21.6 percent, only slightly lower than the U.S. rate. And for the rest of the large developed nations that constitute the G7, the average METR in 2010 was 24.5 percent, slighter higher than the U.S. corporate METR. So while the high statutory rate certainly encourages shifting of reported profits outside the United States, the United States may not be at an extreme disadvantage in attracting corporate capital compared with its largest trading partners.

High corporate tax rates are not the only concern. Persistent uncertainty about the corporate tax code, which creates an uncertain climate for business investment, is a second motivation for corporate tax reform. Over the past dozen years, Congress has enacted an increasing number of temporary business tax provisions. In a few cases, these provisions appear to have been intended as temporary fiscal stimulus. This is true, for example, of bonus depreciation provisions that allow firms to immediately expense a portion of their investment costs. Many other incentives have been enacted on a temporary basis, however, simply because policymakers have not been able to find sufficient budget resources to pay for permanent extension, even for tax incentives as popular as the credit for research and experimentation (often known as the R&D credit) and favorable depreciation rules for small business (Marron 2012). Finding a way to decide which of these provisions to remove and which ones to make permanent – or at least long-lived – is another important goal for tax reform. While a sensible tax reform would surely allow some of these tax benefits to lapse, others (such as the research and experimentation credit) might be extended, which would reduce revenues compared to CBO and OMB projections.

A third set of concerns involves the way that the United States taxes the international income of multinationals that are headquartered here. Currently, U.S. multinationals may defer active profits earned overseas in foreign subsidiaries until the profits are repatriated in the form of dividends to the U.S. parent company. This hybrid treatment of foreign profits satisfies almost no one. Deferral can encourage U.S. multinationals to invest overseas instead of at home and enable tax avoidance by complex schemes that shift reported income to tax havens. But U.S. corporations complain that the repatriation tax keeps their cash locked up overseas and that they are placed at a competitive disadvantage versus multinationals based in other countries that have territorial systems (with full exemption of active income earned overseas). And everyone on both sides of the debate complains about the complexity of international tax rules. The challenging issues raised by international taxation provisions are beyond the scope of this short chapter. But do keep in mind that international reform will almost certainly be part of any broad-reaching discussion of corporate tax reform and, depending on how it is structured, could require offsetting revenue.[4]

2. Carbon revenues and corporate reform

Potential carbon tax revenues

The Congressional Budget Office (2011) presented one carbon pricing option that has since become commonly used in carbon tax analyses. CBO considered a cap-and-trade system for limiting emissions of carbon dioxide and other greenhouse gases. The emissions cap would be set so that allowances would trade at $20 per ton of carbon in the first year and then rise at a nominal rate of 5.6 percent annually (about 3.6 percent annually in real terms, given CBO's long-run projected inflation rate of 2 percent). CBO considered the case where the government would auction all of these allowances, which would be effectively equivalent to a carbon tax of $20 per ton, escalating at 5.6 percent annually.

CBO estimated that this policy would raise about $1.2 trillion over the decade 2012 to 2021, the standard congressional budget window at the time; for simplicity, we assume that it would raise the same annual amounts, starting a year later, for a tax beginning in 2013 and continuing through 2022. That figure reflects the direct revenues from the carbon tax and a partially offsetting reduction in income and payroll taxes.[5]

Reducing the corporate tax rate

In late 2011, the Joint Committee on Taxation (JCT) estimated that reducing the corporate tax rate from 35 percent to 28 percent would lower federal revenues by about $720 billion over 10 years (JCT 2011). Since then, the official budget window shifted and CBO and JCT updated their baseline projections, which now show substantially higher corporate revenue. Adjusting for those changes through August 2012, we estimate that cutting the corporate tax rate from 35 percent to 28 percent would reduce revenues by about $800 billion over the next 10 years and that cutting the rate from 35 percent to 25 percent would reduce revenues by about $1.15 trillion.[6]

Note that these revenue losses are smaller, in relative terms, than the corresponding rate cuts. Cutting the corporate tax rate from 35 percent to 28 percent would be a 20 percent rate reduction, for example, but JCT shows only an 18 percent reduction in revenues. Among other factors, that difference reflects JCT's recognition that lowering corporate income taxes will increase after-tax corporate profits and thus increase the taxable income of individuals, who may receive higher taxable dividends or realize more capital gains.

Making expiring tax provisions permanent

A second goal for many reformers is to find a way to make permanent (or, at least, long-lived) many tax breaks that have recently expired or will do so in the next few years. Table 8.1 lists the 10 largest expiring business provisions as of 2012 (before the passage of the American Taxpayer Relief Act in early 2013, which extended most of them for one year), excluding partial expensing, a stimulus program that we treat as being truly temporary. Two of these expiring provisions are complex rules about taxing the foreign income of domestic multinationals, two offer favorable depreciation allowances, four are incentives for alternative energy sources, one encourages work, and one encourages research and development.

Making all of these business tax provisions permanent would reduce revenues by about $450 billion over the next decade. If some are allowed to expire, however, the overall price tag of making expiring provisions permanent would be less than this $450 billion total.

The revenue loss from extending those provisions that allow corporations to defer or avoid tax on some of their income would also be lower if policymakers

TABLE 8.1 Revenue losses from permanent extension of expiring business tax provisions, 2013–2022

	Expiration	*$ Billions*
Subpart F for Active Financing Income	12/31/11	88
Credit for Research and Experimentation	12/31/11	68
Liquefied Hydrogen Fuel Incentives	9/30/14	65
Alcohol Fuel Tax Credit	12/31/11	59
Section 179 Expensing	12/31/12	36
Depreciation of Leasehold and Restaurant Equipment	12/31/11	19
Payments Between Related Controlled Foreign Corps.	12/31/11	16
Electricity Production Credit for Wind Facilities	12/31/12	12
Biodiesel and Renewable Diesel Credits	12/31/11	12
Work Opportunity Tax Credit	12/31/11	11
Other Expiring Corporate Provisions	Varies	69
Total		**446**

Source: Congressional Budget Office (2012b).

Note: These figures reflect the total revenue impact of these provisions, including both corporate and non-corporate businesses, in current dollars summed over the budget window. These provisions (except the alcohol fuel tax credit) were extended retroactively to 2012 and prospectively until the end of 2013 by the American Taxpayer Relief Act.

reduced the corporate tax rate. (Equivalently, the revenue loss of lowering rates would be smaller if Congress first extended all of these provisions.)

Paying for corporate reform

These figures demonstrate that a carbon tax could finance substantial corporate reforms. A $1.2 trillion carbon tax could pay for reducing the corporate tax rate from 35 percent to 25 percent, for example, or for reducing the rate to 28 percent and permanently extending most expiring business tax preferences.

Indeed, these simple comparisons actually understate the potential because there's no reason to believe a carbon tax should be the only tool for paying for corporate reform. Cutting back on "loopholes" and tax preferences is politically difficult, but that does not mean that policymakers could not do so as part of corporate tax reform. It would be unfortunate if new carbon revenues were used to preserve inefficient or unfair preferences that otherwise could have been eliminated in reform efforts.[7]

But these simple comparisons overstate the potential corporate rate reduction for a separate reason: demands to use the new carbon revenues for other purposes. As discussed in section 3, a pure carbon-for-corporate tax swap would redistribute income from low- and middle-income families towards high-income ones; policymakers may use some carbon tax revenue to offset that rather than to cut corporate taxes. As discussed in section 4, policymakers may also want to dedicate some carbon revenues to deficit reduction. And policymakers may also want to use some of the revenues to finance new spending, whether on climate change adaptation, green energy incentives, or unrelated programs (see Chapters 10 and 12). The net amount of revenue available for corporate tax reform will ultimately depend on all those factors.

3. A distributional mismatch

Distributional issues understandably loom large in tax reform discussions. Citizens and policymakers want to understand who will win and lose from any proposed reforms.

To illustrate the potential distributional effects of carbon-for-corporate tax swap, we consider a stylized carbon tax that would raise revenue equal to 1 percent of Americans' pre-tax income (about $137 billion in 2015). The resulting estimates can then be scaled up or down to match specific carbon tax proposals that, as noted earlier, often include tax rates that increase notably from year to year.

We model the distributional impacts of this stylized tax using the Tax Policy Center (TPC) microsimulation model of the U.S. tax system (see Appendix). The model includes a detailed methodology for analyzing the burden of consumption taxes (Toder et al. 2011). That methodology distributes a consumption tax as a flat rate tax on income (returns to labor and capital plus wage-indexed transfer payments) less the normal return to saving (interest income and a portion of capital gains and dividends) and then reallocates the burden among income groups depending on their relative shares of consumption of those goods and services subject to the tax. That approach reveals that consumption taxes are generally regressive, meaning that the ratio of taxes to income declines as income rises. Consumption taxes are regressive because they exempt much of the income from capital, which is received disproportionately by higher-income households, and because in the long run they reduce real Social Security benefits by lowering real wages.[8]

We model the distribution of a carbon tax the same way as a general consumption tax, except for a relative price effect. The carbon tax raises the relative prices of goods with direct carbon content (e.g., most electric power, gasoline, and home heating fuels) and those with indirect energy content (all goods and services that use carbon-based fuels as a direct or indirect input in production).[9] In general, the consumption of carbon-based products is a larger share of income for low-income groups than consumption generally, so a carbon tax is more regressive than a general consumption tax.[10]

We summarize the model results by examining how the burden of a carbon tax would vary across income groups in 2015, ranging from the lowest fifth (cash incomes less than $20,000 in 2012) to the highest (cash incomes above $108,000).[11] As you might expect, the burden falls highest on low-income households. On average, the carbon tax would place a burden on them – in terms of higher relative prices for goods and services they consume and lower real wages – of about 1.8 percent

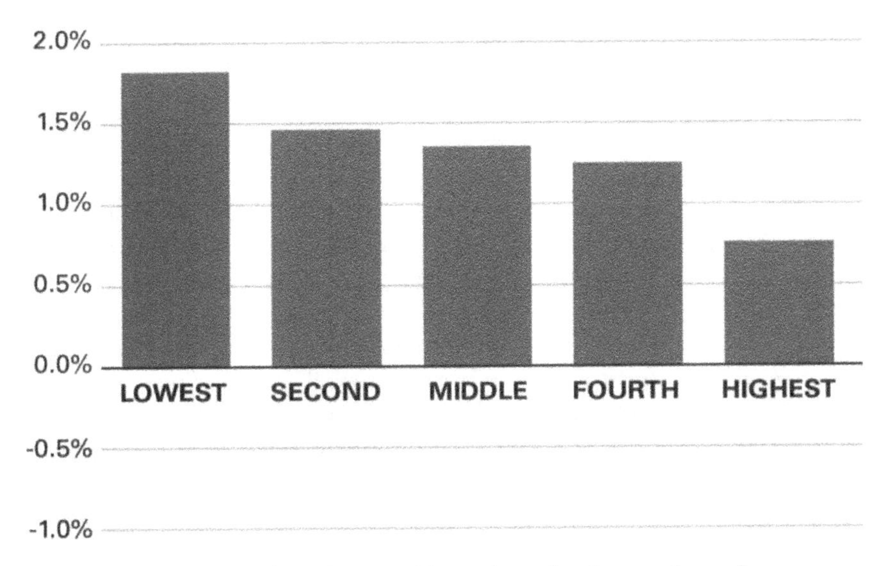

FIGURE 8.1 The burden of a carbon tax Change in tax burden as a share of pre-tax income by income quintile, 2015

Source: Urban-Brooking Tax Policy Center Microsimulation Model and authors' calculations.

Note: Carbon tax chosen so revenues equal 1% of average pre-tax income.

of pre-tax income (Figure 8.1). Each of the three middle-income quintiles – from the twentieth percentile up to the eightieth – similarly bears a burden greater than 1 percent of their pre-tax income, although at a declining rate. Households with the highest incomes, however, bear a burden of less than 1 percent of their income.

In short, a carbon tax is regressive. Although the exact degree of regressivity varies, that finding is consistent with many earlier studies, including Dinan and Lim Rogers (2002), Hassett et al. (2009), Rausch and Reilly (2012), Dinan (2012), and Chapters 6 and 7 of this volume.

Now consider what happens if the carbon tax is paired with a reduction in corporate income taxes. We start with the extreme case in which all carbon revenues are used to reduce corporate taxes.

Corporations are legally liable for the corporate income tax, but economists understand that since a corporation cannot literally have a tax burden, the taxes must lower some peoples' real incomes. The question is which people bear that burden. The classical analysis by Arnold Harberger (1962) found that, under plausible assumptions, suppliers of capital bore about 100 percent of the corporate tax burden, but the burden was spread from corporate shareholders to all capital owners as investors responded to the tax by shifting to non-corporate assets.

More recently, this analysis has been modified in two ways. First, because the corporate income tax is imposed mostly on profits from investments in the United States, capital owners can escape the tax by investing overseas (Harberger 2008, Randolph 2006, Gravelle 2010). Labor is less internationally mobile than capital, so the overseas migration of capital reduces capital per worker and real wages in the United States, shifting some of the corporate tax burden to labor. Second, and working in the opposite direction, a substantial portion of the corporate income tax represents a tax on supernormal profits or economic rents, and that portion falls on owners of corporate equity. The share going to supernormal returns can be approximated by estimates of the share of corporate revenue that would remain if the tax on normal returns were eliminated by allowing expensing of corporate investments (Gentry and Hubbard 1997, Toder and Rueben 2007, Cronin et al. 2012).

TPC has recently reviewed the economic literature on who pays the corporate tax and updated its assumption on who bears the burden of the tax (Nunns 2012). TPC now allocates the tax burden as 20 percent to all capital owners, 20 percent to labor, and 60 percent to corporate shareholders. Because a substantial part of the corporate tax burden falls on capital and high-income people receive a large share of capital income, the burden of the corporate tax is very progressive.

Figure 8.2 shows the net change in overall tax burden from a revenue-neutral carbon-for-corporate tax swap (dark bars). Households at all income levels benefit from the reduction in corporate income taxes. Households in the lowest income quintile, for example, see their net burden decline from 1.8 percent of their pre-tax income to about 1.4 percent. That reflects the portion of the corporate income tax that is borne by workers (and the relatively small portion on investors that fall in that quintile). The benefit of reducing corporate income tax rises as income rises.

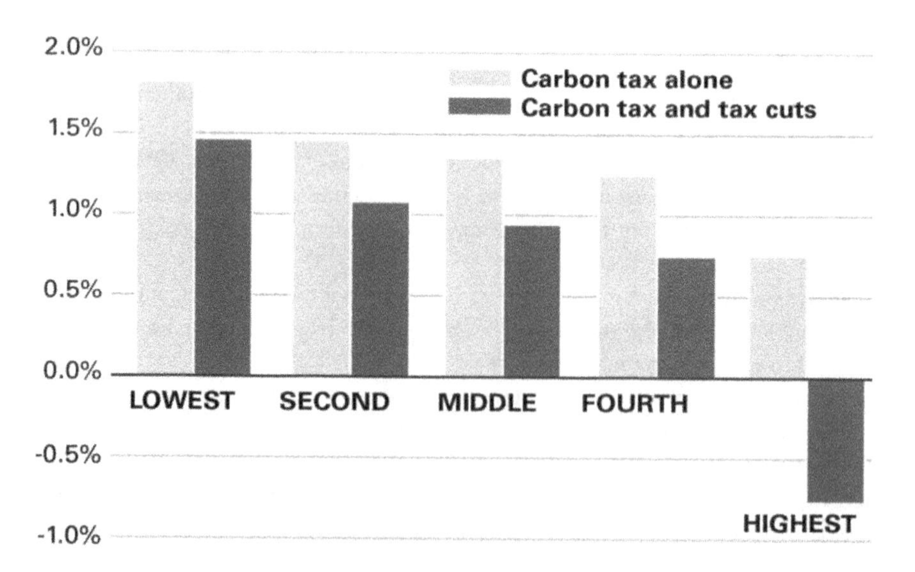

FIGURE 8.2 The net burden of a carbon tax and a cut in corporate income taxes
Change in tax burden as a share of pre-tax income by income quintile, 2015

Sources: Urban-Brookings Tax Policy Center Microsimulation Model and authors' calculations.

Note: The cut in corporate income taxes offsets all carbon revenues.

The bottom 80 percent, by income, all bear a net tax increase from the swap with the burden from the carbon tax more than offsetting the benefit of cutting corporate income taxes.

Households in the top quintile, however, come out ahead on average. For them, the benefit of lower corporate income taxes more than offsets the burden of the carbon tax. On average, they benefit by about 0.8 percent of pre-tax income.

A revenue-neutral carbon-for-corporate tax swap is thus particularly regressive. It introduces a new, regressive tax while cutting an existing progressive one. Such a swap is efficiency enhancing, internalizing an externality while reducing a particularly distortionary tax, but the distributional implications are a concern.

One way to address those concerns would be to earmark a portion of the carbon revenues for lump-sum rebates. For example, one-quarter of the net carbon revenues might be used to pay a fixed rebate to each adult, and half as much for each child. The remaining three-quarters of the carbon revenue would then be used to lower corporate income taxes.

Figure 8.3 illustrates the net effect of this policy combination (dark bars). All income groups continue to benefit from the cut in corporate income taxes, but that benefit is three-quarters its previous size. In addition, all groups now benefit from the lump-sum rebates. Because they are fixed in dollar terms, those rebates are larger as a share of pre-tax income for households with lower incomes. Indeed, households

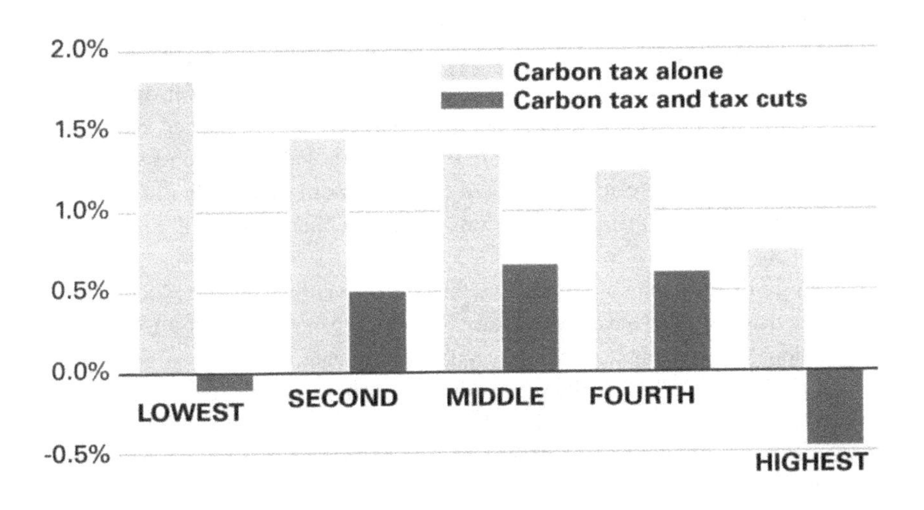

FIGURE 8.3 The net burden of a carbon tax, a cut in corporate income taxes, and a credit Change in tax burden as a share of pre-tax income by income quintile, 2015

Source: Urban-Brooking Tax Policy Center Microsimulation Model and authors' calculations.

Note: The cut in corporate income taxes offsets 75% of carbon revenues and a per capital credit offsets 25%.

in the lowest income quintile would come out slightly ahead, on average, as a result of this policy. Households in the top quintile would also come out ahead although less so than with a pure carbon-for-corporate swap; the benefit they get from the rebate is much less than they would have received from a larger corporate tax cut.

Middle-income households end up bearing an increased tax burden. The corporate tax reduction and the rebate are not enough to offset the additional burden of the carbon tax.

This simple example illustrates a more general conclusion. It is difficult to design a carbon-for-corporate tax swap that would be distributionally neutral (that would be even more true with finer gradations in the analysis). Doing so may require the use of more than one additional rebate policy. For example, using some of the revenues not allocated to corporate tax relief to provide cuts in the payroll tax rate instead of a lump sum rebate would allocate more of the remaining rebate to middle-income households and less to the bottom quintile.

Therefore, if distributional neutrality is an important policy goal or political constraint, policymakers may need to bring additional policy tools or issues into the mix in order to accommodate it. They should recognize, however, that using revenue for lump-sum or other rebates targeted at low-income households rather than corporate tax cuts will reduce the potential efficiency gains of the overall policy package.

4. Room for deficit reduction

Our distributional analysis considered cases in which all net carbon tax revenues were used to reduce corporate income taxes or finance transfers to low-income families. Given America's daunting long-run fiscal outlook, however, we should also consider the possibility of using some carbon revenues for deficit reduction. Because of population aging and rising health care costs, federal spending will likely be larger in the future, relative to the size of the economy, than it has been in the past, even after some entitlement trimming. If so, the United States will need higher revenue (Chapter 1).

One way to combine corporate tax reform with eventual deficit reduction is to recognize that tax policy changes can have different revenue impacts over time. The congressional budget process focuses on 10-year budget windows, so a revenue-neutral tax swap would typically use 10 years of carbon tax revenues to pay for 10 years of corporate tax rate cuts.[12] But that does not mean carbon tax revenues will match corporate tax rate cuts each year. Indeed, the two revenue streams can have quite divergent time paths.

CBO's example of a $1.2 trillion carbon tax, for example, has revenues increasing rapidly over time (Figure 8.4). Revenues start around $90 billion in 2013 and then rise to $150 billion in 2022. That increase reflects the net effect of four forces: (1) the 5.6 percent annual increase in the tax per ton, (2) real economic growth, (3) declining emissions intensity of economic activity, and (4) emissions reductions in response to the tax. CBO projects that the first two factors more than offset the third and fourth, at least over the first decade, so annual revenues rise substantially.

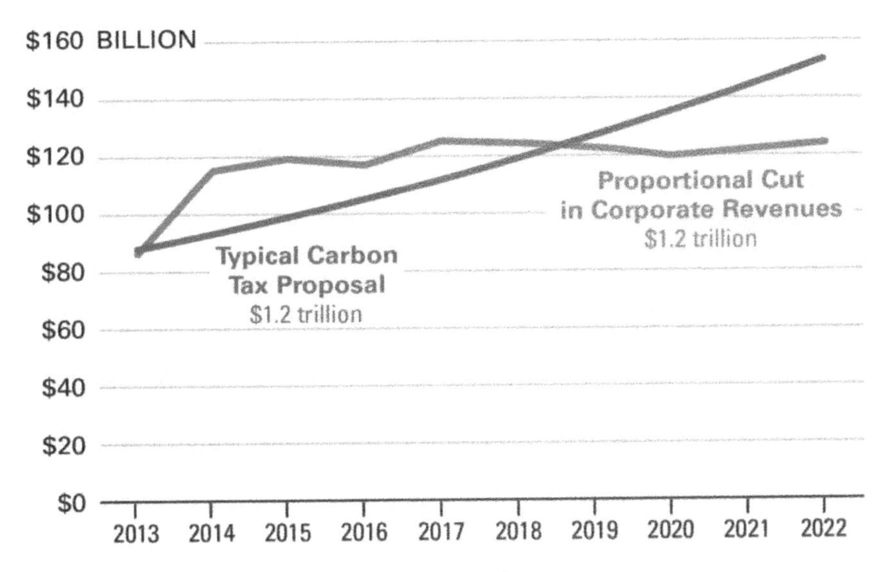

FIGURE 8.4 Carbon and corporate revenues over time

Source: Congressional Budget Office and authors' calculations.

A proportional $1.2 trillion cut in corporate tax revenues over those 10 years would have a much flatter trajectory. Such a cut could reflect a combination of rate cuts and preference extensions. After an early jump (reflecting the economic recovery and expiration of temporary tax cuts), the resulting path of revenues losses would fluctuate closely around $120 billion annually. That path reflects the net effect of three forces: (1) nominal economic growth, (2) a reduction in the share of national income that accrues as corporate profits (which CBO projects will move back towards historical norms after recently expanding), and (3) a declining share of economic activity happening in corporations (as more businesses structure as partnerships, limited liability companies, and other non-corporate forms). CBO projects that the latter two factors roughly offset the first. As a result, corporate revenues remain roughly flat over the decade (and decline relative to the economy).

By the end of the budget window, the carbon tax would be raising about $150 billion annually, while the corporate tax cut would be losing about $120 billion. This "revenue-neutral" tax swap thus generates positive net revenues as we look beyond 10 years. In the eleventh year, for example, it would increase revenues (and thus reduce the deficit) by about $30 billion. That amount would likely grow for a number of years beyond the budget window, although it will start turning around eventually as the carbon tax leads to a substantial cut in emissions.

A carbon-for-corporate tax swap could thus be revenue-neutral inside the official budget window and still contribute to long-term deficit reduction. Of course, the ultimate contribution to deficit reduction would depend on the long-run trajectory of carbon and corporate revenues and policymakers' willingness to let them run their course. Also, if policymakers wanted to accomplish more deficit reduction sooner, they could target revenue neutrality over a shorter window, perhaps five years, with net revenue gains beyond it.

5. Conclusion

America's tax system is broken. It's needlessly complicated, economically harmful, and fails at its most basic task, collecting enough revenue to finance government services. Tax and fiscal reform thus deserves increased attention from policymakers. But current discussions often get bogged down in tired debates and progress has been slow.

Adding a carbon tax to the discussion could help facilitate needed reforms. The environmental argument for a carbon tax is strong, given growing evidence of a changing climate. And the resulting revenue could play a vital role in facilitating tax reform and deficit reduction.

Corporate tax reform could be particularly beneficial, given the potential for reducing economic distortions. But those potential efficiency gains must be balanced with distributional considerations. A carbon-for-corporate tax swap would be highly regressive, increasing tax burdens on low-income families while lowering them on high-income ones. Policymakers may therefore want to pair such a swap with other policies (explored in more detail in Chapter 7) that ameliorate some of the distributional consequences. They may also want to capture some carbon revenues for future deficit reduction.

Taxing carbon, reforming our flawed corporate tax system, and addressing long-term deficits should all be priorities for American policymakers. Combining a carbon tax with corporate tax relief may open up possibilities for coalitions that could overcome the resistance that either policy considered in isolation would engender.

Appendix: TPC's model of the U.S. tax system

The Tax Policy Center has developed a large-scale microsimulation model to analyze the revenue and distributional consequences of the U.S. federal tax system. Based primarily on a public-use sample of information taken from tax returns, the model projects incomes and other tax-related variables for subsequent years and simulates actual and proposed tax laws to estimate and compare their effects on tax units. The model is similar to those used by the Congressional Budget Office, the Joint Committee on Taxation, and the Treasury's Office of Tax Analysis. A detailed description of the model is available online (Urban-Brookings Tax Policy Center 2012). Nunns (2012) explains how the model distributes the burden of corporate income taxes, and Toder et al. (2011) explain how it distributes the burden of consumption taxes.

Notes

* Donald Marron is director of economic policy initiatives at the Urban Institute and former director of the Urban-Brookings Tax Policy Center. Eric Toder is co-director of the Center and an Institute Fellow at the Urban Institute. We thank Jim Nunns for help with revenue calculations, Joseph Rosenberg for preparing the original distributional analysis used in our calculations, Ian Parry for helpful comments, and the Energy Foundation for financial support. The views in this article are those of the authors alone and do not represent the views of the Urban Institute, its board, or its funders.

1. A cap-and-trade system could also raise revenues if the government sells the allowances rather than giving them away for free. For purposes of this chapter, a carbon tax and a comparably-scaled cap-and-trade system with auctioned or sold allowances are essentially equivalent.
2. The researchers interpret this as strong efficiency grounds for reducing corporate taxes. However, they do note an important caveat: "Lowering the corporate tax rate substantially below the top personal income tax rate can jeopardize the integrity of the tax system as high-income individuals will attempt to shelter their savings within corporations" (Johansson et al. 2008, p. 8). That is one of several reasons why comprehensive tax reform – involving both the corporate and personal income tax systems – may ultimately be necessary. Halperin (2009) notes, however, that there are possible measures that could address a disparity between individual and corporate rates, such as limiting the corporate rate reduction to publicly traded corporations or strengthening rules against converting individual to corporate income through the use of "personal service corporations."
3. Because state taxes are deductible from the federal income tax, the combined federal-state rate is less than the sum of the federal and state rates.
4. More fundamental reforms that change the basic structure of the corporate tax are also beyond the scope of this chapter. Some experts have proposed, for example, that reform focus on moving more towards a consumption tax base, allowing immediate expensing of business investment while eliminating interest deductibility, rather than lowering corporate rates (Auerbach 2010). And a long-standing tradition argues for better integrating the corporate and personal income tax systems (U.S. Department of the Treasury 1992). Such fundamental reforms deserve attention, but for the purposes of this chapter we focus on reforms policymakers are currently discussing. And policies that move towards expensing or towards integration of the corporate and personal income taxes are likely to reduce revenue from corporate investments, so they too would require offsetting revenue increases somewhere else in the system.
5. Under the standard revenue estimating assumptions used by the Joint Committee on Taxation and the Congressional Budget Office, changes in tax law are assumed to have no effect on gross domestic product (GDP). With GDP held fixed, any increase in excise taxes must necessarily reduce incomes earned by workers and investors. With lower incomes, revenues from individual income taxes, corporate income taxes, and payroll taxes decline, offsetting about 25 percent of the revenue from the new or increased sales or excise tax (Woodward 2009). This approach differs from that used by other modelers (for example, Rausch and Reilly 2012) who attempt to track how carbon taxes affect overall economic activity.
6. To get these estimates, we scaled up JCT's figures to reflect the increase in projected corporate revenues. At the time of their estimate, 10-year corporate revenues were projected at $3.92 trillion, while in CBO's August 2012 baseline they are $4.36 trillion (CBO 2012a). The specific calculations are: $800 billion = $718 billion × $4.36 trillion / $3.92 trillion, and $1.15 trillion = $800 billion × (35% - 25%) / (35% - 28%).
7. Of particular interest here are existing tax preferences for traditional and alternative energy sources (Marron 2011). By internalizing some external costs of fossil-fuel production and encouraging use of alternatives, a carbon tax would weaken the policy

rationale for tax subsidies supporting alternative energy. Trimming those subsidies (or not extending expiring ones) would thus become more attractive. By the same reasoning, cutting back on subsidies for fossil energy sources would become somewhat less attractive, although the economic arguments against such subsidies would remain.

8. TPC's analysis finds consumption taxes less regressive than would a method that distributed the burden in proportion to the ratios of consumption to income for different groups. The higher regressivity of analyses that rely on snapshot comparisons of annual income and annual consumption reflects in part the mismatch between when income is earned and when it is spent, with individuals with temporarily low (high) incomes consuming very high (low) shares of current income and retirees funding their current consumption from prior years' instead of the current year's income. To prevent this timing mismatch, TPC bases its distribution of a consumption tax on its effects on real incomes, adjusted for effects on the relative prices of goods consumed by households in different income groups.
9. Our analysis considers the long-run, fully phased-in effects. In the short run, a carbon tax will also affect profits of companies and employment and earnings of workers in carbon-intensive sectors, such as coal mining.
10. TPC modeled the burden of a carbon tax as part of a previous exercise to design a deficit reduction plan. Kevin Hassett supplied TPC with the input-output matrix that was used to develop estimates of the indirect effects of a carbon tax on consumer prices.
11. The Tax Policy Center model ranks households by cash income, a broader measure than adjusted gross income that includes other forms of income such as contributions to tax-deferred retirement savings plans, tax-exempt interest, and cash transfers. The income quintiles used in this chapter are based on the estimated income distribution for the entire population in 2013 and contain an equal number of people, not households. The income breaks, in 2012 dollars, are: 20 percent $20,113, 40 percent $39,790, 60 percent $64,484, and 80 percent $108,266.
12. Budget rules sometimes include intermediate targets, for example, not increasing the deficit over five years. Those targets can complicate policy design, but don't weaken our general point here.

References

Auerbach, Alan. 2010. "A Modern Corporate Tax." Paper for the Center for American Progress and The Hamilton Project, December. Washington, DC.

Congressional Budget Office. 2011. *Reducing the Deficit: Spending and Revenue Options.* Washington, DC: Congressional Budget Office.

Congressional Budget Office. 2012a. *An Update to the Budget and Economic Outlook 2012 to 2022.* Washington, DC: Congressional Budget Office.

Congressional Budget Office. 2012b. *Expiring Tax Provisions—August 2012 Baseline.* Washington, DC: Congressional Budget Office.

Cronin, Julie-Anne, Emily Lin, Laura Power, and Michael Cooper. 2012. "Distributing the Corporate Income Tax: Revised U.S. Treasury Methodology." OTA Technical Paper 5. Washington, DC: U.S. Department of the Treasury.

Dinan, Terry. 2012. "Offsetting a Carbon Tax's Costs on Low-Income Households." Working Paper 2012–16. November 13. Washington, DC: Congressional Budget Office.

Dinan, Terry and Diane Lim Rogers. 2002. "Distributional Effects of Carbon Allowance Trading: How Government Decisions Determine Winners and Losers." *National Tax Journal* 55(2): 199–221.

Gentry, William M. and Glenn R. Hubbard. 1997. "Distributional Implications of Introducing a Broad-Based Consumption Tax." In *Tax Policy and the Economy,* Vol. 11, edited by James M. Poterba (1–47). Chicago: University of Chicago Press.

Gravelle, Jennifer C. 2010. "Corporate Tax Incidence: Review of General Equilibrium Estimates and Analysis." Working Paper 2010–03. May. Washington, DC: Congressional Budget Office.

Halperin, Dan. 2009. "Mitigating the Potential Inequity of Reducing Corporate Rates." Occasional Paper. Urban-Brookings Tax Policy Center. July 29. Washington, DC: The Urban Institute.

Harberger, Arnold C. 1962. "The Incidence of the Corporate Income Tax." *Journal of Political Economy* 70(3): 215–240.

Harberger, Arnold C. 2008. "Corporate Tax Incidence: Reflections on What Is Known, Unknown, and Unknowable." In *Fundamental Tax Reform: Issues, Choices, and Implications,* edited by John W. Diamond and George R. Zodrow (283–308). Cambridge, MA: MIT Press.

Hassett, Kevin and Aparna Mathur. 2011. "Report Card on Effective Corporate Tax Rates: United States Gets An F." Washington, DC: American Enterprise Institute.

Hassett, Kevin, Aparna Mathur, and Gilbert E. Metcalf. 2009. "The Incidence of a U.S. Carbon Tax: A Lifetime and Regional Analysis." *The Energy Journal* 30(2): 155–177.

Johansson, Asa, Christopher Heady, Jens Arnold, Bert Brys, and Laura Vartia. 2008. "Tax and Economic Growth." Organization for Economic Cooperation and Development, Economics Department Working Paper No. 620, July. Paris, France: Organization for Economic Cooperation and Development.

Joint Committee on Taxation. 2011. Memorandum by Thomas Barthold. October 27. Washington, DC: Joint Committee on Taxation.

Marron, Donald. 2011. "Energy Policy and Tax Reform." Testimony before the House Subcommittee on Select Revenue Measures and the Subcommittee on Oversight of the Committee on Ways and Means, United States House of Representatives, September 22. Washington, DC.

Marron, Donald. 2012. "The Tax 'Expirers'." Testimony before the House Subcommittee on Select Revenue Measures of the Committee on Ways and Means, United States House of Representatives, June 8. Washington, DC.

Nunns, Jim. 2012. "How TPC Distributes the Corporate Income Tax." Tax Policy Center Report, September 13. Washington, DC: The Urban Institute.

Office of Management and Budget. 2012. *Analytical Perspectives, Budget of the United States Government, Fiscal Year 2013.* Table 17–2: 255–260. Washington, DC: Office of Management and Budget.

Randolph, William C. 2006. "International Burdens of the Corporate Income Tax." Working Paper 2006–09. August. Washington, DC: Congressional Budget Office.

Rausch, Sebastian and John Reilly. 2012. "Carbon Tax Revenue and the Budget Deficit: A Win-Win-Win Solution?" MIT Joint Program on the Science and Policy of Global Change, Research Report No. 228, August. Cambridge, MA: Massachusetts Institutes of Technology.

Toder, Eric. 2012. "International Competitiveness: Who Competes Against Whom and for What?" *Tax Law Review* 65: 505–534.

Toder, Eric, Jim Nunns, and Joseph Rosenberg. 2011. "Methodology for Distributing a VAT." TPC Research Report, April 12. Washington, DC: The Urban Institute.

Toder, Eric and Kim Rueben. 2007. "Should We Eliminate Taxation of Capital Income?" In *Taxing Capital Income,* edited by Henry J. Aaron, Leonard E. Burman, and C. Eugene Steuerle (89–141). Washington, DC: Urban Institute Press.

Urban-Brookings Tax Policy Center. 2012. "The Tax Policy Center Microsimulation Model." http://taxpolicycenter.org/taxtopics/TPC-Model-Overview-2012.cfm. Washington, DC: The Urban Institute.

U.S. Department of the Treasury. 1992. *Integration of the Individual and Corporate Tax Systems: Taxing Business Income Once.* Washington, DC: U.S. Government Printing Office.
Woodward, G. Thomas. 2009. "The Role of the 25 Percent Revenue Offset in Estimating the Budgetary Effects in Legislation." Washington, DC: Congressional Budget Office.

9

CARBON TAXES AND ENERGY-INTENSIVE TRADE-EXPOSED INDUSTRIES

Impacts and options

*Carolyn Fischer, Richard Morgenstern, and Nathan Richardson**

KEY MESSAGES FOR POLICYMAKERS

- A unilateral carbon tax could hurt the international competitiveness of U.S. firms, and resulting production shifts could cause emissions to "leak" outside of the U.S., undermining the environmental benefit of the tax.
- These effects are most pronounced for energy-intensive, trade-exposed (EITE) industries.
- Economic analysis suggests that competitiveness and leakage effects from a modest carbon tax are likely to be very small. But they do exist, and may be politically important.
- Policy options exist for addressing competitiveness and leakage under a carbon tax. None are perfect, but they have important relative advantages and disadvantages.
- Of these policies, output-based rebates and border carbon adjustments are superior to alternatives (carbon tax exemptions or general corporate tax cuts). But between the preferred options, it is unclear which approach is better. Both have important disadvantages.
- It is likely that domestic and international politics would be important factors in determining which of these two options for addressing competitiveness and leakage emerge.

Introduction

An important consideration for any carbon-oriented tax (or regulatory policy) is its effect on the competitiveness of U.S. firms in sectors exposed to international trade. A carbon tax would increase energy costs for U.S. firms, but not their foreign competitors (until and unless those competitors also face a tax or other carbon-pricing policy). Firms in sectors that are both energy-intensive and trade-exposed (EITE) are particularly vulnerable to competitiveness effects. A consequential shift in market share to producers in other nations, particularly those with weak or nonexistent carbon pricing policies, also erodes environmental benefits from a carbon tax, to the extent that emissions are displaced rather than truly abated. This "leakage" effect, along with the loss of U.S. competitiveness itself, is cited by opponents of a carbon tax as a key argument against it.

Most research suggests that these disproportionate effects are relatively modest; however, critics are correct that they are real problems. Furthermore, addressing competitiveness-related leakage need not greatly erode the carbon tax base. To put the scale of the issue into perspective, consider that all industrial-sector CO_2 emissions accounted for less than 14 percent of total U.S. emissions in 2011, up only slightly from the 2009 low but down substantially from 1990 levels of almost 17 percent.[1] Of course, the relevant EITE firms form a smaller subset of that group, since energy use is highly concentrated. As shown in Figure 9.1, three-quarters of manufacturing output is from industries with energy expenditures less than 2 percent of the value of their output (roughly the average across all manufacturing), while industries with energy expenditures in excess of 5 percent of the value of their output account for only one-tenth of the value of U.S. manufacturing output and less than 2 percent of U.S. GDP.

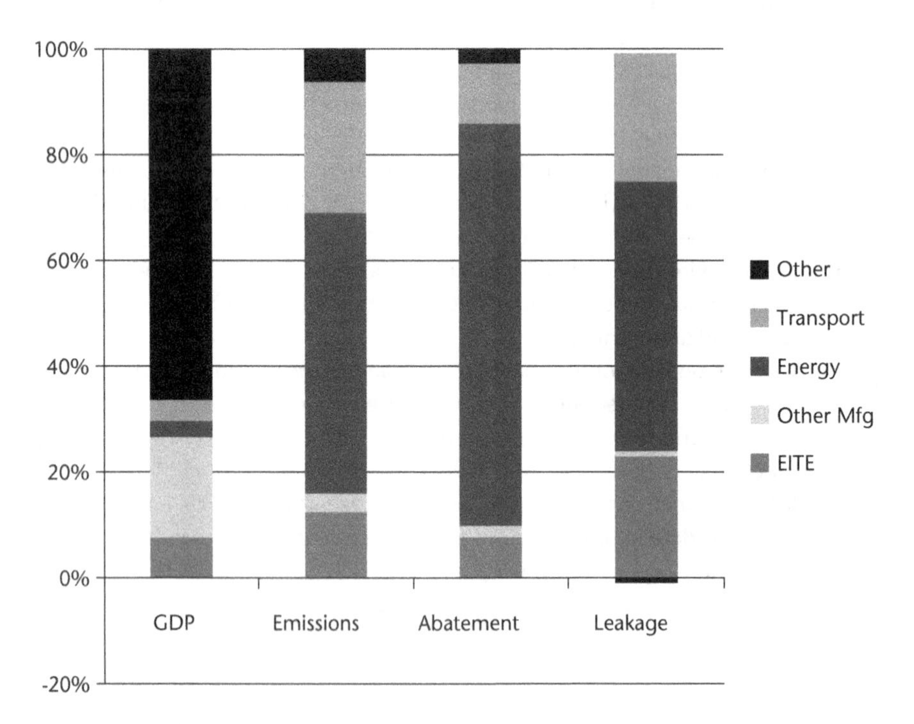

FIGURE 9.1 Share of GDP, emissions, abatement, and leakage by sector

Source: GTAP database, 2004 base year, and simulations from Fischer and Fox (2010).

While emissions shares represent a rough upper bound of the emissions revenues that might be in play, abatement shares (or the reductions that might be in play) are significantly smaller. Due to the abundance of lower-cost emission reduction opportunities in the electric power sector, studies estimate that the share of emissions reductions under an economy-wide CO_2 pricing program coming from the industrial sector would be smaller than that sector's share in nationwide emissions.[2] At the same time, reductions in EITE sectors are predicted to result in disproportionate leakage rates – as much as 3 to 5 times the economy-wide leakage rate.[3]

As a result, some sort of special treatment for the affected sectors may be justified. Clearly, the most effective approach to reduce carbon leakage is to ensure that other countries take comparable carbon pricing action. But until that becomes politically viable, other policy options are available, some of which have been considered by the U.S. Congress in previous proposals for carbon pricing (in the form of cap-and-trade). This chapter examines four potential approaches for addressing competitiveness concerns for EITE sectors in the context of an economy-wide carbon tax. While not exhaustive, these options are commonly discussed in U.S. policy circles:[4]

- partial or full exemption from the carbon tax;
- rebates to firms tied to their output levels;
- border adjustments for energy- or carbon-intensive goods; and
- across-the-board lowering of corporate income tax rates.

The chapter begins with a short discussion of estimates of the size of both leakage and EITE sector competitiveness losses under a carbon tax. Following sections describe policy options for addressing these issues, their relative advantages, disadvantages, and practical design considerations. A final section offers broad conclusions.

The scope of the problem

In order to determine whether special policy treatment for EITE sectors is needed, and if so, to evaluate which policy options are best and how they should be structured, you first have to know the magnitude of the problems you are trying to address. Over the past few years economists have investigated this question and, while their results are sensitive to modeling assumptions, the best evidence suggests that, given a moderate carbon price, both competitiveness losses and emissions leakage are real but modest problems – though higher carbon prices undoubtedly present greater challenges.

Identifying EITE sectors

First, it is important to determine what qualifies as an EITE sector. This is critical for measuring the size of competitiveness or leakage problems, since you need to know where to look for them. But it is also important for policy. If preferential provisions are applied to sectors that are less susceptible to adverse impacts from the carbon tax, they risk weakening the tax's ability to both raise revenue and to reduce emissions. Moreover, the legality of preferential treatment for EITEs depends on their link to environmental goals – namely, mitigating carbon leakage. The less finely such treatment is focused, the more difficult it is to justify it on environmental grounds, increasing legal risk.

Consequently, relatively narrow definitions of vulnerable sectors are preferred. Such definitions should be limited to sectors that meet the twin hurdles of (1) being highly energy-intensive, either as a result of direct combustion of fossil fuel, especially a carbon-intensive fuel such as coal, or from heavy reliance on carbon-intensive electricity as an input to production; and (2) being highly vulnerable to international competition. For example, the American Clean Energy and Security (ACES) Act (the 2009 cap-and-trade bill also known as Waxman-Markey or H.R. 2454) defines EITE sectors as manufacturing industries (excluding refining) that are at least 5 percent energy (or CO_2) intensive *and* 15 percent trade-intensive, *or* 20 percent energy-intensive.[5] According to an analysis conducted by the Environmental Protection Agency, the ACES definition of EITE industries included 44 presumptively eligible of almost 500 manufacturing industries defined at the six-digit level of the North American Industrial Classification System (NAICS).[6] Most of the presumptively eligible industries are within the chemicals, paper, nonmetallic minerals (e.g., cement and glass), or primary metals (e.g., aluminum and steel) sectors. Refineries were not eligible for production rebates under Waxman-Markey, although they were freely granted 2 percent of total allowances without regard to current output, that is, via grandfathering. Other sectors could petition for inclusion. Taken together, these EITE industries account for 12 percent of total manufacturing output and 6 percent of manufacturing employment (half a percent of total U.S. non-farm employment). At the same time, they account for more than two-thirds of direct combustion and process emissions from the manufacturing sector, and slightly more than half when emissions from electricity are also included in the calculations. By way of comparison, the European Union used less stringent criteria for its allocation benchmarking provisions, including a broader group of eligible sectors, albeit with smaller rebates.

Further, with any special tax provision, it is useful to consider the scope for potential administrative complexity. The U.S. Census Bureau reports there were slightly less than 9,200 establishments in 2007 in the presumptively eligible EITE industries under H.R. 2454.[7] In addition, Census reports 189 refiners would be covered under the provision for grandfathered allowances, bringing the total number of establishments to just under 9,400. However, that is likely a substantial overestimate of the number of actual firms that would be affected by any tax mechanism tailored to the EITE sector, as many firms own two or more individual establishments.

Carbon taxes and competitiveness

Recent modeling results have examined the impacts of domestic carbon taxes on industrial output for EITE industries (with their scope defined as in the ACES Act) under two distinct timeframes: the short run, typically less than three years, when it is assumed firms cannot alter production techniques; and the long run, when firms can alter production techniques and make a broad range of new investments in response to the carbon tax regime.

Not surprisingly, the results vary considerably by industry. For the case of a $15/ton CO_2 tax imposed in the U.S. and other Kyoto Annex I (developed) countries, the estimated output decline in EITE industries ranges from less than 0.1 percent to more than 3 percent.[8] Across all such firms, the average output decline is slightly less than 0.5 percent in the short run and about twice that level in the long run (Adkins et al. 2012). Some of the hardest-hit industries include cement, aluminum, lime and gypsum, petroleum refining, and chemicals, rubber, and plastics. These industries operate in highly competitive international markets and generally have difficulty passing along higher costs not also imposed on their competitors. Interestingly, in some industries the long-run output declines are smaller than in the short run, as markets and technologies respond to the new price signals to restore some of the short-run competitiveness losses. The modeling projects that the vast majority of emission reductions achieved by these industries will be from reductions in the emission-intensity of their production (e.g., increased energy efficiency, or shifts to lower-emission production methods), rather than from declines in production associated with increased imports from unregulated countries.

Carbon taxes and emissions leakage

Protecting domestic competitiveness, per se, may confer political benefits, but many attempts to do so run counter to the obligations associated with membership in the World Trade Organization (WTO). However, certain exceptions may be made for measures that are necessary to protect human health or to conserve exhaustible resources, including the global climate.[9] Thus, for international legitimacy, it is important to establish a connection between competitiveness impacts and environmental impacts.

"Emissions leakage" is generally defined as *the increase in foreign emissions that can be attributed to the effects of emissions reduction policies at home.* The "leakage rate" is measured as the increase in foreign emissions as a percentage of (economy-wide) domestic emissions reductions – for example, if a policy led to 100 tons of domestic reductions but 10 additional tons elsewhere (the leakage), it would have a 10 percent leakage rate. Recent studies find overall leakage rates in the range of 5 to 20 percent (Böhringer et al. 2012),[10] although these amounts are sensitive to the size of the international coalition taking action.

Importantly, emissions leakage can arise not only from increases in net exports from nonparticipating countries, but also from the drop in global energy prices that follows the withdrawal of demand for fossil fuels on the part of the regulating countries, inducing nonregulating countries to become more emissions-intensive with cheaper energy. Indeed, most leakage estimated in the widely used global trade models results from the latter mechanism, that is, global energy price induced leakage, which explains why in absolute terms a large amount of leakage is expected in the electricity and transportation sectors, despite the lack of trade in those goods. In the case of a carbon price in the Annex I countries only, model results suggest the countries with the largest percentage increase in their emissions are the oil-exporting nations, then India and China, in that order.[11]

However, leakage rates for certain individual EITE sectors – that is, the increase in emissions by foreign firms in those sectors, relative to the emissions reductions made in those sectors at home – can be much higher than the economy-wide leakage rates. Estimates of leakage for steel and nonferrous metals, for example, can run as high as 50 percent, due to strong competitive pressure and energy price effects for producers of these energy-intensive commodities (Fischer and Fox 2012a).

Put in another perspective, according to simulations by Fischer and Fox (2010), based on 2004 data), EITE sectors in aggregate contribute 7.3 percent of abatement, more similar to their 6.5 percent GDP share than their 10.3 percent share of economy-wide emissions. By contrast, increases in EITE sector emissions abroad represent 23 percent of total leakage. (Figure 9.1 displays the shares of these totals for five broad categories of industries.) Furthermore, while this model finds economy-wide leakage is roughly 8 percent of abatement, leakage rates among the EITE sectors average a much higher 28 percent.

In short, this research indicates that the emission reductions from a U.S. carbon tax are likely to be eroded somewhat by leakage. The much higher rates of leakage in EITE sectors, due in large part to the competitiveness losses discussed above, strengthen the case for targeted policies aimed at reducing the competitiveness impacts of a carbon tax on these sectors. The next section discusses some such policy options.

Description of alternative policy options

Partial or full exemption from the carbon tax

One straightforward option for addressing competitiveness concerns is to exempt certain industries from the broader GHG-reduction policy, in whole or in part. Conceptually, the mechanics of actually providing exemptions are relatively easy. If downstream entities – primarily energy users – are subject to the tax, exempt firms in EITE sectors would have reduced obligations (or perhaps none at all) to cover their emissions.[12] For a tax scheme focused on upstream entities (i.e., energy suppliers), exemptions could be provided to downstream firms indirectly via a procedure of credits based on their emissions or fuel use. The credit could come in the form of a tax credit or rebate, or a tradable allowance that could be sold to firms with positive tax liabilities. However, this method would undermine some of the attractiveness of a carbon tax on fuel suppliers as it would require not only a finding of eligibility but also calculations of exempt portions of fuel payments.

The principal advantage of exemptions is that they can be used to protect vulnerable firms or industries in a convincing and targeted way, potentially making it more politically feasible to adopt a higher overall carbon tax rate on domestic emissions.

But exemptions tend to be an inefficient and costly means of addressing competitiveness and leakage issues. This is primarily because exempting certain firms or sectors would almost certainly leave untapped some relatively inexpensive options for cutting emissions in these sectors. If all emissions from the H.R. 2454 presumptively eligible industries were exempt without compensating increases on other sources of emissions, both emission reductions and revenues would decline by about 8–10 percent, depending on the treatment of emissions associated with electricity use.

This approach may also raise equity concerns: if some industries or firms are exempt from participating, a greater burden would be placed on the remaining nonexempt industries. As a result, exemptions would likely increase the total, economy-wide cost of achieving a given emissions target, unless leakage effects are very strong.[13]

The energy (Btu) tax proposed by the Clinton administration in 1993 is a cautionary lesson concerning the political hazards of exemptions. At that time, many firms and industries made claims of business hardship. As a result, the final House legislation included a long list of exemptions added at the request of members or recommended by the administration. Ultimately, of course, the Btu tax was defeated in the Senate and the policy was never implemented – in part because of the effectiveness and equity concerns raised by the exemptions.

Rebates tied to output levels

A less-obvious but possibly more effective policy remedy for loss of EITE sector competitiveness is a rebate tied to firms' domestic output. Such an approach would reduce the adverse competitiveness impacts of the carbon tax, while maintaining the incentives the tax creates to reduce the carbon intensity of production (the most important source of emissions reductions for these sectors).

The ACES Act offers a starting point for illustrating how output-based rebates have been proposed in practice. That legislation would have created a national cap-and-trade system for greenhouse gases, under which covered emitters would have been required to surrender an allowance for each ton of carbon they emit. The limited number of such allowances (the cap) and their ability to be traded gives them a market value, imposing a cost on carbon emissions. This cost has similar competitiveness effects on firms in EITE industries as a carbon tax.

The ACES Act addressed this concern by allocating allowances to firms in EITE industries based on their output levels for an initial period.[14] Crucially, firms did not receive block allocations simply by doing business in such an industry. Instead, these rebates would have been updated on the basis of recent output levels. If a firm in an EITE sector increases its output, it gets a bigger allowance allocation, *regardless of its emissions*. If the firm is able to increase its output without increasing its emissions – that is, if it is able to become less carbon-intensive – all the better, since under a cap-and-trade system it can sell its excess allowances.

The rebate thus functions like a subsidy to output of EITE firms, offsetting some or all of their costs of the remaining emissions embodied in that output. Under ACES, sector-wide average carbon intensity was chosen as the benchmark in rebate calculations.[15] This allocation is relatively generous, as it implies that a large share of firms in each sector will receive net subsidies. In contrast, the European Union chose a benchmark of the performance of the top 10 percent of firms (i.e., those with lowest emissions intensities) in a sector.

Figure 9.2 shows the estimated output losses for some of the industries most affected by the ACES proposal, with and without rebates and over both short- and long-run horizons. Except for the long-run impacts on the petroleum and coal industries, a large portion of the estimated output losses are offset by the output-based rebates in ACES. At the same time, there is also a small amount of overcompensation in several industries.

Output-based rebates can be designed in conjunction with a carbon tax to work much like those under cap-and-trade. Several potential approaches offer similar economic incentives, but the legal treatment might differ. Importantly, the WTO Subsidies Code prohibits specific subsidies for industry and does not allow for environmental exceptions (unlike for tariffs, discussed in the next section). One

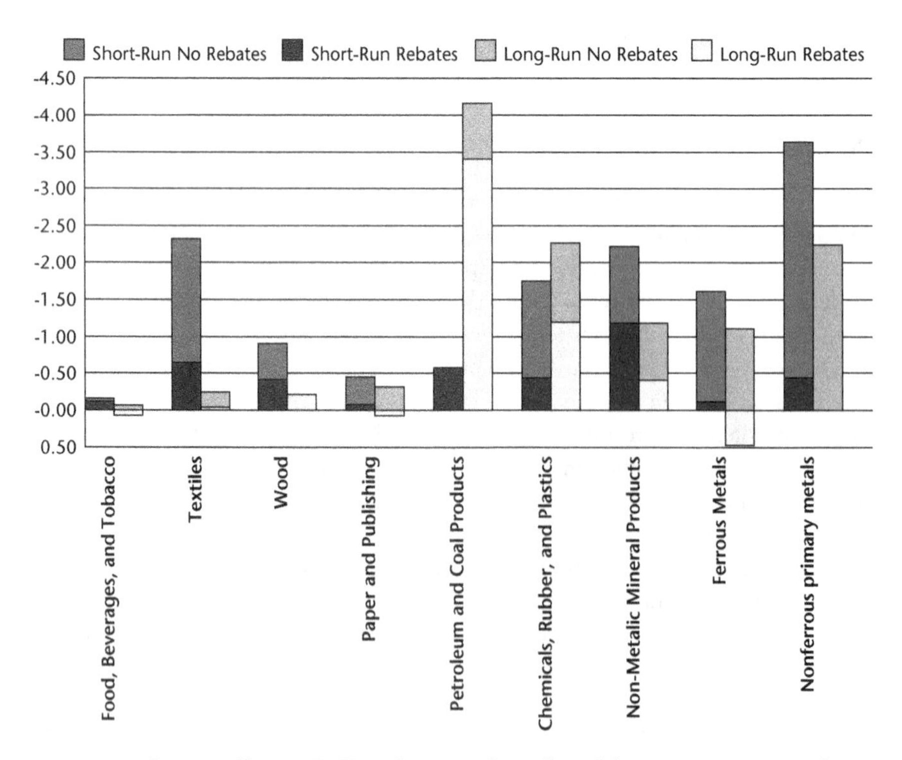

FIGURE 9.2 Output effect on EITE industries of a unilateral \$15/ton CO_2 tax, with and without rebates

Source: Adkins et al. (2012).

approach would simply offer firms in EITE sectors direct rebates, based on the carbon tax rate multiplied by a sector benchmark emissions rate and their output. This is relatively simple, but seems likely to be viewed by the WTO as an illegal subsidy.

Another option would be to embed the rebate into the carbon tax in the determination of the tax base for EITE sectors; that is, the carbon tax liability would be for emissions above a level equal to the benchmark rate multiplied by the firm's output. With this option, especially if the benchmark is generous, the tax would have to be refundable (allowing for negative tax payments) to preserve the incentive effects for relatively clean firms. A final (hybrid) option is to exempt EITE sectors from the carbon tax, but then to regulate them separately with a distinct market-based climate policy, like a tradable performance standard,[16] along with measures to ensure the price follows the carbon tax.[17] For example, if the performance standard is not very tight, allowing EITE firms to sell excess credits to firms subject to the carbon tax would preserve abatement incentives for the former and reduce the tax liability of the latter. If the performance standard is more stringent, allowing EITE firms the alternative compliance method of paying the tax would ensure that EITE carbon prices do not rise above the carbon tax rate.[18]

Reducing adverse competitiveness impacts of the carbon tax while preserving (most of) the incentives to reduce emissions are the key advantages of output-based rebates versus the simple carbon tax exemption for EITE industries. However, both options reduce net government revenues, by the extent of the emissions payments either rebated or forgiven; the amounts thus depend on the generosity of the rebate, but could range to 8 or 10 percent.[19]

An additional disadvantage of the rebate mechanism is its complexity. In addition to the challenge of accurately and equitably identifying EITE industries (a problem faced by all competitiveness policies), under output-based rebates the government must devise a dynamic rebate system and track the output of all firms in EITE industries. For example, information on current and historic output levels would be required, imposing an additional (though probably modest) reporting burden over and above that already required by the carbon tax. For some industries, it will be difficult to define appropriate output metrics across the varied products (e.g., chemicals). Presumably, the Internal Revenue Service, in cooperation with other agencies, would need to develop the needed metrics and collect the relevant data from the affected parties.

Perhaps the most important disadvantage of output-based rebating is its potential to distort firm production decisions in ways that yield unintended consequences. The rebate does discourage consumers from diverting their demand to imported alternatives, but it also discourages them from seeking less emissions-intensive alternative products/services, including conservation practices. In other words, output-based rebates may encourage production of the very energy-intensive materials that a carbon tax is designed to discourage. The costs of these distortions tend to grow as more trading partners adopt carbon pricing, and can soon outweigh the benefits from limiting any remaining carbon leakage (Böhringer et al. 2011; Fischer and Fox 2012b).

Border carbon adjustments

Border carbon adjustments (BCAs) address the leakage and competitiveness impacts of a carbon tax by ensuring that consumers face consistent carbon pricing signals, regardless of the country of origin for energy-intensive manufactured products. Import adjustments would require importers to pay an equivalent carbon tax, based on the actual or estimated emissions embodied in the product. Full border adjustment would add relief for exports in order to implement a fully "destination-based" carbon tax. However, most proposals (including the ACES Act) address only imports.[20] Recent studies indicate that the additional leakage reduction benefits of export adjustments are small, relative to those from import adjustments (Böhringer et al. 2012).

BCAs entail several important design issues: To which products should they apply? How are embodied emissions calculated? Should adjustments be made to the BCA for carbon regulations in the country of origin? Which exporting countries should be exempt? As trade-related measures, BCAs are controversial and should be designed with an eye toward maintaining compliance with WTO obligations and the principle of "common but differentiated responsibilities" enshrined in the UNFCCC (see Cosbey et al. 2012). Indeed, only mitigating carbon leakage is likely to be a legally permissible justification for interfering with international trade; protecting competitiveness contradicts WTO principles.

Scope of application: Like the other policy options, a BCA should only be applied to qualified EITE sectors. This narrow scope ensures that the benefits of reduced leakage outweigh the substantial administrative costs. It also is more likely to be legal under trade law – or, specifically, to be found to justify a GATT Article XX exemption from nondiscrimination obligations.

Defining embodied emissions: The definition of embodied emissions determines not only the extent to which consumers will pay for carbon embodied in imports, but also the extent to which foreign producers have an incentive to reduce their emissions. This incentive only occurs if producers can influence their calculated carbon content. Basing the BCA on actual embodied emissions would thus create some incentive for exporters to reduce their emissions; however, such calculations may be complex, costly, and legally challenging. Ideally, all firms would accurately report their embodied emissions, credibly verified by a third party. In practice, a default measure of embodied emissions is necessary for cases when actual data are either unavailable, unreliable, or prohibitively costly to obtain. Regardless of this benchmark, foreign producers should be afforded the opportunity to demonstrate if their emissions are actually lower. Although it raises compliance costs, this opportunity may be essential for WTO compatibility; it also can preserve some incentives for firms that can become relatively clean.

Setting a high default benchmark – for example, at the level of embodied emissions of a "worst technology" producer in the foreign country – would create a powerful incentive to report actual embodied emissions. Some producers, especially those in developing countries, would probably need technical assistance to measure their emissions, or else a high benchmark would unduly penalize them. Another option is to set the benchmark at the average emissions from the sector in the country of export, although sufficient data to calculate this average may not be available. A third benchmark option is to use domestic (that is, U.S.) industry average emissions. This would avoid any discrimination by country of origin, possibly increasing the legal and political palatability of BCAs.[21] But purely domestic-based benchmarks may systematically underestimate embodied emissions if domestic sectors are relatively clean. Indeed, any benchmark reduces incentives for more carbon-intensive producers, that is, those well above the benchmark, to cut their emissions, since marginal reductions have no effect on the BCA.

A reasonable compromise would be to adopt technology-based benchmarks for direct emissions related to production processes (since these are likely to be reasonably consistent across countries) and country-specific emissions factors for indirect emissions related to power generation (since this type of reporting already occurs to the UNFCCC). This option seeks to balance administrative burden and accuracy (Cosbey et al. 2012).

Adjustments to the embodied carbon tariff: The BCA should account for any output-based rebates offered to domestic industries, to ensure equal treatment with foreign producers. The BCA should also account for carbon prices paid in the exporting country; of course, this calculation may not be so simple, as corresponding tax breaks (like OBR) can undermine the price impact. A BCA need not attempt to account for the costs of non-price-based carbon regulation, as the intent is not to adjust for abatement costs that reduce emissions but rather for the cost of payments for the embodied emissions that remain.

Exemptions: One type of adjustment is simply to exempt imports from certain countries. The goals of such exemptions should be to minimize administrative costs, ensure the environmental effectiveness of the carbon tax, and foster compatibility with the UNFCCC goals of "common but differentiated responsibilities" (CBR). For example, Cosbey et al. (2012) recommend exemptions for the following types of countries:

- those with an effective national emissions cap, and for sectors with an effective sectoral cap (as this by definition prevents leakage);
- least developed countries (for CBR compliance); and
- those from which imports in these sectors or goods fall below a de minimus level.

Offering any sort of exemption entails a need for trans-shipment provisions – without them, goods could be produced in countries subject to BCA, then shipped to the U.S. via exempt countries to avoid BCA. The resulting complexity necessitates a high threshold of applicability that probably excludes manufactured goods and covers only a small number of commodities.

If these design issues can be appropriately addressed, the principal advantage of BCA is the preservation of carbon price signals for consumers. The disadvantages relate to administrative complexity and an arguably more combative approach to leakage, from the perspective of international relations.

Lower corporate income taxes

U.S. corporate income taxes are higher than those in most other developed economies, and many U.S. firms, including those in EITE sectors (see Chapter 8), argue that these tax rates put them at a competitive disadvantage with foreign rivals. Accordingly, some have suggested that the U.S. should broadly reduce corporate taxes, possibly as part of a revenue-neutral policy package which would include a carbon tax.

Relative to a carbon tax alone, broad corporate tax reductions would not only make U.S. companies, including those in EITE industries, more competitive, but might also broadly reduce incentives to shift production to other jurisdictions.[22] Overall corporate tax bills might be higher or lower relative to today under a policy package including both a carbon tax and corporate tax cuts, depending on the size of the corporate tax cut relative to the carbon tax, and the energy intensiveness of the firm. But corporations' tax bills would generally be lower (and their competitiveness greater) under such a package than under an equivalent carbon tax alone. And since firms would be paying the carbon tax, they would still have an incentive to make their production less carbon-intensive.[23]

However, broad corporate tax reductions are a blunt instrument for addressing competitiveness concerns associated with the carbon tax, as they make no distinction between EITE and non-EITE firms. Many firms liable for U.S. corporate taxes are not in EITE sectors, and there is therefore little need for policies to shield them from the effects of a carbon tax. Even small reductions in corporate tax rates will have significant effects on the revenue the federal government receives from such taxes, but only a small part of both the change in tax receipts and increased competitiveness would occur in the EITE sectors.[24] Indeed, effective U.S. corporate tax rates vary widely across sectors and firms, based on differences in tax breaks, foreign tax liabilities, and also profitability.[25] Corporate taxes are based on profits, so a lower rate only has a significant effect in sectors with significant profits. If firms in EITE sectors are already struggling, the benefits of a lower tax rate are limited; in fact, it can reduce the benefits from operating losses they would carry forward.

As a result, it is difficult to justify such across-the-board corporate tax reductions purely based on their ability to reduce negative effects on EITE firms. If one thinks that U.S. federal corporate taxes are currently set about right, then cutting those rates to address EITE competitiveness and leakage concerns arising from a carbon tax is an expensive, poorly directed policy. Alternatively, if you think that federal corporate taxes should be lower for reasons unrelated to a carbon tax, you might favor a revenue-neutral package that trades lower corporate taxes for a carbon tax. However, if the rationale for reducing corporate taxes is to improve the competitiveness of U.S. firms in EITE industries, a carbon tax erodes some or all of these benefits for those sectors. This puts us back where we started – one of the policy options discussed above is still needed. Cutting broad corporate taxes even more deeply could achieve the desired competitiveness benefits for firms in EITE sectors. But those cuts would be deeper than necessary to achieve the target competiveness boost for the larger group of trade-exposed (but not energy-intensive) firms (or the desired economy-wide efficiency improvements of lower corporate taxes). In short, the cart would be driving the horse.

A strong business advocate might argue that nothing is wrong with this approach – that lower corporate taxes are always a good thing. But the mismatch between broad corporate tax reductions and the problem of EITE sector competitiveness under a carbon tax exists whatever level of general corporate taxes one considers ideal. Once corporate taxes have been set at this ideal point, further cuts to address EITE sector competitiveness overshoot the target for other firms. In the extreme case, if one believes the correct corporate tax rate is zero, further cuts to benefit EITE sectors are not possible (setting aside negative taxes – that is, direct subsidies to U.S. firms – as both politically implausible and a clear violation of trade law).

In other words, the general competitiveness of an economy relative to other countries is a different (and possibly much larger) issue than the competitiveness of EITE sectors under a carbon tax. A general reduction in corporate taxes helps with the former, but only indirectly with the latter. And any help it would provide must be weighed against other uses for corporate tax revenues – cuts in other, other distortionary taxes, deficit reduction, infrastructure or technology investment, or anything else worthwhile that the government may do. More targeted policies for addressing EITE competitiveness would almost certainly have a smaller revenue impact.

One important advantage of broad corporate tax reductions, however, is that they could not be construed as breaches of international trade law. Nothing in U.S. WTO obligations restricts the freedom to set domestic corporate tax rates. Measures that specifically target EITE sectors are much more likely to trigger legal scrutiny.

At the same time, lobbyists for EITE sectors may argue that under a carbon tax + corporate tax cut package, their industries are still disadvantaged relative to other industries, since all industries benefit roughly equally from the tax cuts, but EITE sectors are disproportionately harmed by the new carbon tax. As the 1993 Btu tax example mentioned above illustrates, heavy lobbying is likely under any new tax proposal, and this may be a superficially compelling equity argument. However, firms making this argument are complaining about their tax position relative to firms in other industries – that is, firms against which they do not compete. EITE firms' domestic competitors would be similarly (though not necessarily identically) affected by a carbon tax. And as noted above, EITE firms might be more or less competitive with foreign firms in the same industry, but would be better off than they would be under a carbon tax alone.

Comparing the options

The goal of a carbon tax would be to raise revenue for public finances while reducing an important distortion to the economy, namely, the lack of a price reflecting the damages from greenhouse gas emissions.

Both exemptions and output-based rebates would forego part or all of the revenue that might be raised from a carbon tax on those sectors. Since EITE sectors in the U.S. account for perhaps 10–15 percent of emissions (depending on how broadly or narrowly the category is defined), the foregone revenue would be significant but not extraordinary. At the same time, with border adjustments, the revenue consequences would be minor, particularly if tariff revenues are returned to exporting countries (as suggested by some), but even if retained domestically, since imports represent only a share of production in EITE industries. A corporate income tax reduction, on the other hand, would absorb the large part (or all) of the carbon tax revenue.

In terms of cost-effectiveness, a broad literature has analyzed the design of climate policies in the context of unilateral or Annex I actions. It is well understood that the optimal policy requires uniform emissions pricing across sectors and countries, ensuring that marginal abatement costs are equal.[26] When such coverage cannot be complete, so-called second-best policies have been considered as complements to emissions pricing among the covered sectors.

Although analysis has not yet been done for the effects of corporate tax reforms on EITE industries, it is clear that the least efficient of the options discussed is setting lower carbon prices (that is, lower carbon tax rates) for the EITE sectors or exempting them altogether. The reason is that it forgoes some important incentives for those sectors to reduce their emissions. Still, when leakage effects are very strong, differentiated emissions pricing can lower the costs (both domestically and globally) of meeting a given global emissions reduction target, by shifting more of the burden toward sectors with less leakage potential. Although some studies find substantial justification for emissions price differentiation to deter leakage, from a global perspective, using these differentiated prices only slightly lowers the economic cost of unilateral climate policy versus uniform pricing.[27]

Unlike exemptions, output-based rebating retains the incentive effects of the carbon price on encouraging firms in EITE industries to find cost-effective ways to reduce their emissions intensity. At the same time, like exemptions, rebates help keep product prices from rising, and thus discourage substituting toward uncovered alternatives as a means of reducing emissions. However, this de facto subsidy also discourages conservation as a means of reducing emissions, resulting in an efficiency trade-off. In fact, it may actually encourage production of the energy-intensive products. While this may be attractive to some firms, it may be inconsistent with the overall aims of the carbon emissions reduction policy. Thus, this treatment is only recommended for sectors in which a large share of the production loss that might be associated with a carbon tax would arise from import displacement rather than reduced domestic consumption. Furthermore, it may only be recommended while the group of countries with carbon regulation is relatively small: as the climate coalition grows larger, the costs of the lost conservation incentives ultimately outweigh the benefits of reducing leakage (Böhringer et al. 2011).

BCAs tend to be significantly more cost-effective than these other options, since they not only address competitiveness-related leakage by discouraging carbon-intensive imports from nonparticipating nations, but also encourage those same countries to adopt a carbon pricing regime. In addition, they also preserve the incentives for consumers to find less carbon-intensive products. They are, however, the most politically controversial option and face significant legal and administrative barriers.

A few economic studies compare BCAs with output-based rebates and exemptions. In combination with a carbon tax, all the options promote domestic production to some extent and reduce leakage. However, none would necessarily reduce *global* emissions in a given sector: while they reduce emissions abroad, they expand domestic firms' emissions. The net effect depends on the relative responses of domestic and foreign producers to price changes and their relative emissions intensities. Fischer and Fox (2012a) find that for most U.S. sectors, full border adjustment would be most effective at reducing global emissions, but output-based rebates can be more effective than import adjustments (particularly when those are limited by WTO law to a weaker, nondiscriminatory standard) at reducing emissions leakage and encouraging domestic production (see Figure 9.3).

Figure 9.3 also illustrates the point that while these policies can address some of the leakage related to competitiveness impacts, they do not eliminate all leakage. That is because the carbon tax also causes global energy prices to fall, encouraging more carbon-intensive fossil fuel use among foreign EITE sectors.

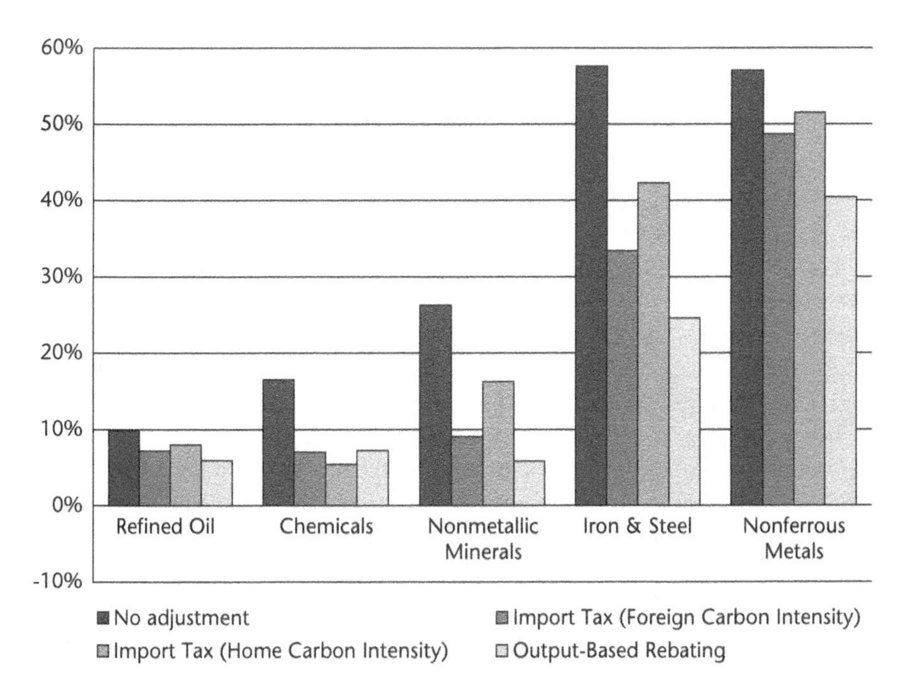

FIGURE 9.3 Carbon leakage rates by sector and policy option

Source: Based on Fischer and Fox (2012a), which assumes that a \$14/ton CO_2 carbon tax is applied to energy-intensive industries.

Of course, the effect of policies on leakage rates is distinct from the effect on economic well-being, which recognizes the benefits of emissions reductions as well as the costs of mitigation. Many trade models have compared these policies in a cost-effectiveness framework that holds global emissions (and thus the environmental benefits) constant, requiring the domestic carbon price to adjust in response to anti-leakage measures. From a global cost-effectiveness standpoint, border adjustment for imports achieves most of the benefits associated with full border adjustment, and results in lower costs regardless of coalition size. However, BCAs also entail the largest shifting of the burdens of climate regulation toward developing and emerging economies (Böhringer et al. 2012).[28] At the same time, requiring that the tariff revenues be returned to exporting countries can allow developing countries overall to enjoy net benefits from the policy (Fischer and Fox 2012b). When the regulating coalition is sufficiently small, output-based rebating offers a middle ground of improving cost-effectiveness without shifting the burden excessively (see also Böhringer et al. 2011).

Conclusions

Critics point to competitiveness and leakage as major problems that would undermine the effectiveness of a carbon price. While economic modeling indicates that the magnitude of these impacts may not be as severe as these critics suggest, they are legitimate concerns that merit attention.

There are, however, policy options available for mitigating these twin problems. At least three specific policy options focused on EITE industries exist: sector-specific exemptions, output-based rebates similar to the ones advanced in the ACES Act, and border carbon adjustments. Broad corporate tax reform not aimed exclusively at EITE industries also addresses loss of competitiveness, and could be viewed as a crude pro-competitiveness/anti-leakage policy.

As shown in Table 9.1, some of these policies are likely to be more effective than others. Broad-based corporate tax reform, while attractive for many reasons, is not an effective mechanism for addressing the disproportionate impacts imposed on EITE industries by a carbon tax. Exempting firms from carbon taxes altogether is also a fairly blunt instrument that can be both inefficient and inequitable.

Two other policies are relatively attractive, although each has its own advantages, disadvantages, and challenges. BCAs offer the potential to address both competitiveness and leakage concerns most directly, although their design poses many challenges and their use may be constrained by a variety of legal problems. Output-based rebates are more effective in addressing competitiveness than leakage impacts, could be implemented simultaneously with the carbon tax itself, and might be more likely to survive legal challenge. But they are relatively complex to administer and might be more prone to unintended consequences from firms' production decisions. For either policy, the devil is largely in the details. But both appear to offer viable methods for addressing legitimate concerns about the adverse impacts of a U.S.-only carbon tax.

TABLE 9.1 Pros and cons of alternative policies

Policy	*Pro*	*Con*
Tax Exemption EITE firms exempt from some/all of the carbon tax	Simple Political resonance	Reduces revenues Undercuts incentives to reduce emissions
Output-based Rebates EITE firms subsidized based on their output	Maintain incentives to cut emissions	Complex Reduces revenues
Border Carbon Adjustments Imports taxed/exports subsidized for EITE sectors	Directly targets leakage/ competitiveness	Legally vulnerable Internationally controversial Complex
General Corporate Tax Cuts Lower corporate taxes for all firms	None	Poorly targeted to EITE sectors Reduces revenues

Notes

* This paper was presented at an October 23, 2012, workshop at Resources for the Future "Roundtable Discussion on the Economics of Future US Fiscal and Carbon Policy" and at the November 12, 2012, conference "The Economics of Carbon Taxes" sponsored by the American Enterprise Institute, Brookings Institution, IMF, and Resources for the Future.

1. EPA, "Draft Inventory of U.S. Greenhouse Gas Emissions and Sinks: 1990–2011," Washington, DC, February 13, 2013.
2. For example, the U.S. Energy Information Administration (EIA) in its 2009 analysis of H.R. 2454 (see below) estimated the industrial sector would contribute 9.7 percent of total reductions in 2020 and 5.5 percent in 2030. Reflecting recent declines in natural gas prices and the corresponding increase of gas use in the electric power sector, today's outlook probably calls for slightly greater shares from the industrial sector, although EIA analysis is not yet available on this issue.
3. Fischer and Fox (2010).
4. Other less commonly discussed options include cuts in employer payroll taxes or subsidies to energy-efficient technologies for EITE industries.
5. Energy intensity is defined as the share of total costs that is from energy use (combustion and electricity). Trade intensity is defined as exports plus inputs divided by the sum of the value of shipments and imports.
6. U.S. Environmental Protection Agency 2009.
7. U.S. EPA, 2009.
8. Output losses in the case of unilateral taxation would be somewhat higher.
9. These exceptions are described in Article XX of the General Agreement on Tariffs and Trade (GATT).
10. For a coalition of Kyoto Annex I (developed) countries, excluding Russia, with a reduction target of 20 percent below 2004 emissions; the corresponding carbon prices range from \$15 to \$60 per ton CO_2. Studies of similar unilateral U.S. policies find slightly larger leakage rates (Fischer and Fox 2012b).
11. Adkins et al. 2012, p. 43.
12. Sweden has adopted this approach. Its industries are exempt from 80 percent of the CO_2 tax.
13. See, for example, Böhringer et al. (2010); Böhringer et al. (2012); Fischer and Fox (2012b).
14. The freely allocated allowances were designed to start phasing down in 2025, and to end completely by 2030.

15. For example, CO_2 emissions per ton of product, although the precise definition would be left to EPA.
16. Under a tradable performance standard, regulators would set a benchmark carbon intensity or other measure of efficiency on a sector-by-sector basis. Firms that do better than this standard would receive credits tradable to firms that do not reach it.
17. A feebate is sometimes mentioned as another possible option, although it is almost identical to a carbon tax combined with output-based rebates. The principal distinction is that with a feebate you could have a lower tax rate by blending exemptions and output-based rebates, likely resulting in lower abatement incentives.
18. For example, the Alberta intensity-based emissions trading scheme allows the compliance option of contributions to the Climate Change and Emissions Management Fund at a price of CND 15 per metric ton of CO_2e.
19. Rebates in excess of emissions liabilities on average would likely run afoul of WTO law on subsidies.
20. For one thing, the legal basis for export rebates is shakier than that for import adjustments. Rebates fall under the WTO Subsidies Code, which does not have an environmental exception akin to Article XX in the GATT, which might allow for discrimination among "like" products if necessary in order to protect health or conserve natural resources.
21. Opinions vary as to whether BCAs that differentiate across countries can be compatible with WTO obligations to treat imported goods no less favorably than "like" domestic products and to accord Most Favored Nation Treatment to all WTO members. Exceptions might be made for policies deemed necessary to preserve the environment, as long as they are the least trade restrictive option.
22. On the other hand, if lower corporate taxes also mean similarly reduced taxes on multinational corporations' repatriated foreign income, there might be little or no additional incentive to avoid shifts of production (and therefore leakage) that a carbon tax would promote.
23. Carbon and corporate tax interactions could be complex. One relatively straightforward observation is that carbon tax payments (both direct and indirect) are partly corporate-tax-deductible, depending on their effect on corporate profits. Market power, the availability of substitutes, and other factors determine the extent that firms will be able to pass carbon tax costs on to consumers or be forced to absorb them with corresponding reductions in profits and, therefore, corporate tax payments.
24. Of course, it is possible to cut corporate taxes only for EITE sectors. Doing so is broadly similar to a carbon tax plus rebates tied to output levels (discussed above), except that it is achieved via a different part of the tax code and is only useful to firms that are making profits (and therefore paying corporate tax).
25. See, for example, http://dealbook.nytimes.com/2012/12/11/not-all-companies-would-welcome-a-lower-tax-rate/
26. Of course, this assumes an otherwise ideal world, barring additional major distortions.
27. In one analysis, second-best carbon prices for EITE sectors were found to be roughly 40 percent lower than for other sectors overall (Böhringer et al. 2010).
28. See also the full special issue in *Energy Economics* (December 2012).

References

Adkins, L., R. Garbaccio, M. Ho, E. Moore, and R. D. Morgenstern. 2012. Carbon Pricing with Output Based Subsidies: Impact on U.S. Industries over Multiple Timeframes. Discussion Paper 12–27. Washington, DC: Resources for the Future.

Böhringer, C., E. Balistreri, and T. F. Rutherford, 2012. "The Role of Border Carbon Adjustment in Unilateral Climate Policy: Insights from a Model-Comparison Study." Discussion Paper 12–54, The Harvard Project on Climate Agreements, Cambridge, MA.

Böhringer, C., J. C. Carbone, and T.F. Rutherford. 2012. Unilateral Climate Policy Design: Efficiency and Equity Implications of Alternative Instruments to Reduce Carbon Leakage. *Energy Economics* 34(Supplement 2): S208–S217 (December).

Böhringer, C., C. Fischer, and K.E. Rosendahl. 2011. Cost-Effective Unilateral Climate Policy Design: Size Matters. Discussion Paper 11–34. Washington, DC: Resources for the Future.

Böhringer, C., A. Lange, and T.F. Rutherford. 2010. Optimal Emission Pricing in the Presence of International Spillovers: Decomposing Leakage and Terms-of-Trade Motives. Working Paper 15899. April. Cambridge, MA: National Bureau of Economic Research.

Cosbey, A., S. Droege, C. Fischer, J. Reinaud, J. Stephenson, L. Weischer, and P. Wooders. 2012. A Guide for the Concerned: Guidance on the Elaboration and Implementation of Border Carbon Adjustment. ENTWINED Report. Stockholm.

Fischer, C., and A.K. Fox. 2010. On the Scope for Output-Based Rebating in Climate Policy: When Revenue Recycling Isn't Enough (or Isn't Possible). Discussion Paper 10–69. Washington, DC: Resources for the Future.

Fischer, C., and A.K. Fox. 2012a. Comparing Policies to Combat Emissions Leakage: Border Tax Adjustments versus Rebates. *Journal of Environmental Economics and Management* 64(2): 199–216 (September).

Fischer, C., and A.K. Fox. 2012b. Climate Policy and Fiscal Constraints: Do Tax Interactions Outweigh Carbon Leakage? *Energy Economics* 34(Supplement 2): S218–S227 (December).

U.S. Environmental Protection Agency. 2009. The Effects of H.R. 2454 on International Competitiveness and Emission Leakage in Energy-Intensive Trade-Exposed Industries: An Interagency Report Responding to a Request from Senators Bayh, Specter, Stabenow, McCaskill, and Brown. Washington, DC: EPA. Available at: www.epa.gov/climatechange/economics/pdfs/InteragencyReport_Competitiveness&EmissionLeakage.pdf

10

THE ROLE OF ENERGY TECHNOLOGY POLICY ALONGSIDE CARBON PRICING

Richard G. Newell

KEY MESSAGES FOR POLICYMAKERS

- In the context of a future carbon tax or market-based trading system, advanced energy technologies offer the opportunity to significantly reduce costs and expand options for meeting emissions reduction targets.
- There are two specific market problems to consider in the realm of GHG-relevant technology innovation: the environmental externality of global climate change and issues related to the market for innovations.
- This chapter presents a two-part strategy to climate innovation that directly confronts these two problems: (1) establish a price on GHG emissions through a carbon tax or market-based trading system, supplemented by permanent tax credits for all (not just energy-related) R&D; and (2) increase public funding for basic strategic research inspired by critical, climate-related needs.
- This approach seeks balance by increasing both the demand for and the supply of GHG-reducing innovations. Pursuing either demand- or supply-side measures alone, but not both, will likely lead to higher costs or missed opportunities in achieving emissions targets.
- The innovation strategy also aims to harness the power of private sector incentives for societal gain, recommending that the direct governmental research role complement, rather than substitute for, activities commonly undertaken by industry. This strategy emphasizes those aspects of the overall innovation process that the private and public components of the system are best oriented toward advancing.

Introduction and conceptual background

Substantial reductions in U.S. greenhouse gas (GHG) emissions will require large-scale innovation and adoption of GHG-reducing technologies throughout the U.S. economy, including technologies for increased energy efficiency, renewable energy, nuclear power, and carbon dioxide (CO_2) capture and storage. In the context of a policy regime that places a price on GHG emissions – either in the form of a carbon tax or market-based trading system – advanced energy technologies hold the potential to significantly lower costs and expand options for meeting GHG mitigation goals. While the importance of new technology in solving the climate problem is widely understood, there is considerable debate about what specific public policies and programs are necessary to bring about these technological changes as effectively and efficiently as possible.

The proposed climate technology innovation strategy presented in this chapter is based on the simple principle that, within a market-based economy, success is maximized if policies directly address specific market problems. In addressing such problems, policies should be designed to harness the power of private sector incentives for societal gain, and the direct governmental research role should be designed to complement, rather than substitute for, activities commonly undertaken by industry.

In the context of GHG-relevant technology innovation, there are two principal market problems. First and foremost, there is the environmental externality of global climate change. If firms and households do not have to pay for the climate damage imposed by their GHG emissions, then these emissions will be too high, and demand for GHG-reducing technologies will be too low. In turn, there will be insufficient incentive for companies to invest in mitigation technology research and development (R&D), because there will be little market demand for any potential innovations that result. A market-based emissions policy that places a price on GHGs is widely accepted to be a cost-effective response to this issue.

Second, there are problems specific to the market for innovations – not just with respect to climate, but more broadly.[1] Knowledge, just like a stable climate, is a public good. It is well known that individual companies cannot capture the full value of investing in innovation, as that value tends to spill over to other technology producers and users, thereby diminishing individual private incentives for R&D.[2] This problem tends to worsen the more basic and long term the research (and may be exacerbated by technology transfer to other countries without sufficient intellectual property protection). Well-targeted policy that boosts climate technology innovation therefore has the potential to lower the overall cost of attaining long-term climate goals.

The proposed strategy thus has two main parts to directly confront these two market problems: (1) establish a price on GHG emissions through a carbon tax or market-based trading system, supplemented by permanent tax credits for all (not just energy-related) R&D; and (2) increase public funding for basic strategic research inspired by critical, climate-related needs. The total revenue required for these purposes would be on the order of $10–15 billion per year.

Taking these parts together, the strategy seeks to increase both the demand for and the supply of GHG-reducing innovations in a balanced way – one that emphasizes those aspects of the overall innovation process that the private and public components of the system are best oriented toward advancing. R&D push without the pull of demand is like pushing on a rope: ultimately having little impact.[3] In fact, ratcheting up R&D and other technology policies in an attempt to compensate for insufficiently stringent emissions policy can dramatically raise the cost of achieving a given amount of GHG mitigation. Conversely, market demand-pull without supportive R&D policies may miss longer-term opportunities for significantly lowering GHG reduction costs and expanding opportunities for greater GHG mitigation. A coupled "emissions price plus R&D" strategy, as suggested here, offers the best opportunity for mitigating GHG emissions at the lowest possible cost to society.

The remainder of the chapter will consider each stage of the technology innovation process separately – research and development, demonstration, and deployment – elaborating on the innovation strategy proposed above, and highlighting in particular its ability to address the unique challenges and opportunities at each point of the innovation process.

Research and development

Both parts of the innovation strategy work toward the advancement of climate mitigation technology, although each targets a different part of the economy.

First part of the innovation strategy: Stimulating private sector R&D

The first part of the innovation strategy seeks to harness the power of private sector investment. Industry is central to the U.S. innovation system, performing 71 percent and funding 62 percent of total U.S. R&D (Figure 10.1). The single most important part of solving the climate technology problem is therefore to address the GHG externality through emissions pricing. The emissions price attaches a financial cost to GHGs and – just as people will consume less of something that carries a price than they will of something given away for free – will induce households and firms to buy technologies with lower GHG emissions (such as lower-emission power plants and more efficient cars and appliances). In turn, the emissions price creates a demand-driven, profit-based incentive for the private sector to invest in R&D and other innovative efforts to bring new, lower-cost, climate-friendly technologies to market. In all, the GHG price helps to stimulate progress at multiple stages of the innovative technology process: basic and applied research, development, demonstration, and deployment (Box 10.1) (demonstration and deployment to be discussed in greater detail below).

Emissions pricing is not the only important tool necessary to achieve climate mitigation goals, however. Although private sector incentives for innovation are supported by intellectual property protection, secrecy, and other means, there is still a substantial portion of the benefits of innovation that cannot be captured by

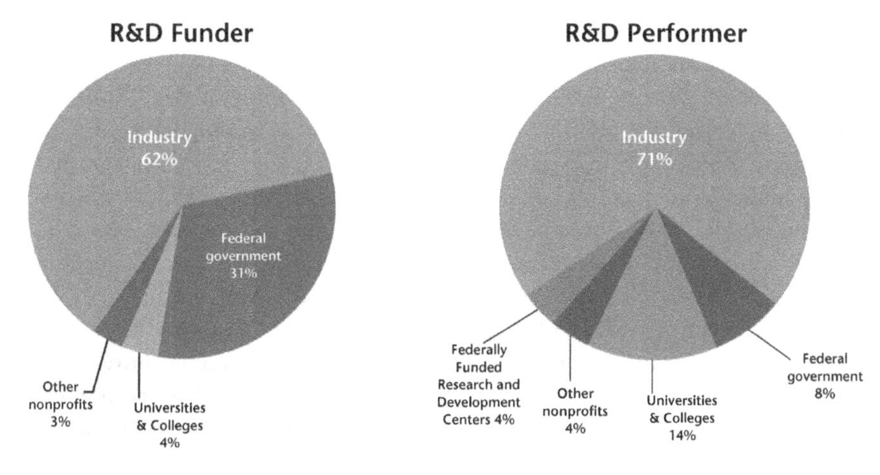

FIGURE 10.1 U.S. R&D expenditures by funder and performer

Source: 2009 data from National Science Board (2012), Table 4–3.

BOX 10.1 STAGES OF R&D: COMMON DEFINITIONS

R&D. According to international guidelines for conducting R&D surveys, R&D comprises creative work "undertaken on a systematic basis to increase the stock of knowledge . . . and the use of this stock of knowledge to devise new applications" (OECD 2002, p. 30).

Basic research. The objective of basic research is to gain more comprehensive knowledge or understanding of the subject under study without specific applications in mind. Although basic research may not have specific applications as its goal, it can be directed in fields of present or potential interest. This is often the case with basic research performed by industry or mission-driven federal agencies.

Applied research. The objective of applied research is to gain knowledge or understanding to meet a specific, recognized need. In industry, applied research includes investigations to discover new scientific knowledge that has specific commercial objectives with respect to products, processes, or services.

Development. Development is the systematic use of the knowledge or understanding gained from research directed toward the production of useful materials, devices, systems, or methods, including the design and development of prototypes and processes.

Source: National Science Board (2008), p. 4.9.

innovating firms, as new innovations build on existing knowledge and the benefits of new technology are passed onto consumers. This leads to a generic argument in favor of R&D tax incentives to boost the level of all (energy- and non-energy-related) private sector R&D.

Section 41 of the U.S. Internal Revenue Code allows taxpayers to claim credits against corporate and individual income tax for extra expenditures on research and experimentation above a defined baseline. Known informally as the "R&D tax credit," it was originally added to the tax code in 1981 as a temporary measure. It has been renewed 14 times since then – sometimes retroactively and/or after lapses – and under current law expired on December 31, 2013, though it is expected to be renewed for 2014. Given that R&D efforts typically span multiple years, this kind of uncertainty makes long-range research planning based on tax considerations difficult. As such, encouraging increased private sector R&D by making the R&D tax credit permanent would bolster private incentives for innovation that would be induced by the emissions price, and would improve innovation incentives more generally. In recent years tax expenditures for the R&D tax credit have been in the range of $8–9 billion annually, with forward-looking estimates of about $10 billion per year to make the credit permanent.

It is difficult to pin down exactly how much and what type of innovation is likely to be generated by a GHG emissions price bolstered by the R&D tax credit, but the innovation is sure to come from a wide array of businesses currently engaged in the development and use of energy producing and consuming technologies, including:

- Provision of electricity and transportation services.
- Agro-biotech sector (assuming there are incentives for CO_2 sequestration in forests and other biomass sinks).
- Companies that produce and consume other non-CO_2 GHGs (e.g., chemical companies).
- Less obvious sectors such as information technologies (e.g., in the context of energy management and conservation).

Second part of the innovation strategy: Public support for strategic basic research

Although basic and applied research is critical to the innovation process, more than three-quarters of industrial R&D is instead focused on development. In contrast, universities, other nonprofits, and federal labs perform the vast majority of basic research (about 80 percent), more than half of which (approximately 53 percent) is funded by the federal government.[4] Although it may have low short-term returns to individual firms, basic research can have high returns to society over the long run by building the intellectual capital that lays the groundwork for future advances in technology. In this way, universities, nonprofits, and federal labs play a complementary role to industry in the innovation system, so there is a need for policy that will supplement industrial R&D with more basic research relevant to lowering the cost of GHG mitigation and meeting other energy policy goals.

The second part of the climate innovation strategy addresses this need by proposing to gradually increase federal spending for energy R&D. But how high should federal energy R&D budgets go? While ideally one might like to optimally determine and allocate the federal R&D budget across the wide variety of funded areas – thorough detailed evaluation of the technical opportunities, cost of research efforts, likelihood of success, and ultimate economic and social payoff of research – this is not realistic and may not even be practically possible. Nonetheless, a significant expansion of well-directed energy R&D funding is warranted based on scientific opportunities for advance, plausible assumptions about the rate of return on such spending and other factors, to roughly $8 billion per year over the next several years, or roughly a 40 percent addition in energy R&D from 2012 levels. A gradual and sustained ramp-up is preferable to a dramatic spike in R&D spending due to the nature of the R&D process, which involves the training of scientists and engineers and their gradual movement into the innovation system, where they then require ongoing support. To avoid crowding out of other beneficial R&D (i.e., diverting engineers and scientists away from non-energy sectors), this funding should therefore be phased in.

Other studies have recommended substantially higher levels of R&D funding, including a 2010 study by the President's Council on Science and Technology, or PCAST, that recommended double this amount – $12 billion for R&D and $4 billion for large-scale demonstration and deployment annually – although without a specific scale-up period. The PCAST study also recommended that the additional funds ($10–11 billion annually in their case) come from new revenue streams, such as carbon pricing, rather than traditional federal appropriations.

This funding should place a priority on strategic basic research inspired by critical needs arising from efforts to develop new and improved GHG mitigation technologies. The concept of strategic basic research emphasized here is close in spirit to Stokes's notion of use-inspired basic research, which (unlike pure basic research) is inspired by the desire to develop improved technology, but (unlike pure applied research) also seeks to develop improved fundamental understanding (see Figure 10.2). A few examples include:

- Direct conversion of solar energy to electricity and chemical fuels.
- Understanding of how biological feedstocks are converted into portable fuels.

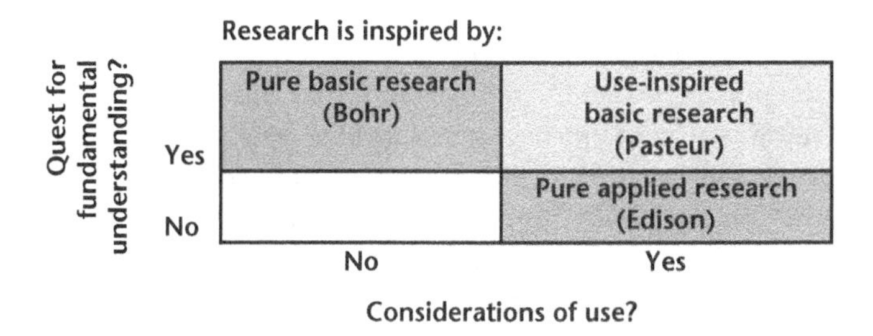

FIGURE 10.2 An alternative conception of research: "use-inspired" basic research

Source: Based on Stokes (1997).

- New generation of radiation-tolerant materials and chemical separation processes for fission applications.
- Addressing fundamental knowledge gaps in energy storage.

At the same time, this funding should also invest in training the next generation of scientists and engineers. This would tend to imply prioritizing additional funding to universities and colleges, which accounted for only about 22 percent of DOE energy-related R&D funding obligations in 2009, while government laboratories represented nearly two-thirds of funding obligations in that year.[5] During the past two administrations a number of innovative new R&D programs have been developed to increase the application of basic scientific research to energy problems, including the Energy Frontier Research Centers and Bioenergy Research Centers.

In order to encourage exploration of novel, emergent, or integrative concepts for addressing climate change, there should also be a program of exploratory research that pursues transformational technologies that may not fit well within existing basic or applied research programs. The Advanced Research Projects Agency-Energy, or ARPA-E, is the natural foundation on which to further build that program, where appropriate in cooperation with other key funding agencies such as the National Science Foundation, the Department of Defense, and the Department of Agriculture. ARPA-E currently includes programs in the areas of electrical energy storage, microorganisms and plants for liquid transportation fuels, and innovative materials and processes for carbon capture, to name a few.

Improved research communication, coordination, and collaboration

In the context of increased funding for strategic basic research, perhaps the most important additional recommendation is to improve processes for communication, coordination, and collaboration between the U.S. Department of Energy (DOE) and the private sector and within DOE among the basic research programs in the Office of Science, the applied energy research programs within the DOE program offices (fossil fuel, nuclear, renewables, end-use efficiency, and electricity delivery), and ARPA-E. The first *Quadrennial Energy Technology Review*, published in 2011, was a positive step in this direction and will require continued attention from the point of view of implementation.

In addition, the President's Council on Science and Technology has recommended to broaden this process beyond DOE through a Quadrennial Energy Review process that would engage the many other agencies with active energy policy roles, including but not limited to the Departments of Agriculture, Commerce, Defense, Energy, Interior, Transportation, State, and Treasury, and the Environmental Protection Agency. Increased resources need to be tied to an effective strategy for managing and coordinating climate mitigation technology research to ensure these funds are employed efficiently. The PCAST study and others (see Suggestions for Further Reading) also made a number of useful recommendations for improving federal management of energy innovation, both within DOE and across the federal decision-making complex.

Inducement prizes

Finally, a portion of the proposed public funds should be used to supplement the traditional research contracts and grants structure with inducement prizes (sometimes called challenges) that provide financial rewards for achieving significant advances in climate mitigation innovation. Prizes of this type can help focus research efforts on clearly defined objectives, instill a sense of urgency and competition, and engage a broad set of innovators. In contrast to other instruments such as research contracts, grants, and R&D tax credits, prizes have the attractive incentive property of targeting and rewarding innovation outputs, rather than inputs: the prize is paid only if the objective is attained. Prizes or awards can also help to focus efforts on specific high-priority objectives, without specifying how the goal is to be accomplished.

Because prize competitors select themselves based on their own knowledge of their likelihood of success – rather than being selected in advance by a research manager – prizes can also attract a more diverse and potentially effective range of innovators from the private sector (e.g., industry or individual entrepreneurs), universities, and other research institutions. Building on recent experience gained at the federal level – including at DARPA, NASA, and the DOE – the DOE (and possibly other relevant agencies) should continue an experimental series of prizes over the next several years employing a small fraction of the additional recommended funds, with additional funds thereafter in accordance with an evaluation of the first phase. For example, in the past several years DOE has launched a number of inducement challenges including:

- The Bright Tomorrow Lighting Prize for energy-efficient lighting;
- SunShot prize for low-cost rooftop solar PV;
- Apps for Vehicles Challenge to spur and highlight innovations from vehicle-generated open data; and
- Apps for Energy prizes for software that helps utility consumers make informed decisions.

Inducement prizes are not suited to all innovation goals, however, as they tend to require well-defined and measurable objectives that can be stated in advance.

Demonstration

Overall, public funding for precommercial research tends to receive widespread support among experts based on the significant positive spillovers typically associated with the generation of new knowledge. Agreement over the appropriate role of public policy in technology development tends to weaken, however, as one moves from support for R&D to support for large-scale demonstration projects, and to widespread commercial deployment.

Technology demonstration projects – which seek to prove the viability of new technologies at commercial scale – occupy a middle ground between R&D

and deployment. Arguments for public support of technology demonstration projects tend to point to the large expense; high degree of technical, market, and regulatory risk; and inability of private firms to capture the rewards from designing and constructing first-of-a-kind facilities. Most compelling from an economic perspective, there may be knowledge generated in the process of undertaking first-of-a-kind demonstration projects – which can help improve the design of future technology, lower technical risks, and serve as a basis for well-designed regulations – but profits from this knowledge may not be appropriable by individual firms.

Conversely, caution is required because, despite good intentions, the most infamous failures in government energy R&D funding (e.g., the Synthetic Fuels Corporation, Clinch River Breeder Reactor) tend to be associated with large-scale demonstration projects – using up large portions of limited R&D budgets in the process (see Suggestions for Further Reading). The experience with the FutureGen Initiative for clean-coal power tends to reinforce this perspective, although that project is still under development.

In sum, while it should not be the focus of climate mitigation innovation investments by the public sector, there may be a compelling rationale for well-designed public support for a limited number of first-of-a-kind mitigation technology demonstration projects, so long as the purpose is the generation of substantial new knowledge (as opposed to meeting production or deployment targets).

Deployment

The first part of the innovation strategy, pricing GHG emissions through a carbon tax or market-based trading system, would provide direct, cost-effective, and technology-neutral financial incentives for the deployment of GHG-reducing technologies. Beyond GHG emissions pricing, however, there are other technology policies – divided into the broad categories of standards and subsidies – that might also be enacted to aid in the deployment of climate-related innovations. If such additional policies are pursued alongside GHG pricing, they should be targeted at addressing market problems other than emissions reductions per se – and thus can be viewed as a complement to, rather than a substitute for, an emissions pricing policy.

There are several specific market problems to which technology deployment policies could be efficiently directed, if the benefits of practicable policies were found to justify the costs in particular circumstances. These market problems include:

- impediments to cost-effective energy-efficiency investment decisions;
- knowledge spillovers from learning during deployment;
- different information available to project developers versus lenders;[6]
- network interactions in large integrated systems, such as with transportation fuels, vehicles, and fueling infrastructure; and
- incomplete insurance markets for liability associated with specific technologies.[7]

Although market problems are often cited in justifying technology deployment policies, such policies in practice may go further in promoting particular technologies than a response to a particular market problem may require. Therefore, while conceptually sound rationales may exist for implementing these policies, in practice one must evaluate whether, as actually proposed and implemented, they would provide a cost-effective addition to market-based emissions policies. Critics point out that deployment policies intended to last only during the early stages of commercialization and deployment often create vested interests that make the policies difficult to end.

As complements to a market-based trading system, technology deployment policies will tend to lower the allowance price associated with achieving a given emission target, rather than producing additional emissions reductions below the cap. As complements to a GHG tax, such policies will tend to increase the total amount of emissions reductions achieved by a given tax.

Technology standards and subsidies. Technology standards and subsidies can be viewed as different means to achieve the same ends (such as increased energy efficiency and/or greater reliance on renewable energy, for example). Just as there are important differences between an emissions-trading program and an emissions tax, however, standards and subsidies tend to differ in terms of who bears the cost, how their impact evolves over time, and what kinds of outcomes they guarantee.

Cost distribution. Regarding distributional consequences, the cost of imposing a standard tends to fall primarily on households and firms in the regulated sector. By contrast, the cost of providing subsidies tends to fall on taxpayers more generally – unless those subsidies are funded directly by energy-related fees (e.g., electricity ratepayer surcharges). However, this distinction can also be altered somewhat through self-financing mechanisms such as "feebates" (for example, to promote improved automobile fuel economy, subsidies for efficient vehicles could be funded by fees on inefficient vehicles – see Chapter 12).

Evolution over time. Different deployment policies also have different dynamic properties. The incentives generated by standards are typically more static in the sense that industry has no reason to exceed the standard, which eventually becomes less binding as technology matures (of course, as technology improves, policymakers may also respond by raising standards). Fixed subsidy levels, on the other hand, may continue to provide incremental deployment incentives, depending on the payment structure. If the rationale for a particular deployment policy is early technology learning, however, designing technology deployment policy instruments to automatically lose their impact over time as the technology matures could very well be desirable.

Variable outcomes. Standards tend to guarantee that specific technologies will be deployed in a certain quantity (or as a minimum share of the market) or that certain performance criteria will be achieved, but leave the cost of achieving the standards uncertain. Technology subsidies, on the other hand, pin the incremental cost spent on technology to the level of the incentive and leave uncertain how much deployment (or what level of performance) will be achieved at that cost. Ceilings (and floors) on credit prices within a tradable standards system can blur these distinctions.

Designing for flexibility and efficiency. As with emission standards, the cost-effectiveness of technology-oriented standards can be increased by incorporating flexibility mechanisms such as credit trading, banking, and borrowing. Likewise, tendering or reverse auctions – whereby the government has a competitive bidding process for the provision of technology such as renewable electricity generation – can help facilitate cost competition by making subsidy recipients bid for the minimum subsidy needed to deliver a specified quantity of new technology. This approach can help reduce the cost of technology deployment over time by ensuring that a given expenditure of public resources produces the maximum amount of deployment (or conversely, that a given deployment target is achieved at the lowest possible cost to taxpayers). This is particularly true if the approaches involve a differentiated rather than uniform bid structure.

Other aspects of deployment policy. Finally, a number of other polices may be critical in helping certain GHG-reducing technologies compete effectively to potentially gain a foothold in the marketplace. The successful deployment of new technologies often requires better information and verification methods; infrastructure planning, permitting, compatibility standards, and other supporting regulatory developments; and institutional structures that facilitate technology transfer, such as rule of law, judicial or regulatory transparency, intellectual property protection, and open markets. A balance must be struck, however, between enabling technologies to compete and constructing policies that preferentially support specific technology options or systems.

Conclusion

The purpose of this chapter is to outline how a well-targeted set of climate policies, including those targeted directly at science and innovation, could help lower the overall costs of mitigation. It is important to stress, however, that poorly designed technology policy will raise rather than lower the societal costs of climate mitigation. To avoid this, government support should emphasize areas that are least likely to be undertaken by the private sector, assuming that industry will face substantial incentives in the form of a market-based price on GHG emissions. As discussed, this would tend to emphasize strategic basic and applied research that advances science in areas critical to climate mitigation. In addition to generating useful results, such funding also serves the critical function of training the next generation of scientists and engineers for future work in both the private and nonprofit sectors.

Climate technology policy must complement rather than trying to substitute for emissions pricing. On the research side, R&D without market demand for the results is like pushing on a rope and would ultimately have little impact. On the deployment side, technology-specific mandates and subsidies may have some emission reduction benefits, but will tend to generate those reductions in a relatively expensive, inefficient way relative to an economy-wide emissions price. The scale of the climate technology problem and our other energy challenges requires a solution that is as cost-effective as possible.

Notes

1. These issues persist despite intellectual property and other protections; refer to the section below on "Research and Development" for a more detailed discussion.
2. Some incentives, however, may go the other way. For example, it is possible that multiple firms competing against one another may over-invest in a single technology or set of technologies from a broader societal point of view. The sense among experts, however, is that on net positive spillover concerns dominate, leading generally to under- rather than over-investment in innovation.
3. In some cases, such as where existing products are made more energy-efficient (e.g., vehicle fuel efficiency improvements), demand for new technologies is generated from the prospect of fuel savings alone. Nonetheless, this demand incentive is still insufficient if fuel prices do not reflect the GHG externality.
4. Based on 2009 statistics from National Science Board (2012).
5. Statistics were calculated from data in: National Science Foundation, National Center for Science and Engineering Statistics, Federal Funds for Research and Development, FY 2009-11, Table 16. For the purposes of these statistics, "energy-related R&D" comprises Electricity Delivery and Energy Reliability, Energy Efficiency and Renewable Energy, Fossil Energy, Nuclear Energy, and the Office of Science within the U.S. Department of Energy.
6. Loan guarantee programs – which involve government guarantee of loans at low financing rates – may be conceptually justified if informational asymmetries exist in credit markets for relevant technologies. On the other hand, loan guarantees create implicit subsidies; as such, their benefits should justify their costs. Because loan guarantees insulate projects (at least in part) from default risk, they can incentivize developers to take on riskier projects while doing less than they should to guard against preventable risks.
7. There may be a rationale for establishing a joint insurance pool or limiting liability for certain technologies like carbon storage if there is insufficient availability of private liability insurance or there are substantial potential difficulties in assigning liability. On the other hand, liability protection provides a form of implicit subsidy by insulating parties from potential damages caused by their technologies. Thus, if designed poorly they may reduce incentives for those parties to take appropriate actions to mitigate risks where possible.

References and suggestions for further reading

Henderson, R.H. and R.G. Newell. 2011. Introduction and Summary. In *Accelerating Energy Innovation: Insights from Multiple Sectors.* Edited by R.H. Henderson and R.G. Newell. Chicago: University of Chicago Press for the National Bureau of Economic Research, 1–24.

National Science Board. 2008. *Science and Engineering Indicators.* Arlington, VA: National Science Foundation.

National Science Board. 2012. *Science and Engineering Indicators 2012.* Arlington VA: National Science Foundation (NSB 12–01).

Newell, Richard G. 2007a. Climate Technology Research, Development, and Demonstration: Funding Sources, Institutions, and Instruments. Issue Brief 9 in *Assessing U.S. Climate Policy Options,* ed. Raymond J. Kopp and William A. Pizer. Washington, DC: Resources for the Future.

Newell, Richard G. 2007b. Climate Technology Deployment Policy. Issue Brief 10 in *Assessing U.S. Climate Policy Options,* ed. Raymond J. Kopp and William A. Pizer. Washington, DC: Resources for the Future.

Newell, Richard G. 2008. *A U.S. Innovation Strategy for Climate Change Mitigation.* Washington, DC: The Hamilton Project (Brookings Institution).

Newell, R.G. 2009. International Climate Technology Strategies. In *Post-Kyoto International Climate Policy: Implementing Architectures for Agreement*. Edited by J.E. Aldy and R.N. Stavins. Cambridge: Cambridge University Press, 403–438.

Newell, R.G. 2011. The Energy Innovation System: A Historical Perspective. In *Accelerating Energy Innovation: Insights from Multiple Sectors*. Edited by R.H. Henderson and R.G. Newell. Chicago: University of Chicago Press for the National Bureau of Economic Research, 25–48.

Newell, R.G. and N.E. Wilson. 2005. Technology Prizes for Climate Change Mitigation. RFF Discussion Paper 05–33. Washington, DC: Resources for the Future.

Organisation for Economic Co-operation and Development (OECD). 2002. Proposed Standard Practice for Surveys on Research and Experimental Development (Frascati Manual). Paris: OECD.

President's Council of Advisors on Science and Technology (PCAST). 2010. Report to the President on Accelerating the Pace of Change in Energy Technologies through an Integrated Federal Energy Policy. Washington, DC: Executive Office of the President.

Stokes, Donald E. 1997. *Pasteur's Quadrant: Basic Science and Technological Innovation*. Washington, DC: Brookings Institution Press.

U.S. Department of Energy (DOE). 2011. Quadrennial Technology Review 2011. Washington, DC: DOE.

U.S. Department of the Treasury. 2011. Investing in U.S. Competitiveness: The Benefits of Enhancing the Research and Experimentation (R&E) Tax Credit. A Report from the Office of Tax Policy, March 25, 2011.

11

MIXING IT UP

Power sector energy and regional and regulatory climate policies in the presence of a carbon tax

*Dallas Burtraw and Karen L. Palmer**

KEY MESSAGES FOR POLICYMAKERS

- A carbon tax creates long-run incentives for innovation and is expected to be the most efficient instrument available for reducing emissions.
- However, given that a carbon tax would provide a valuable revenue source, and the institutional structure of decision making in implementing a tax, it is uncertain whether the tax would be set efficiently with respect to climate policy goals and updated effectively. Congress should therefore preserve a role for an expert agency (such as EPA) to revise the tax level periodically based on new information.
- If an effective carbon tax were in place, policies that impose costs on the federal government such as tax credits that promote new technology can be scaled back. However, even if a tax is used efficiently, it may not work exactly as anticipated in the conventional economic model. If successful, carbon taxes could make other regulations such as technology standards or EPA's authority to regulate under the Clean Air Act irrelevant; if not these regulations might provide important benefits. Potential redundancy may impose unnecessary costs and this should be balanced against the appeal of a portfolio approach in the face of uncertainty about the effect of any individual policy.
- The preemption of state and local actions seems especially poorly advised because climate change is a problem fundamentally characterized by the incentive for free riding. The only justification for preemption is where these policies interfere with other constitutional protections for business, which is unrelated to the introduction of a carbon tax.

- Climate policy should and will continue to be a complex mix of regulations at various levels of government even with a carbon price. Nonetheless it is important for local, state, and national governments to consider the interactions of other policies with a carbon tax and the potential for unanticipated consequences.

Introduction

Federal budget deficits in the United States have reached record levels in recent years. During the recession, the federal budget deficit spiked to $1.1 trillion in fiscal year 2012, roughly 30 percent of total government spending or 7 percent of GDP. Although the deficit is now falling (not least because of the sequester), over the longer term the fiscal outlook is set to deteriorate alarmingly (see Chapter 1). US policymakers are therefore in the midst of an ongoing debate about how to address these budget challenges and rationalize the federal tax system and one of the options under discussion is a carbon tax. A carbon tax of, say, $25 per ton of carbon dioxide (CO_2) could contribute as much as $125 billion per year in additional revenue.

Unlike other deficit-reduction options (like raising taxes on labor and capital income or cutting spending), a carbon tax has many virtues as an environmental policy. If applied economy-wide, it would address emissions from all sectors, and encourage producers and consumers to make efficient choices across fuels and technologies for energy production and use. Unlike a cap-and-trade program (at least one without price stability provisions), a carbon tax also provides cost certainty to regulated sectors of the economy. And if the revenues from a tax are used efficiently, most obviously to reduce the need for burdensome taxes on other sectors (either now or, through deficit reduction, in the future), a carbon tax is a highly cost-effective instrument for reducing CO_2 emissions.

Yet, would a carbon tax necessarily be an efficient policy in practice?

A prerequisite for this outcome is that the tax, ideally, would be scaled to match the environmental damages from CO_2 emissions. This is challenging from a conceptual perspective, given that damage assessments are much disputed and between reports issued in 2010 and in 2013 an interagency working group has increased the estimate for 2020 by 50 percent, to a central case estimate of about $40 (US Government 2010, 2013; Johnson and Hope 2012). In this evolving scientific and economic context, a $25 per ton value might be a reasonable lower bound value to aim for (based on US Government 2010). Further, as suggested in our opening paragraph, the current political relevance of a carbon tax comes from its potential contribution to revenue and broader tax reform, which casts some doubt (along with some other considerations discussed below) on whether the tax would be set and updated efficiently from an environmental perspective. There is also some consideration that the incentives introduced by a carbon tax may reach all decision makers in the economy. To the extent that a carbon tax in practice might fall short of achieving environmental goals, there is a potential role for other policies – like policies to promote renewables and energy efficiency and planning for land use infrastructure – to achieve additional emissions reductions.

Other policies might also have a role if they address a variety of other factors (again discussed below) that could prevent sufficient investment in clean technologies, even if a carbon tax were set appropriately from an environmental perspective.

This chapter offers a way to think about when it makes sense for other regulations (with a focus on the power sector) at both the national and state/local level to coexist with a carbon tax. In fact there is much confusion over this issue. Some have argued that if the country enacts a carbon tax, then other policies to help mitigate greenhouse gas emissions and encourage the deployment of renewable energy and other clean technologies are unnecessary.[1] And others have suggested policy swaps with the withdrawal of current and preclusion of future regulations of carbon emissions as a way to gain political support within the Congress for establishing a carbon tax (Gayer 2012; Fraas and Richardson 2013).

Our conclusion is more nuanced. A well-designed and well-implemented carbon tax is likely to generate the lion's share of environmental improvements, not least because a carbon price is imperative for creating long-run incentives for innovation and efficiency across-the-board. Nevertheless, we expect limitations in the practical implementation of a carbon tax, leaving room for the coexistence of other policies at the state and local levels and regulations at the national level. Ultimately, each of these policies and their interactions with a carbon tax deserves its own analysis on a case-by-case basis.

We begin in the next section by reviewing the conceptual economic approach to analyzing the efficiency of environmental regulations. We next discuss the interaction of a carbon price with other energy and environmental policies at the national level, with the Clean Air Act, and with regulations at the subnational level. We then offer a conclusion.

The conceptual context for a national carbon tax policy

Economic advice about the design of climate policy typically builds on the suggestion that a single policy objective (e.g., reducing emissions) requires a single policy instrument. This idea is used to assert that well-designed climate policy would therefore use *only one* policy instrument and that the use of more than one policy to address the same objective cannot reduce, and may well increase, costs. For example, if the climate problem is due to society's overconsumption of fossil fuels because CO_2 emissions from their combustion are not priced, then it follows that the solution should be to introduce a price on emissions. Coupling this policy with another one that constrains consumer choices would raise the cost of reducing emissions if consumers were denied choices that might be highly valued in some situations or if it would direct consumers to choices that reduce emissions at a higher cost than would result if they were to make decisions based on the carbon tax alone.

According to this logic, the only justification for multiple policies (that is, policies in addition to a carbon tax) is the existence of multiple problems. For instance, if the benefit of one firm's research and development (R&D) into cleaner technologies accrues to other parties, then that firm might invest less in R&D than would be desirable from society's perspective. In this case, efficient climate policy might combine a carbon price aimed at reducing emissions with other instruments to promote R&D (see Chapter 10).[2]

To take another example in the energy context, one might observe that consumers exhibit myopic decision making that places too much emphasis on the short-term cost, forgoing options that save energy and money in the long run (Allcott and Wozny 2012). Such myopic decision making might be true in a variety of consumer decisions such as telephone and internet services, and usually economists pay little attention to small consumer mistakes (*caveat emptor*). However, in the energy context consumer mistakes are especially relevant because of associated environmental consequences. Again, a carbon price by itself might not invoke the consumption and investment behavior that would reduce emissions at the least cost. This myopic decision making might provide justification for additional policy such as performance standards that restricted the set of choices for consumers or other policies that nudged their behavior toward cost-minimizing choices (Gillingham and Palmer 2013).

We are interested also (and perhaps more important) in another reason policy might not achieve its intended policy outcome – the role of institutions. Several institutional settings are relevant here: (i) the apparatus that would implement a carbon price, (ii) the interaction of climate policy with other energy policies, (iii) existing EPA regulatory authority under the Clean Air Act, and (iv) the federalist structure of government in the US under which a price-based policy or other regulatory approaches would take effect. Each of these settings provides a potential justification for combining a carbon price with additional policies.

Imagining the implementation of a carbon price

With rare exceptions, economic models implicitly ascribe the governance of a price on carbon to a unitary government actor. As illustrated in Figure 11.1, that actor

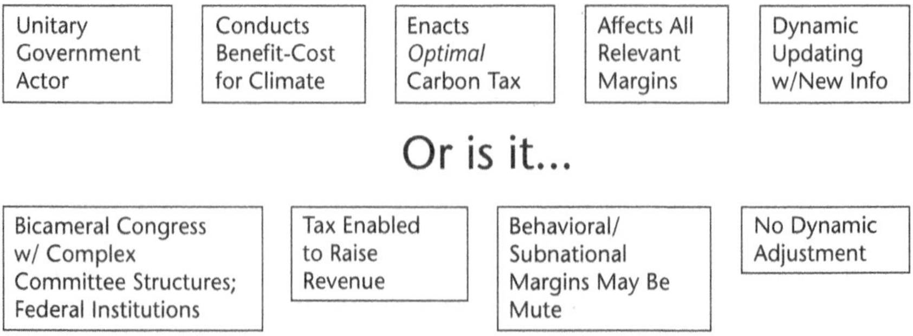

Unitary Government Actor	Conducts Benefit-Cost for Climate	Enacts *Optimal* Carbon Tax	Affects All Relevant Margins	Dynamic Updating w/New Info

Or is it...

Bicameral Congress w/ Complex Committee Structures; Federal Institutions	Tax Enabled to Raise Revenue	Behavioral/ Subnational Margins May Be Mute	No Dynamic Adjustment

FIGURE 11.1 Mental models of implementation of a carbon tax

would balance benefits and costs to identify and implement an efficient carbon tax (from an environmental perspective). Presumably the tax affects emissions throughout the economy, and consumption and investment adjust accordingly (myopic decision making mentioned above notwithstanding). When new scientific or economic information becomes available, the model assumes the optimal carbon tax is adjusted in a timely way.

Realistic expectations about how a carbon price would be implemented depart importantly from this view, however, for several reasons.

First, both the level and structure of a carbon price would be decided in a bicameral Congress with a complex committee structure and with the overhang of required presidential approval. The primary function of the relevant committees is to raise revenue, not determine a level of taxation that achieves an environmental goal with respect to determination of the level of carbon emissions.

As regards the tax level, how would this be determined? Legislative staff on budget and revenue committees might lack the expertise to evaluate the link between a carbon price and associated emissions, and associated complexities such as the effect of a tax on international competitiveness or the level of commitment to reducing emissions that has been made by other countries. The committee might base the tax level on technical information such as the "social cost of carbon" recommended by US Government (2010; 2013). However, this expectation is not fully consistent with the perspective of contemporary advocates for a carbon tax, who generally suggest it is politically possible because of its ability to raise revenue. With this as the enabling justification for a carbon tax, it may be unlikely that it also would be optimized to satisfy a specific social goal that is generally not the domain of the relevant congressional committees.[3] Economic analysis suggests that the carbon tax should generally be set to reflect environmental damages, with other revenues needs met through broader fiscal instruments (e.g., Goulder 2002), but only sometimes is this advice taken in practice.

And tax revenue measures are typically complicated, with provisions to protect national and special interests. There also tend to be many exemptions, suggesting the implementation may not be uniform throughout the economy and a price signal may not equate the marginal cost of emissions reduction opportunities across different agents in the economy, although the possibility for exemptions might be reduced with an upstream carbon charge levied on fuel suppliers – coal processors, refineries, and so on (Chapter 3).

A second reason why the tax may not be set efficiently is distributional considerations. Tax economists typically relegate the remediation of distributional effects directly to distributional policies and seek to design a tax system that is as efficient as possible; however, in practice policy coalitions form to balance multiple objectives. A price on carbon has distributional outcomes that can be offset by other energy policies such as renewable energy targets, which tend to lower market prices of power (see below).[4] Up to a point, energy efficiency and renewable policies also have weaker effects on energy prices than carbon taxes (as the latter involve the pass through of carbon tax payments in higher prices). This may also explain the coexistence of a carbon price (below its efficient level) and other energy policies even when those other policies are less environmentally effective.

Third, as noted above, concerns about impacts on the overall economy or competitiveness may be another reason why the tax level might be insufficient from an environmental perspective, or applied unevenly across the economy. One reason might be to lessen the loss of economic activity associated with introducing a new tax on top of already existing taxes. However, these concerns can be at least partly addressed through using some of the carbon tax revenues to lower tax burdens elsewhere in the economy, or use of additional measures to help energy-intensive, trade-exposed industries (see Chapter 9).

Fourth, the mental model illustrated in Figure 11.1 also assumes that the carbon price would automatically be adjusted to assimilate new scientific information. However, it is hard to imagine how this would occur, since the tax level and perhaps a time path for automatic annual adjustments is established by congressional action and revisiting the technical basis for this legislative decision might be just as fraught or flawed as the initial implementation of a tax.[5] One approach to assimilating new information might be to delegate authority to adjust the carbon tax to an expert agency such as EPA (Fraas and Richardson 2013; Burtraw 2013). However, the authority to tax is the exclusive domain of Congress, and this is one power the legislative branch guards jealously and rarely delegates.

In sum, economic theory tells us that a carbon tax is a relatively efficient instrument. However, an efficient instrument does not guarantee an efficient policy. The outcome depends on how the instrument is used, and that depends importantly on the institutional context for policymaking. The implementation of a carbon price seems likely to be imperfect suggesting a potential role for other policies to help mitigate carbon emissions.

National power sector policies with a carbon tax

National energy policy focused on the electricity sector is multifaceted with several policies that are motivated in part by a desire to reduce greenhouse gas emissions. Chief among these are policies to encourage the adoption of energy efficient appliances and equipment and policies to encourage the use of renewables and other non- or low-CO_2-emitting sources of electricity generation. If a carbon tax were imposed, should these policies be continued, and if so, in what form and to what extent?

The answers to these questions depend on several considerations. The first is whether the carbon tax is set in a way that fully internalizes into private decisions the environmental damages associated with CO_2 emissions. The second is whether there are other market barriers preventing sufficient deployment of renewables or consumer investments in energy efficiency (despite the carbon tax) that these policies could address. The third consideration is how these policies promoting renewables and efficiency interact with each other and with the carbon tax. A fourth consideration is distributional concerns, which, as noted above, might explain the coexistence of a carbon price with other energy policies.

We first discuss these issues in the contexts of renewables policies and then energy efficiency. Then we turn briefly to the potential interactions between carbon taxes and other environmental concerns stemming from electricity generation with coal and from nuclear power.

Renewables policies and motivations

Knowledge spillovers provide a reason why markets may not lead to sufficient development and deployment of renewable electricity technologies. These spillovers can occur during research and development (e.g., when some firms copy or imitate for free new technologies developed by others or use information about those technologies to further their own research programs) and deployment (e.g., when some firms benefit from the experience of others in learning how to use a new technology more efficiently). Policies that provide incentives for renewables investment and generation directly encourage learning and indirectly provide incentives for R&D by increasing the returns to deployment of renewables technologies resulting from successful R&D.

The main federal policy mechanisms for this purpose are tax incentives. For a number of renewable technologies, these incentives take the form of a production tax credit that is applied to each kWh generated by the facility for the first 10 years of operation. In early January 2013, Congress extended the production tax credit to apply to eligible generators that have begun construction by the end of 2013.[6] The tax credit varies between 2.2 cents per kWh for wind, geothermal, and closed-loop biomass and 1.1 cents per kWh for others including small hydro, wave, and tidal energy and landfill gas. (These credits are roughly comparable to the 1.4 cent and 2.8 cent per kWh relative cost advantage that renewables would have to natural gas and coal, respectively, under a $25 per ton carbon tax.)

Notably the production tax credit is lower for the more nascent technologies such as wave and tidal energy than it is for the more mature technologies such as wind power, which may be a poor match to where the greater benefits from technological learning might reside. In addition to the production tax credit, there is also an investment tax credit that applies to solar and small wind facilities.[7] Overall the fiscal costs of these incentives are relatively modest; according to the Congressional Budget Office, the total cost to the government of the renewable production and investment tax credits is estimated to be about $2.5 billion in fiscal year 2013 and another $2.6 billion for the grants to renewables in lieu of tax credits (Dinan 2013), which totals to roughly 4 percent of the revenue expected from a $25 per ton carbon tax.

A renewable portfolio standard (RPS) focuses on mandating renewables production instead of on reducing their costs. An RPS specifies the minimum share of electricity sales that must be produced using qualified renewables. RPS policies typically are accompanied by a credit trading provision allowing retail electric utilities that do not generate enough renewable electricity themselves to comply with the standard by purchasing credits (from others that exceed the standard). Currently

29 states and the District of Columbia have state-level RPS policies in place, and a federal RPS was proposed as a part of the Waxman-Markey climate cap-and-trade legislation passed by the US House in 2009.

An RPS is a less cost-effective approach to reducing CO_2 emissions than pricing CO_2 emissions directly (Palmer et al. 2010; Palmer et al. 2011) for several reasons. First, the RPS does not reduce emissions outside the power sector. Second, the RPS does not raise electricity prices the way a tax would and thus does little to encourage conservation; in some instances it can lead to lower electricity prices than with no policy (Fischer 2010). Third, an RPS does not differentially disadvantage fossil technologies in relation to their emissions intensity. Palmer et al. (2010) show that a Clean Energy Standard, an alternative policy that seeks to encourage a wider array of non- and low-emitting generation technologies such as nuclear and natural gas, could be a more effective and more cost-effective approach to reducing CO_2 emissions than an RPS. President Obama and former Senator Jeff Bingaman have both proposed Clean Energy Standards (CES) that are targeted to provide substantial reductions in CO_2 emissions (Mignone et al. 2012).

Do these renewable or clean energy technology policies make sense in the presence of a carbon tax? The answer depends importantly on the size of the spillovers related to technological learning.

One study, which uses estimates of technology learning effects embodied in the Energy Information Administration's National Energy Modeling System (NEMS) model, suggests that the level of an RPS or a renewables subsidy justified by learning spillovers, in the presence of a carbon tax, is much lower than levels specified in current policies, implying that some scaling back of these policies, or even a complete phase-out, would likely be warranted.[8] Their work suggests that a substantial gap between the adopted carbon tax and the optimal carbon tax would be necessary to motivate even modestly ambitious renewables tax credits or RPS targets.

To the extent these other policies are retained, there are ways in which they might be improved. Most obviously, greater certainty about the future course of the renewable tax credit will create a more stable investment climate and help to prevent large fluctuations in renewables investment that have occurred in response to periodic lapses in that policy. If a national RPS is instituted alongside a carbon tax, an alternative compliance payment that effectively caps the price of tradable renewable energy credits (preferably one scaled to the environmental damages from CO_2) could help to contain the cost an RPS.

One aspect of learning that has not been considered in the literature and that could be very important as renewables become more prevalent is learning about better ways to integrate intermittent renewables into the grid as their share of the generation mix grows. Policies that provide incentives to expand transmission capacity between locations with substantial renewable resources and locations with high demand for electricity should help to facilitate integration, as would increasing the ability of energy conservation and load management programs to be more responsive to price fluctuations in wholesale electricity markets.

Energy efficiency policies and motivations

Historically policies to promote energy efficiency have been motivated by the apparent *energy efficiency gap* – that is, the empirical observation that consumers fail to adopt energy efficient technologies that appear to more than pay for the upfront investment costs in terms of expected energy savings over the life of the technology (Jaffe and Stavins 1994). Some have suggested that closing the gap through widespread adoption of more efficient technologies could produce substantial reductions in CO_2 emissions essentially for free, or even with negative cost (McKinsey & Company 2009).

Explanations for this gap are wide ranging and have differing implications for policy. Some explanations point to underestimates of costs due to a failure to account for some categories of costs; for example, some consumers' reluctance to switch to compact fluorescent lighting may be partly due to a reduction in the perceived quality of lighting when compared with incandescent bulbs. Other factors, including differences in energy use and opportunities for energy savings across customers and the value of waiting for uncertainty about future energy prices to be resolved before making irreversible investments in more efficient appliances or building improvements, also suggest that the efficiency gap has been overstated (Murphy and Jaccard 2011). Collectively, these arguments indicate that the role for energy efficiency enhancing policy may be more circumscribed than studies based on engineering costs have suggested (Allcott and Greenstone 2012).

However, other explanations suggest that there may be a role for policy to encourage greater adoption of energy-efficient durable goods. For example, in rented properties there is typically a mismatch between those paying energy bills and those responsible for efficiency upgrades, which may cause underinvestment. Even homeowners may be reluctant to make investments in enhancing energy efficiency if they don't believe they can recover that investment upon sale of their house. Policies that make energy cost information more transparent in housing transactions could address that concern. In general, which explanations are most relevant for explaining the gap likely varies across energy users and energy uses, and sorting them out continues to be a topic in need of research.

Motivated by both climate and energy efficiency gap considerations, there are numerous federal policies currently in place to promote energy efficiency including mandatory standards, mandatory and voluntary information programs, and financial incentives. Federal appliance standards currently apply to more than 50 categories of appliances and equipment, ranging from air conditioners to light bulbs, and Parry et al. (2010) estimate that 60 percent of total electricity consumption is associated with durables that are potentially subject to minimum efficiency standards. Federal rules also require Energy Guide labels that display expected annual energy use and costs for many consumer durables.[9] To encourage manufacturers to make products that are more energy efficient, EPA runs a voluntary product certification program known as Energy Star, for which the top 25 percent most efficient products of a particular type are eligible. The US Department of Energy also offers free energy use assessments to small and medium-sized manufacturing firms through its Industrial Assessment Centers program. For many years the federal government acting through the states has offered weatherization assistance to low-income families, and there are tax incentives for certain types of efficiency enhancing upgrades to homes.

Should policies to promote energy efficiency be continued in the presence of a carbon tax, and if so, which ones? Again, the answer depends in part on whether the carbon tax is high enough, or comprehensive enough, from an environmental perspective. It also depends on the size and causes of the energy efficiency gap. Information programs can help address lack-of-information problems; up to a point, efficiency standards may be more effective and even more cost-effective than a carbon tax if consumers are systematically failing to account for future savings in their actions. Targeting of policies toward affected populations and particular market failures will tend to raise overall cost-effectiveness relative to less targeted policies. For example, limiting subsidies for efficient appliances or building shell enhancements to rental units.

Assuming that carbon taxes are effective at reducing CO_2 emissions, research suggests that efficiency policies designed to address underinvestment in energy efficiency could be made less stringent than if their aim were to achieve substantial reductions in CO_2 emissions. Fischer et al. (2012) show that an energy efficiency subsidy alone set to achieve a 20 percent reduction in CO_2 emissions from the electricity sector is roughly three times as stringent as an energy efficiency policy designed to offset an assumed 10 percent efficiency undervaluation by consumers in the presence of an optimal carbon tax and other policies to address renewables. This is true in part because imposing a tax on carbon will lead to higher electricity prices (due to the pass-through of carbon tax costs) and, therefore, stronger incentives for conservation. Higher rates of undervaluation would suggest a greater role for efficiency policy to deal with undervaluation alone. Parry et al. (2010) find the literature on implicit discount rates suggests an undervaluation anywhere between 0 and 65 percent. Given the uncertainty about the extent of consumer undervaluation of energy efficiency, it is difficult to make strong statements about the appropriate stringency of efficiency policies with or without a carbon tax in place. Fischer and coauthors also show that interactions between renewables policies and energy efficiency policies will matter for the prescribed level of each type of policy necessary to fully address the relevant issues. By introducing renewables that have low or zero operating costs, the RPS tends to shift out the electricity supply curve and lower the market price of electricity in competitive markets, which in turn raises consumption levels. The flip side is that subsidizing energy efficiency if consumers do not undervalue it will result in less adoption of renewables and thus exacerbate the learning externality. Careful consideration of these interactions is an important part of energy policy design with or without a carbon tax in place.

Emissions from coal and nuclear power

Historically, the electricity sector has been a major source of emissions of criteria air pollutants such as SO_2 and nitrogen oxides (NO*x*) and of air toxics such as mercury. Over the past 23 years, as a consequence of the SO_2 cap-and-trade program under

Title IV of the Clean Air Act Amendments of 1990 and subsequent regulations including the NO*x* Budget Program, the Clean Air Interstate Rule, and the recently adopted Mercury and Air Toxics Standard (MATS), substantial reductions in these emissions have been achieved. For example, when MATS is fully implemented in 2015–2016, emissions of SO_2 are expected to be reduced to roughly 2.3 million tons per year, well below the 8.95 million ton annual cap introduced by the 1990 amendments. Because the SO_2 cap is no longer binding, imposing a carbon tax would result in further reductions in SO_2 emissions as electricity generation shifts away from coal to greater reliance on natural gas and renewables. Estimates of the marginal health and environmental benefits of reducing SO_2 emissions are varied, ranging from $1,640 per ton (Muller and Mendelsohn 2012, 2000 year dollars), to $1,800–$4,700 (Banzhaf et al. 2004, 1999 dollars). The EPA recently used an estimate of $29,000 in the Eastern states and $8,300 in the Western states (US EPA 2011, 2007 dollars). A recent analysis of carbon pricing in the electricity sector (Paul et al. 2013) finds that a CO_2 tax of $43 per ton in 2020 leads to a 2.1 million ton reduction in annual SO_2 emissions from the electricity sector. Using EPA estimates for the Eastern states, where most of these reductions will occur, the ancillary benefits from reductions in SO_2 would approach $60 billion. All of these estimates suggest that any ancillary SO_2 reductions that result from a carbon tax will have a positive net benefit to society, thus reinforcing the case for a carbon tax.

A carbon tax will also improve the relative cost of nuclear power compared to coal or natural gas. Whether this change in the economics would lead to more investment in nuclear plants is an open question, not least because the tsunami in Japan and damages to the Fukushima nuclear plant there have raised concerns about domestic safety that will take effort to resolve. Also, the US federal government's failure to come up with a long-term solution for storage of spent nuclear fuel means that waste storage in hardened casks on the site of existing nuclear plants is the default strategy for dealing with this highly hazardous material. This suggests that the development of future nuclear power plants over the next decade may be limited to available space at the sites of existing plants where there is experience in dealing with waste and an existing risk profile.

EPA authority to address greenhouse gas emissions under the Clean Air Act

What should become of EPA's authority to regulate CO_2 under the Clean Air Act if a carbon price is introduced? There are two aspects to this question. First is the role of EPA in the administration of the carbon price, and second is EPA's independent development of regulations affecting emissions from specific sets of sources.

EPA's role in the administration of a carbon price

Does EPA have a role in the administration of a carbon price? An important lesson from the flagship SO_2 trading program is the need to preserve a role for an expert agency in updating the program. Congress proved unable to update the level of the SO_2 emissions cap based on new technical information. This might not be surprising, given the specific knowledge necessary to evaluate and remain current with new economic and scientific information. On the other hand, executive branch agencies are organized with the purpose of maintaining technical expertise; for environmental matters this responsibility falls to EPA.

New information might concern the cost of achieving emissions reductions, which would be very relevant in determining the level of the emissions cap under a trading program.[10] However, the economically efficient level of a carbon tax does not depend on the total costs of emissions reductions; it depends only on the environmental benefits per ton of reductions.

A second possibility is that new scientific information might reveal that benefits of emissions reductions are greater or less than initially thought. Such a finding would not be reached quickly; it involves a scientific assessment of the expected damages from emissions, which would be incorporated into estimates of the social cost of carbon and which would take time to assess. To accomplish this assessment and translate findings into an updated regulation requires a role for an expert agency, a role that is built into regulation under the Clean Air Act and is needed also under a carbon tax. Guidelines for this process are not obvious because Congress is protective of its ability to set taxes. If the objective is to set the tax efficiently from an environmental perspective (leaving aside the reasons – noted above – why this may not be the case in practice), then an ongoing oversight role for an expert agency would appear crucial to the efficient implementation of a carbon price.

Greenhouse gas regulations under the Clean Air Act

In 2007 the US Supreme Court affirmed the authority of EPA to regulate greenhouse gases under the Clean Air Act.[11] Subsequently EPA made a formal, science-based determination that greenhouse gases are dangerous to human health and the environment, which compels the agency to develop regulations to mitigate that harm. Important regulations have already been finalized in the mobile source sector and for the construction permitting of new and modified stationary emissions sources (new source review). Draft new source performance standards have been proposed for new fossil-steam electricity generators, and regulations for other sectors are expected to follow. Most important, the Obama administration has signaled its intent to move forward with performance standards for existing stationary sources that might be flexible, allowing averaging across facilities. With the introduction of a price on carbon, should these regulations and EPA's authority to regulate be withdrawn?

We have argued already that a carbon price is not guaranteed to affect decisions at every margin in the economy, potentially leaving cost-effective emissions reduction efforts unrealized. And we argued that the carbon price might not be set at a level that is fully efficient. These possibilities would appear to provide justification for continuation of EPA's authority to regulate.

However, *if* a carbon price is accomplishing its intended environmental goal, then what would be the consequence of coincident regulation under the Clean Air Act? This is a bit of a tautology, because if one policy is defined to be successful it would seem that only bad things could happen by adding to it, that is, fixing something that is not broken. To be sure, Clean Air Act regulation will likely be less efficient than a price-based policy because it is likely to miss many opportunities for cost-effective emissions reductions where regulators lack specific information about different technological options for compliance among the regulated sources. In principle, a price on carbon would be expected to invoke cost-effective differences in investment among these sources. Further, regulations are likely to emerge slowly; and when regulations overlap precisely with a carbon price and are calibrated to the same outcome the regulations are likely to be irrelevant. However, this question might resemble the parable of the hare and the tortoise. If a price-based policy performs as one would hope, it will win the race. If the price-based policy stagnates, the regulatory approaches might become relevant. In other words, regulation could be structured such that it would be relevant only if a price-based policy fails to perform, which happened in the first grand experiment introducing a price on SO_2 emissions. From this perspective one might conclude that the regulatory authority under the Clean Air Act might do no harm and might provide important benefits.

There is a nagging concern that redundancy could impose anachronistic measures that would accomplish few emissions reductions but perhaps impose high costs. In some cases, regulation may be counterproductive. For example, technology standards that are differentiated by vintage may require higher efficiency and raise the cost of new investment, resulting in the delayed retirement of older, dirtier facilities (Gruenspecht 1982; Maloney and Brady 1988; Stavins 2006). Patino Echeverri et al. (2013) examine such regulation in the context of a CO_2 performance standard for new power plants and find that an inflexible standard could result in greater cumulative emissions due to the delay in new investment in more efficient generators that would otherwise occur. In its proposed performance standard for new fossil-steam power plants, EPA has attempted to address this problem by introducing a 30-year averaging rule, which would allow new generators to exceed the performance standard as long as retrofits were made by the tenth year of operation allowing the standard to be achieved in the long run. Nonetheless, this illustrates that regulations may impose compliance requirements that are less efficient than if firms were left to respond to the price signal. In this eventuality, redundant regulation might introduce unnecessary costs that undermine efforts to address climate change.

The concern about wasteful redundant regulation might be allayed somewhat by the language of the Clean Air Act that is relevant to the regulation of CO_2 from existing stationary sources. EPA has decided to use Section 111 for this purpose. Unlike other portions of the act that regulate human health effects of pollution, this section has a cost test that requires the agency to take into consideration the effect of regulation on the remaining useful life of a facility. The technical preparation of a regulation is required to address this criterion, providing some protection against regulation that would not achieve meaningful emissions reductions or would do so at high costs. In addition, although the schedule often has not been met, this section of the act calls for the regular review of performance standards, which provides a forum to address the efficiency of existing regulations.

Subnational climate policies to address greenhouse gas emissions

Subnational levels of government have a vital role to play in the societal response to climate change because decisions about the infrastructure that will define social opportunities for the next century reside not primarily at the federal level, where a tax policy might take shape, but at the state and local levels (Shobe and Burtraw 2012). This is where decisions are made that govern industrial operations, siting and permitting, residential land use, building codes, and transit modes and patterns.

In a unitary model of government, the introduction of a price signal is assumed to be transmitted instantly to decision makers at all levels of government so that permitting, land use planning, and other functions of government adjust accordingly. If the introduction of a price signal would be transmitted effectively to all levels of the society, then additional efforts by subnational levels of government would be expected to be redundant at best and inefficient in general because they would lead to different effective marginal costs of emissions reductions across the economy (Goulder and Stavins 2011).

But in fact there is little research to indicate how well this would occur. There are many reasons to think that price signals *may not be* transmitted efficiently through levels of government. Local levels of government do not respond to short-run price signals. Their decisions are accountable to incumbent landowners who value consistency of new construction with the existing architecture, an interest that heavily influences local zoning decisions. In anticipation of this incongruity, the Waxman-Markey proposal included specific incentives to motivate state and local government actions to develop more energy-efficient transportation and land use policies.

Without addressing the potential inconsistency, one might also observe that while there may be a tendency for local jurisdictions to respond slowly to price signals, state governments have often been leaders in the introduction of policies to reduce carbon emissions. These jurisdictions have initiated energy efficiency standards for household appliances and vehicles and emissions standards for vehicles, which subsequently were taken up at the national level, and they have unique purview in the development of building standards. In addition to state-level renewable portfolio standards and energy efficiency programs, 10 states have adopted emissions cap-and-trade programs for CO_2.

The introduction of a national price on CO_2 interacts with these existing activities in sometimes unavoidable ways. The introduction of an emissions cap and trade program would effectively and automatically preempt these efforts because marginal efforts in one locale (or by one individual) to reduce emissions would free up allowances under the cap that could be used in other jurisdictions (Burtraw and Shobe 2009). Goulder and Stavins (2011) refer to this as 100 percent leakage. In contrast, the introduction of a price on carbon through a carbon tax would preserve the additionality of emissions reduction measures by subnational levels of governments or individuals. Emissions reductions achieved through various means will not affect the level of the tax. Consequently, the marginal incentives to reduce emissions as a result of the carbon price would not be diluted.

Incentive-based policies interact in another unavoidable way. Each policy introduces an implicit price, and for multiple policies, those prices interact. There are many possibilities, but one example would be the existence of a cap-and-trade program at the state level interacting with a carbon tax at the national level. In this case the introduction of a national tax would lead the state allowance price to fall. (The marginal cost to reduce emissions would equal the sum of the state allowance price and the national carbon price.) This would affect the availability of revenue generated at the state level. Perversely, even taking steps toward implementation of a tax at the national level would affect decisions about mitigation and banking at the state level in anticipation that the value of banked allowances would be reduced after a national tax was adopted (Stavins 2007). Firms that are holding emissions allowances would suffer a loss of value associated with that asset. To arrest this problem, national policymakers might consider compensation for the loss in value of banked allowances under state programs.

In summary, it may be beneficial from a national perspective to have subnational governments and individuals taking measures to reduce emissions beyond what would be incentivized by a price on carbon alone. Subnational initiatives might provide additional incentives for investments and behavior where the transmission of market price signals is incomplete. Inevitably policies at the national and subnational level will interact, and consideration should be given to how that will unfold.

However, occasionally attention is given to the question of preempting policies at the state and local level. The answer hinges on whether in any circumstance there is a *national interest* in preventing state and local governments from doing more than national policy requires of them to reduce their contributions to a global externality. If subnational policies yield innovative outcomes, those innovations can be expected to spread to other jurisdictions. If subnational efforts are redundant or inefficient, they impose costs on those jurisdictions while they would lower costs or yield benefits for citizens elsewhere. It is hard to construct a justification for preemption of subnational policies.

Conclusion

A carbon tax is expected to be the most efficient instrument available for reducing emissions. However, a carbon tax is politically plausible because of its ability to generate revenue to contribute to the substantial revenue needs of the federal government. In the face of this priority and given the institutional structure of decision making in implementing a tax, we question whether a tax is likely to be set at an efficient level with respect to climate policy goals. If it is set efficiently initially, it is unclear how it would adjust to assimilate new scientific and economic information.

A carbon tax may be an efficient instrument, but it may not be used efficiently. To do so, we argue, requires a role for an expert agency to set the tax and adjust it to reflect new information.

Even if a tax is used efficiently, it may not work as described in the conventional economic model. In particular, it may not, and we think it most certainly will not, affect all relevant margins of decision making in the economy from consumer behavior to the decisions of state and local governments.

The preemption of state and local actions seems especially poorly advised. Climate change is a problem fundamentally characterized by the incentive for free riding. The notion of preempting voluntary actions by subnational jurisdictions to address the global externality would seem to lack a rationale other than its potential interference with other constitutional protections for business. Keohane and Victor (2010) argue that at the international level, solutions to such coordination problems lie in small groups of relevant countries finding commitments that provide incentives for all participating countries to continue to comply with the agreement. The outcomes of such cooperation efforts are likely to be decentralized complexes of networked institutions rather than integrated, hierarchical treaties that govern a coherently defined issue area. We observe the same phenomenon in microcosm happening in domestic policy formation. This is not to say that a price on carbon is not useful or ultimately imperative. We believe it is; however, we differ with the premise of some writers. A price is a necessary policy to address the challenge of climate change, but it is not sufficient.

An array of policies also may offer the advantage over a single, integrated policy regime by enabling the flexibility to address related issues such as other greenhouse gases and the ability to adapt over time. Several policies, including regulation under the Clean Air Act, renewable and clean energy technology standards, tax incentives to promote technologies, and other examples, are meaningfully within the domain of national government, even if they also exist in some form at the subnational level. In an era of budget deficits and constrained resources for federal agencies, targeting activities in an efficient manner is essential. If an effective carbon tax were in place, policies that impose costs on the federal government such as tax credits that promote new technology can be scaled back. Other regulations such as technology standards may be irrelevant if the tax leads to expected changes in emissions. Managers at agencies such as EPA would be likely to divert resources to other priorities and slow development of climate-driven regulations. Nonetheless, these potentially redundant regulations might do no harm and might provide important benefits. A portfolio approach offers diverse measures in the face of uncertainty about the effect of any individual policy, and for this reason it may have intuitive appeal to policymakers.

Finally, we note that public opinion appears to rest solidly in favor of a portfolio approach. Krosnick and MacInnis (2013) report that large majorities of Americans have endorsed a variety of policies to reduce greenhouse gas emissions, such as those we discuss at length, and this support has been consistent for many years. Public support for these policies is sensitive to their cost; however, the public continues to prefer mandated emissions reduction policies over price-based approaches.

For these many reasons, there is no basis for automatic preemption of the many existing climate-related policies in the presence of a carbon tax, but it is nonetheless important for local, state, and national governments to consider the interactions of other policies with a carbon tax and the potential for unanticipated consequences. They will often interact in unanticipated ways that in fact may raise the cost of achieving climate policy goals. To understand how this may occur and how it should be managed requires a realistic characterization of the institutions and behaviors that shape climate policy.

Notes

* Financial support for this research was received from RFF's Center for Climate and Electricity Policy. Direct correspondence to burtraw@rff.org and palmer@rff.org.

1. "To a first approximation, raising the price of carbon is a necessary and sufficient step for tackling global warming. The rest is at best rhetoric and may actually be harmful in inducing economic inefficiencies" (Nordhaus 2006).
2. The idea that a policy problem should be addressed with only one policy instrument is often associated with Tinbergen (1952), but actually he did not make that point. "Tinbergen's rule" prescribes that the number of policy instruments cannot be *less* than the number of policy goals. For example, a single monetary policy rule cannot achieve simultaneous targets with respect to employment and inflation. In energy or environmental policy, the implication of Tinbergen's rule would be that *at least one* policy would be required to address climate-related objectives; that is, "both energy and environmental goals need to be broken down into actionable targets, and there must be at least one policy instrument for each target" (Knudson 2009).
3. On which side of the technically efficient level might a carbon tax fall? The absence heretofore of a price on carbon (or any price on most other environmental externalities) suggests that pricing would likely understate the efficient level, although it could go either way. As a general lesson in political economy, Olson's (1965) *Logic of Collective Action* suggests the costs of externality policy would likely be concentrated on a small number of interests, thereby focusing political opposition to the tax, while the benefits would likely be diffuse (accruing mostly to future citizens, who are not part of the contemporary political economy).
4. See Hirth and Ueckerdt (2012).
5. The inability of Congress to update the emissions cap in the sulfur dioxide (SO_2) cap-and-trade program initiated by 1990 legislation amending the Clean Air Act provides an instructive example. Five years after the act was passed, environmental benefits were estimated to be 10 times as high as initially expected, and in turn these estimates were again dramatically higher in subsequent studies (see Portney 1990, Burtraw et al. 1998; Chestnut and Mills 2005), yet Congress did not respond by altering program stringency. A recent examination (Fowlie 2013) of the evolution of estimates of the environmental benefits of reducing carbon emissions suggests that these estimates have also increased substantially in recent years, as also evident in revisions to the US government estimates of the social cost of carbon (US Government 2010, 2013).
6. Prior incarnations of the law required that a facility be operational before the expiration date of the tax credit in order to be eligible for the credit.
7. Since 2009 it has also been possible for any facility that is eligible for the production tax credit to opt for using the investment tax credit instead.
8. Policy interactions are a greater concern when carbon emissions are regulated using a cap-and-trade approach as in that case and RPS will tend to lower the costs of CO_2 emissions allowances and thus lower the cost of generating with emitting sources relative to a cap by itself (Böhringer and Rosendahl 2010; Fischer and Preonas 2010). This

price lowering effect does not occur with a carbon tax, so marginal incentives to reduce emissions are not diluted.

9. Energy Guide labels are required for refrigerators, freezers, water heaters, dishwashers, washing machines, room air conditioners, central air conditioners, heat pumps, furnaces, and boilers.
10. A revolutionary feature of market-based policies compared to traditional, prescriptive approaches is the ability to discern the change in costs because information about changes in marginal costs of emissions reductions is instantaneously summarized in the market price of an emissions allowance. In a cap-and-trade program a price floor or ceiling allows the supply of allowances to automatically respond to this information (Burtraw et al. 2010; Fell et al. 2012).
11. *Massachusetts v. EPA*, 549 US 497 (2007).

References

Allcott, Hunt, and Michael Greenstone. 2012. Is There an Energy Efficiency Gap? *Journal of Economic Perspectives* 26(1): 3–28.

Allcott, Hunt, and Nathan Wozny. 2012. Gasoline Prices, Fuel Economy, and the Energy Paradox. NBER Working Paper Number 18583.

Banzhaf, H. Spencer, Dallas Burtraw, and Karen Palmer. 2004. Efficient Emission Fees in the US Electricity Sector. *Resource and Energy Economics* 26(3): 317–341.

Böhringer, C., and K. E. Rosendahl. 2010. Green Promotes the Dirtiest: On the Interaction between Black and Green Quotas in Energy Markets. *Journal of Regulatory Economics* 37(3): 316–325.

Burtraw, Dallas. 2013. The Institutional Blind Spot in Environmental Economics. *Dædalus* 142(1): 110–118.

Burtraw, Dallas, Alan J. Krupnick, Erin Mansur, David Austin, and Deirdre Farrell. 1998. The Costs and Benefits of Reducing Air Pollutants Related to Acid Rain. *Contemporary Economic Policy* 16: 379–400.

Burtraw, Dallas, Karen Palmer, and Danny Kahn. 2010. A Symmetric Safety Valve. *Energy Policy* 38(9): 4921–4932.

Burtraw, Dallas, and William Shobe. 2009. State and Local Climate Policy under a National Emissions Floor. RFF Discussion Paper 09–54. Washington, DC: Resources for the Future.

Chestnut, Lauraine G., and David M. Mills. 2005. A Fresh Look at the Benefits and Costs of the US Acid Rain Program. *Journal of Environmental Management* 77(3): 252–266.

Dinan, Terry. 2013. Federal Financial Support for Fuels and Energy Technologies, Testimony before the Subcommittee on Energy, Committee on Science, Space and Technology, US House of Representatives, March 13.

Fell, Harrison, Dallas Burtraw, Richard D. Morgenstern, and Karen L. Palmer. 2012. Soft and Hard Price Collars in a Cap-and-Trade System: A Comparative Analysis. *Journal of Environmental Economics and Management* 64(2): 183–198.

Fischer, Carolyn. 2010. When Do Renewable Portfolio Standards Lower Electricity Prices? *Energy Journal* 31(1): 101–120.

Fischer, Carolyn, Richard Newell, and Louis Preonas. 2012. Environmental and Technology Policy Options in the Electricity Sector: Interactions and Outcomes. RFF manuscript. Washington, DC: Resources for the Future.

Fischer, C., and L. Preonas. 2010. Combining Policies for Renewable Energy: Is the Whole Less Than the Sum of Its Parts? *International Review of Environmental and Resource Economics* 4(1): 52–92.

Fowlie, Meredith. 2013. 400 ppm and the Rising Cost of Climate Change, Energy Economics Exchange blogpost, UC Energy Institute at Haas, May 20. http://energyathaas.wordpress.com

Fraas, Art, and Nathan Richardson. 2013. Should the Clean Air Act Be Traded for a Carbon Price? Washington, DC: Resources for the Future.

Gayer, Ted. 2012. Linking Climate Policy to Fiscal and Environmental Reform. In *Campaign 2012: Twelve Independent Ideas for Improving American Public Policy,* edited by Benjamin Wittes. Washington, DC: Brookings Institution Press, 189–197.

Gillingham, Kenneth, and Karen Palmer. 2013. Bridging the Energy Efficiency Gap: Insights for Policy from Economic Theory and Empirical Analysis. RFF Discussion Paper 13–02. Washington, DC: Resources for the Future.

Goulder, L.H. 2002. *Environmental Policy Making in Economies with Prior Tax Distortions.* Northampton: Edward Elgar.

Goulder, L.H., and R. N. Stavins. 2011. Challenges from State-Federal Interactions in US Climate Change Policy. *American Economic Review* 101: 253–257.

Gruenspecht, Howard K. 1982. Differentiated Regulation: The Case of Auto Emissions Standards. *American Economic Review* 72(2): 329–332.

Hirth, Lion, and Falko Ueckerdt. 2012. Redistribution Effects of Energy and Climate Policy: The Electricity Market. FEEM Working Paper No. 82. Milan: Fondazione Eni Enrico Mattei.

Jaffe, A., and R. Stavins. 1994. The Energy Efficiency Gap: What Does It Mean? *Energy Policy* 22(10): 804–810.

Johnson, Laurie T., and Chris Hope. 2012. The Social Cost of Carbon in US Regulatory Impact Analyses: An Introduction and Critique. *Journal of Environmental Studies and Science,* DOI: 10.1007/s13412-012-0087-7.

Keohane, Robert O., and David G. Victor. 2010. The Regime Complex for Climate Change. Discussion Paper 2010–33. Cambridge, MA: Harvard Project on International Climate Agreements.

Knudson, William A. 2009. The Environment, Energy, and the Tinbergen Rule. *Bulletin of Science, Technology & Society* 29(4): 308–312.

Krosnick, Jon A., and Bo MacInnis. 2013. Does the American Public Support Legislation to Reduce Greenhouse Gas Emissions? *Dædalus* 142(1): 26–39.

Maloney, M., and G.L. Brady. 1988. Capital Turnover and Marketable Pollution Rights. *Journal of Law and Economics* 31(1): 203–226.

McKinsey & Company. 2009. Pathways to a Low-Carbon Economy: Version 2 of the Global Greenhouse Gas Abatement Curve. Washington, DC: McKinsey & Company.

Mignone, Bryan K., Thomas Alfstad, Aaron Bergman, Kenneth Dubin, Richard Duke, Paul Friley, Andrew Martinez, Matthew Mowers, Karen Palmer, Anthony Paul, Sharon Showalter, Daniel Steinberg, Matt Woerman, and Frances Wood. 2012. Cost-Effectiveness and Economic Incidence of a Clean Energy Standard. *Economics Energy and Environmental Policy* 1(3): 59–86.

Muller, Nicholas, and Robert Mendelsohn. 2012. *Using Marginal Damages in Environmental Policy: A Study of Air Pollution in the United States.* Washington, DC: AEI Press.

Murphy, Rose, and Mark Jaccard. 2011. Energy Efficiency and the Cost of Greenhouse Gas Abatement: A Comparison of Bottom-Up and Hybrid Models for the US. *Energy Policy* 39: 7146–7155.

Nordhaus, William D. 2006. *A Question of Balance: Weighing the Options on Global Warming Policies.* New Haven, CT: Yale University Press.

Olson, Mancur. 1965. *The Logic of Collective Action: Public Goods and the Theory of Groups.* Cambridge, MA: Harvard University Press.

Palmer, K., A. Paul, M. Woerman, and D. Steinberg. 2011. Federal Policies for Renewable Electricity. *Energy Policy* 39(7): 3975–3991.

Palmer, K., R. Sweeney, and M. Allaire. 2010. Modeling Policies to Promote Renewables and Low Carbon Sources of Electricity. Backgrounder. Washington, DC: Resources for the Future.

Parry, Ian W. H., David Evans, and Wallace Oates. 2010. Are Energy Efficiency Standards Justified? Discussion Paper 10–59. Washington, DC: Resources for the Future.

Patino-Echeverri, D, Burtraw, D, and Palmer, K. "Flexible mandates for investment in new technology." *Journal of Regulatory Economics* (2013): 1–35.

Paul, Anthony, Blair Beasley, and Karen L. Palmer. 2013. Taxing Electricity Sector Carbon Emissions at Social Cost. Discussion Paper 13-23. Washington, DC: Resources for the Future.

Portney, Paul R. 1990. Economics of the Clean Air Act. *Journal of Economic Perspectives* 4(4): 173–181.

Shobe, William M., and Dallas Burtraw. 2012. Rethinking Environmental Federalism in a Warming World. *Climate Change Economics* 3(4): 1–33

Stavins, Robert. 2006. Vintage-Differentiated Environmental Regulation. *Stanford Environmental Law Journal* 25(1): 29–63.

Stavins, Robert. 2007. Comments on the Recommendations of the Market Advisory Committee to the California Air Resources Board, Recommendations for Designing a Greenhouse Gas Cap-and-Trade System for California, June 15.

Tinbergen J. 1952. *On the Theory of Economic Policy,* 2nd ed. Amsterdam: North Holland.

US EPA (Environmental Protection Agency). 2011. *Regulatory Impact Analysis for the Final Mercury and Air Toxics Standards.* Washington, DC: EPA. www.epa.gov/ttn/ecas/regdata/RIAs/matsriafinal.pdf

US Government. 2010. Technical Support Document: Social Cost of Carbon for Regulatory Impact Analysis under Executive Order 12866. Interagency Working Group on Social Cost of Carbon. Washington, DC: United States Government.

US Government. 2013. Technical Support Document: Technical Update of the Social Cost of Carbon for Regulatory Impact Analysis under Executive Order 12866. Interagency Working Group on Social Cost of Carbon. Washington, DC: United States Government.

12

IMPLICATIONS OF CARBON TAXES FOR TRANSPORTATION POLICIES

Ian Parry and Kenneth A. Small

KEY MESSAGES FOR POLICYMAKERS

- Even if a carbon tax were introduced, motor fuels in the United States would still be undercharged *on average* from the perspective of reflecting broader adverse side effects, or "externalities" (notably congestion, accidents, and local pollution), in fuel prices.
- Ideally, these other externalities would be more effectively reduced through a variety of per-mile tolls (e.g., that vary across time of day and region to address congestion), with fuel taxes retained only to reflect carbon damages.
- A combination of carbon tax and mileage taxes aimed at externalities would make a part of current policy toward fuel economy standards redundant. The remaining part would depend on beliefs about the importance of reducing dependence on petroleum markets (a diminishing concern) and about the extent of apparent consumer undervaluation of fuel economy in making vehicle purchases.
- Certain existing policies, such as regulations on emission of local pollutants, have a valuable role to play even if transportation taxes are reformed to better target externalities. The case for transit fare subsidies is largely dependent on other factors, as subsidies produce relatively modest carbon benefits.
- Appropriately scaled taxes on the carbon content of motor fuels could go a long way toward covering current and growing shortfalls in socially desirable highway funding, although they would not fully fund perceived infrastructure needs. This means that using revenues for

purposes with a high social value may be made especially simple, and politically attractive, by simply using the revenues to augment highway budgets.

- Taking into account the fiscal dividend from gasoline taxes reinforces the economic case for higher tax rates. But mileage taxes would be more stable as a source of highway funding than carbon taxes, as they avoid the erosion of the tax base due to rising fuel economy. By encouraging more efficient use of roads, mileage-based taxes would also improve the productivity of highway investments.

Introduction

This policy note considers what an economically efficient federal transportation policy would look like if carbon emissions from motor fuel consumption were priced through the type of carbon tax motivated in other chapters of this volume. Two broad sets of issues arise.

First, what are the implications for other major transportation policies – such as fuel taxes, fuel economy standards, and support for public transit – that are rationalized, at least in part, on climate grounds? In particular, what role remains for existing policies aimed at problems such as road congestion and local pollution?

Second, what are the implications of the revenues that would be raised from the carbon taxes applied to motor fuels? Use of these revenues for transportation needs would greatly alter the debate about how to fund federal infrastructure projects. This issue is coming to a head as real revenues raised per vehicle mile traveled are steadily eroding due to rapidly rising fuel economy (see Figure 12.1) and to the failure to adjust nominal fuel tax rates for inflation.[1]

This combination of a carbon tax and turmoil in infrastructure finance creates an opportune time to thoroughly re-evaluate transportation policies and highway finance. Roads are steadily becoming more clogged, after a brief respite during the recession, even while some other transportation-related concerns – local pollution, traffic accidents, dependence on foreign oil – are becoming less acute (Figure 12.2). Meanwhile, the backlog of perceived infrastructure needs continues to grow while real fuel tax revenues per vehicle mile fall, the gap being only partially filled by politically volatile appropriations from general revenues (CBO, 2012, Figure 1).

This chapter considers the two sets of issues in turn, focusing most attention on the first.

Reforming motor vehicle policies in light of carbon pricing

We start by briefly describing the main "externalities" (adverse side effects) of motor vehicle use that will remain to be addressed if carbon is priced. We then consider how well they could be addressed by fuel taxes alone. Following that we

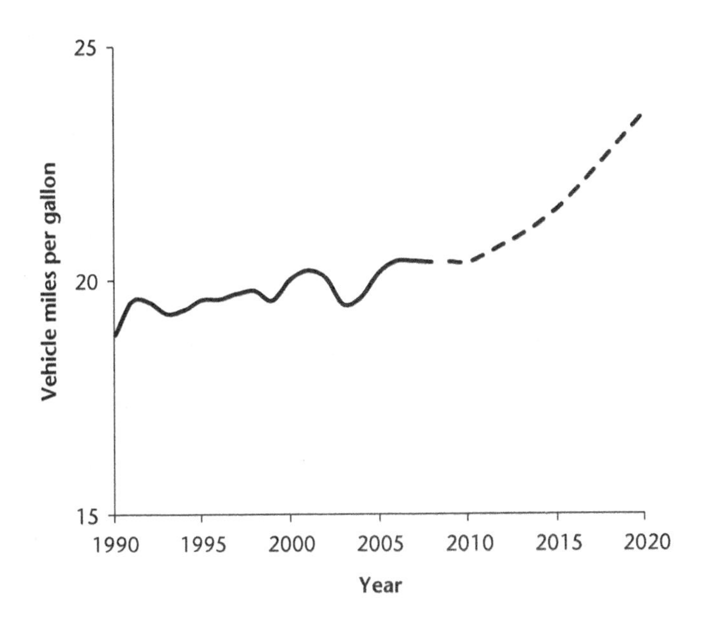

FIGURE 12.1 Fuel economy trends for (on-road) light-duty vehicles

Source: BTS (2012), Tables 4.11, 4.12, EIA (2012), Table 41.

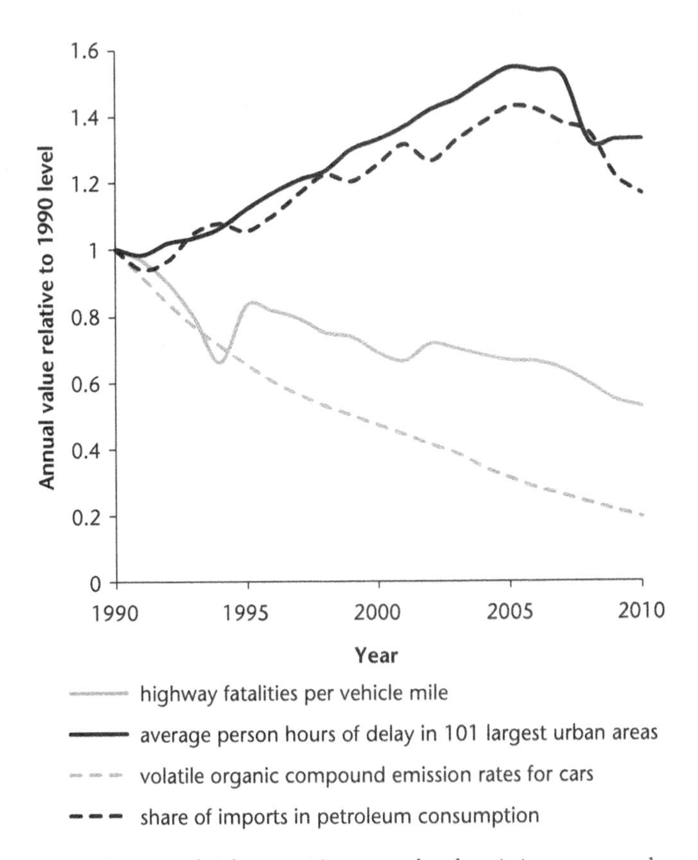

FIGURE 12.2 Trends in travel delays, accident rates, local emissions rates, and petroleum imports

Source: BTS (2012), Tables 1.69, 2.18, and 4.47; EIA (2013).

discuss the role of fuel economy standards in the presence of carbon pricing. Then we turn to various types of per-mile tolls that could alleviate some of the (non-climate) externalities more effectively than fuel taxes. Finally, we briefly discuss the implications of tax reform for a potpourri of other existing policies like road building, support for transit, and local emissions standards.

Transportation-related externalities (for light-duty vehicles)

For our purposes, a negative externality (or external cost) occurs when individual drivers do not take into account costs they impose on others from their own use of fuels and vehicles. Externalities provide a rationale for government policy interventions, one form of which is to effectively reflect these broader social impacts in the prices or costs individuals face. Even if strong policies are put in place to control the adverse effects of carbon emissions, there are at least three other major remaining externalities from use of light-duty vehicles that need to be addressed: air pollution, congestion, and traffic accidents.

Local (or conventional) air pollution

Long-standing pollution control measures, especially those deriving from the Clean Air Act, recognize several key air pollutants from motor vehicle emissions: especially fine particulates, volatile organic compounds, nitrogen oxides, carbon monoxide, and sulfates (the latter almost entirely from diesel vehicles). Most of the damages are from health effects. These arise largely from particulates and ozone, both of which are formed from atmospheric reactions as well as (in the case of particulates) direct emissions. Emission rates have been declining dramatically (Figure 12.2) due to ever more stringent regulations for new vehicles, and this trend will continue as the vehicle fleet turns over. An assessment by NRC (2009) valued the remaining environmental impacts by quantifying the effect of emissions on air quality, the extent of population exposure, the extensive evidence on health impacts, and evidence on people's willingness to pay to reduce health risks. Roughly speaking, they put environmental damages at 1.3 cents per vehicle-mile for the average (on-road) light-duty vehicle (after updating for inflation to 2010), or about 27 cents per gallon at today's average fleet fuel economy of about 20 miles per gallon (see Figure 12.1).

Congestion

Generally, growth in vehicle miles driven has outstripped growth in road capacity for decades.[2] Although individual motorists bear the costs of road delays, they do not take into account their impact on adding to congestion, thereby increasing delays for other road users. The resulting externality, per extra mile of driving, has been inferred from relationships between speed and traffic flows and from measurements of how people value time lost to congestion (often found to be half the market wage or more). Obviously, congestion is very specific to location and time of day: even at a modest level of aggregation, measured external congestion costs per mile vary from zero for free-flowing roads up to around 25–35 cents per vehicle mile for peak travel in large urban centers in the United States (Parry 2009, Table 2). When averaged across different regions and time of day, the congestion externality is perhaps 6.5 cents per vehicle mile.[3]

Traffic accidents

Although highway fatality rates declined by two-thirds between 1980 and 2010 (BTS, 2012, Table 2.18), the annual total costs to society from traffic accidents is large plausibly on the order of $400 billion a year, or 2.5 percent of GDP (Small and Verhoef 2007, pp. 100–101).

But the nature of the externality from traffic accidents is complex, for several reasons. First, some but not all accident risks are taken into account by individuals in their decisions about when and where to drive and in what type of vehicle. Drivers may well consider the risk of injuring themselves in accidents involving only their own vehicle, but not necessarily the risks to pedestrians and other drivers except insofar as it affects drivers' personal liability and/or insurance rates – a highly uncertain prospect. The risk imposed on other drivers is also complex. On average it may be quite small – while the frequency of collisions (per vehicle mile) rises with more vehicles on the road, their average severity falls as people drive slower and more carefully in heavier traffic. What remains very clear is that drivers in heavier vehicles, such as pickup trucks and SUVs, create sizable risks for those in lighter vehicles.

In addition, some costs from injuries are borne by insurance companies and the government, but typically only the monetary medical costs which are just a fraction of individuals' willingness to pay to avoid injuries and fatalities. Studies that attempt to decompose all these types of costs suggest that motorists impose a cost of around 4.0 cents on other individuals and third parties on average for each extra mile of driving.[4]

Other side effects from motor vehicle use

Another significant externality is additional road maintenance caused by traffic, insofar as it exceeds fees paid by those vehicles. However, this is almost entirely caused by heavy-duty trucks rather than light-duty vehicles since road wear is a rapidly escalating function of the vehicle's axle weight. Policymakers have also been concerned about energy security; but recently the share of imports in the nation's oil consumption has declined (Figure 12.2) and is set to decline further with continued expansion of domestic oil shale reserves. In any event, the implications of energy security for fuel conservation policies are opaque, as discussed later.

Summary

The left-hand bar of Figure 12.3 provides a summary of the relative importance of the externalities discussed above, by comparing them all on a per-gallon basis (i.e., by assuming that an extra gallon of gasoline use leads to 20 additional miles of driving). We include carbon dioxide (CO_2) based on damage of $25 per metric ton and 0.009 tons of CO_2 generated per gallon of fuel. Most noteworthy is that unless CO_2 damage is several times larger than this estimate,[5] congestion and accidents easily dominate it in terms of costs of motor vehicle use. This provides an immediate clue that for motor vehicles, the "side effects" of climate policy may be more important than the direct effects, perhaps not surprisingly given that motor vehicles have strong and very immediate effects on matters of great concern in everyday life – time use and risk to health and safety.

Corrective fuel taxes

Suppose, for the moment, that gasoline taxes are initially the only available fiscal instrument to address the above externalities (as emphasized later, other instruments are much better tailored to addressing some of the externalities). If a new charge on carbon emissions is then introduced, what does this imply for the economically efficient level of gasoline taxes?

There is a standard formula in the literature for assessing the appropriate level of such "corrective" taxes (see Appendix). Based on this formula, the right-hand bar in Figure 12.3 summarizes the contribution of different externalities to the efficient gasoline tax.

Carbon contributes 23 cents per gallon, identical to its external cost as shown in the left-hand bar in the figure.

Local pollution contributes 14 cents per gallon, only half the value of its external cost. This is because, as discussed in the Appendix, roughly speaking there are no local pollution benefits from improvements in vehicle fuel economy in the presence of binding emission rate standards – pollution only falls if the amount of driving is reduced. Because about half of reductions in gasoline usage come from fuel economy improvements rather than reduced driving, the local pollution benefits per (tax-induced) gallon reduction in gasoline are therefore diluted. (However, there is an important caveat related to fuel economy regulation here, as noted below.)

The same dilution applies for congestion and accidents – as a first pass, these externalities depend only on amount of driving, not fuel consumption. Again the congestion and safety benefits per gallon of fuel reduction are less than their average external cost per gallon of fuel use. Nonetheless, these externalities are large, and so still contribute an estimated 50 and 37 cents per gallon, respectively, to the efficient fuel tax.

Overall, the estimated corrective fuel tax is $1.23 per gallon, more than three times the current tax level, which is $0.40 per gallon ($0.184 at the federal level plus on average $0.218 at the state level).

Estimates of the efficient tax on diesel fuel used by heavy-duty trucks are in the same ballpark as those for efficient gasoline taxes. For the year 2007, Parry (2011) estimated it at $1.15 per gallon – though the relative contribution of different externalities to the efficient tax is somewhat different than for gasoline. In fact at

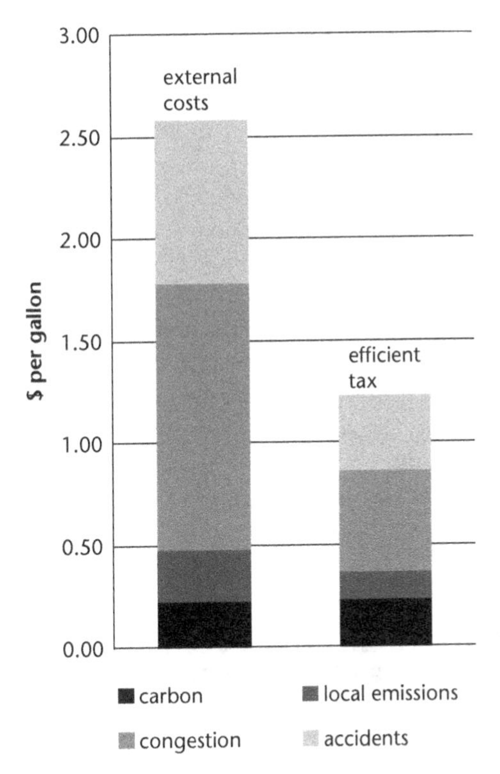

FIGURE 12.3 External costs and efficient fuel tax for light-duty vehicles (expressed in year 2010 dollars per gallon of gasoline)

Source: Costs imposed on others from local emissions, congestion, and accidents, per extra vehicle mile, are discussed in the text. These are converted into per-gallon costs (for the left-hand bar) assuming a gallon results in 20 extra vehicle miles driven. Carbon costs equal an assumed damage of $25 per metric ton of CO_2 multiplied by emissions of 0.009 metric tons per gallon of gasoline (Davis et al., 2012, Table 11.11). See Appendix for details of the efficient tax computation (for the right-hand bar).

first glance we might expect the efficient tax to be higher as one extra vehicle mile by a truck adds more to externalities than one extra car mile (e.g., trucks take up more road space and drive slower, thereby adding more to congestion, and they also cause road damage). But this is offset because trucks have much lower fuel economy than light-duty vehicles, which means that a higher per-mile cost can translate into roughly the same per-gallon cost.

In short, the case for taxing fuel goes far beyond climate damage due to carbon emissions. Indeed, even if the climate damages in Figure 12.3 were incorporated into a broad-based carbon tax without reducing the current fuel tax, the latter still would not be nearly high enough to account for the other externalities described here. Furthermore, as discussed in Box 12.1, the corrective fuel tax estimates in Figure 12.3 may be on the low side in some regards: for example, in the presence of stringent fuel economy standards, fuel taxes behave more like mileage-based taxes which would justify higher levels. Thus, if fuel taxes were the only available instrument for dealing with these externalities, we would want to raise them further based on those grounds alone.

Fuel economy standards

Rather than fuel taxes, the centerpiece of ongoing efforts to reduce motor fuel use is a set of progressively escalating fuel economy standards.[6] We examine the case for continuing such a policy if carbon emissions were effectively priced through another means. Whether a fuel economy policy would be redundant appears to depend mainly on two factors: energy security and the so-called energy paradox.

As regards energy security, to the extent that the national interest is served by reducing our reliance on petroleum – especially imported petroleum – fuel economy standards are one way to address that need. While it is difficult for economists to provide guidance on the appropriate stringency of fuel economy standards to address this issue (Box 12.1), fuel economy standards appear to have considerable political acceptability.

BOX 12.1 WHY CORRECTIVE FUEL TAX ESTIMATES MIGHT BE UNDERSTATED

For several reasons, the corrective fuel tax estimates in Figure 12.3 might be understated.

They ignore the possibility of an energy security benefit from reducing gasoline consumption, which might arise from reduced macroeconomic vulnerabilities to oil price shocks or from less reliance on oil-exporting nations whose international objectives may be contrary to ours. But the nature of this benefit is often difficult to state rigorously: at least some of the risks from oil price shocks should already be taken into account in firm and household decisions, and the influence of specific exporting nations is limited by the fact that oil prices are determined in world markets. To the extent any benefits have been measured, however, they seem to be fairly modest in magnitude: see, for example, Brown and Huntington (2013) on the macroeconomic vulnerabilities. Moreover, concerns about energy security are diminishing as domestic oil shale reserves are coming online.

Our corrective fuel tax estimates also ignore the possibility of a safety bonus for other road users to the extent that high fuel taxes encourage some people to shift to smaller and lighter vehicles to economize on fuel, thereby reducing the damage caused to other vehicles during collisions. Nor do they consider the dramatic, recently promulgated, ramping up of fuel economy standards through 2025. These standards will mute the tendency of motorists to respond to fuel taxes by buying more fuel-efficient cars, since they are already mandated to do so. Thus, a greater portion of a given tax-induced fuel reduction will in future come from people driving less, thereby magnifying the benefits (per gallon of fuel reduced) of reduced local pollution, congestion, and accidents. In other words, this trend will cause the effect of the fuel tax on mileage-based externalities to be less "diluted," in the terminology of the text, so that the tax will come to resemble more closely a tax on vehicle miles.

Finally, assuming that underfunding of infrastructure will not be eliminated quickly, the already large congestion externality is likely to grow larger still, adding to the appropriate level of the corrective fuel tax.

Just as for CO_2 emissions, however, a broader tax – namely, one applied to all petroleum products – would be more effective than a policy aimed only at new-vehicle fuel economy, because it would exploit some additional opportunities for reducing oil consumption. In particular, the broader tax would discourage motor vehicle usage – with the favorable impacts on congestion, accident, and local pollution already discussed. It would also promote conservation of other oil-based products. Furthermore, to the extent that oil imports, rather than oil consumption, is the problem, a better targeted policy would be an oil import fee if it could be made compatible with trade agreements.

The second factor, however, could justify a regulatory policy even if there were no feasibility limit to the fuel tax. This factor is the potential for market failure in energy markets if consumers inadequately consider the lifecycle fuel-saving implications when choosing among vehicles with different fuel economies. There is tantalizing but not definitive evidence that this is the case. Engineering assessments suggest a range of fuel-saving technologies that would yield lifecycle fuel savings – discounted at market rates – in excess of the costs from incorporating them into new vehicles. At the same time, many though not all empirical studies have found that consumers implicitly apply high discount rates in their market choices, perhaps failing to account for as much as two-thirds of the true private value of future savings in fuel costs when making their purchases. The observation that such seemingly "negative cost" technologies are not always adopted is known as the "energy paradox."[7]

But as discussed in Box 12.2, there is much dispute about the reasons for the energy paradox, and whether it warrants government policy intervention. Parry et al. (2014) suggest that, even under generous assumptions about the size of any market failure, the energy paradox by itself probably does not fully justify the aggressive ramping up of standards recently promulgated through 2025. In any case, this ramping up would become partially redundant in the presence of a carbon tax.

Aside from policy stringency, the structure of the CAFE program has recently undergone some other changes, most notably provisions that promote credit trading and that link standards to individual vehicle size, which raise further issues. But since they are tangential to carbon policy, we confine our discussion of them to Box 12.3.

BOX 12.2 THE ENERGY PARADOX CONTROVERSY

Numerous explanations have been proposed to explain the energy paradox (Helfand and Wolverton 2011). Most of them do indeed imply a market failure. Consumers may have limited information, or limited ability to calculate future fuel costs from the information they have. There is evidence, for example, of the incorrect belief that increasing efficiency from say 30 to 31 miles per gallon (mpg) provides the same future fuel savings as increasing it from 20 to 21 mpg. Or consumers may have more vehicle traits to consider than they can process, and so omit fuel economy. Or they may be mistrustful of claimed fuel cost savings, doubtful about future fuel prices, or short-sighted in their assessment of the future. Informational inefficiencies in used car markets could perpetuate such short-sightedness by not permitting people to reap the full advantage of more fuel-efficient cars in their sale prices upon trade-in or sale. Moreover, consumers may be subject to borrowing constraints causing them to underinvest in energy-saving technologies relative to what would be desirable from society's perspective.

Other explanations, however, do not imply a market imperfection. For example, the observed reluctance of consumers to pay for vehicles with higher fuel economy may reflect their awareness of possible undesirable side effects, such as reduced acceleration or greater likelihood of needing repairs. Such "hidden costs," if real, would then create an additional cost of a policy mandating high fuel economy.

The appropriate policy response depends therefore not only on the size of the energy paradox (if it indeed exists at all), but on the reason for it. If it is large and is caused by one of the factors involving a market failure, some policy intervention can be justified even aside from any environmental costs associated with energy use. Indeed, the official regulatory impact analysis of currently adopted US standards for model years through 2025 can be read as implying that whether current efficiency standards are worth their cost depends more on the energy paradox than on environmental and energy security concerns. At any rate, there is an urgent need for more definitive evidence about the extent and exact nature of any energy paradox.

BOX 12.3 RECENT STRUCTURAL REFORMS TO THE CAFE PROGRAM

On the plus side, manufacturers now have greater flexibility to trade fuel economy credits among themselves, across different periods of time, and across their car and light truck fleets. This improves the cost-effectiveness of the program as, for example, manufacturers can go beyond the standard in years of high gasoline prices when consumers are more willing to buy fuel-efficient vehicles, and use those banked credits in low-fuel-price periods when it would otherwise be costly to meet a rigid fuel economy standard.

On the debit side, standards now vary inversely with a vehicle's size (or footprint), meaning that manufacturers can effectively relax their average CAFE requirements by shifting to larger-size vehicles. Removing this perverse incentive would improve the cost-effectiveness of the program for a given overall fuel economy improvement, while also alleviating the risk to other road users posed by larger-size vehicles.

An alternative to fuel economy standards is a type of financial incentive known as a "feebate": a fee charged or rebate given for the purchase of a motor vehicle proportional to the difference between its rated fuel consumption per mile and some arbitrarily chosen standard (see e.g., Greene et al. 2005a, b and Small 2012 for analyses). In theory, these incentives can be chosen to give results equivalent to those of a fuel economy standard; the similarity is even closer if one recognizes that a manufacturer subject to an efficiency standard has an incentive to adjust its prices to encourage purchase of fuel-efficient vehicles at the expense of fuel-inefficient vehicles. However, because feebates involve potentially significant financial flows, their implementation requirements and political implications may be quite different from those of a fuel economy standard.

Mileage-based taxes

Our earlier discussion makes the case that a higher fuel tax – perhaps much higher – continues to be justified on the basis of other externalities besides CO_2 emissions provided the fuel tax is the only available means to tighten current control of these externalities. But in fact a number of other policies are often proposed for this purpose.

Most externalities are much more closely related to number of miles driven than to fuel consumed. Therefore, it is natural that policies to discourage motor vehicle use are often considered as front-line policies to address motor-vehicle-related externalities. Here we discuss a novel class of these policies, involving per-mile tolls of various kinds.

In principle, a tax on vehicle miles traveled (VMT) is the most effective disincentive to use motor vehicles. Substantial interest in VMT taxes has emerged recently, usually as a replacement for fuel taxes and largely driven by fiscal considerations (see below). The state of Oregon has led the way with extensive experiments examining implementation strategies based on tax collection at time of refueling, while the United Kingdom and the Netherlands have seriously considered (though not yet implemented) nationwide VMT taxes based on Global Positioning System (GPS) technology.

While a VMT tax would be a large improvement over fuel taxes for controlling most externalities, it is still a relatively blunt instrument. Each of these externalities can vary widely by time, place, and other circumstances. Therefore, more effective policies would vary per-mile tolls accordingly.

For congestion, this reasoning leads to a per-mile charge for vehicles driving on busy roads, with the charge varied by location (high for Los Angeles, New York, Washington, DC, zero for rural areas, etc.) and rising and falling during the course of the rush hour. This policy, known as "congestion pricing," exploits multiple possibilities for drivers to alter behavior in order to alleviate congestion. Examples include moving trips away from the peak of the rush hour, encouraging alternate modes (e.g., carpools, public transit, walking, bicycling), reducing trip-making (e.g., via telecommuting or combining trips), shifting to less congested routes, changing job or residential locations, and other strategies that analysts may not even have thought of. Some form of, albeit very limited, congestion pricing has been implemented in Singapore, London, and Stockholm. Partial versions exist in the United States in the form of "value pricing," usually meaning express lanes that are free to certain users (e.g., carpools) and available at a charge to others (Poole 2012).

For accidents, per-mile charges could be scaled according to the extent of external accident risk: higher for higher-risk drivers (based, perhaps, on their rating factor as determined by insurance companies) and higher for vehicle classes that pose greater risks for other drivers and third parties. A start toward such a system has been made privately in the form of insurance rates based on the number of miles driven as well as the usual ratings factors. A government-imposed version has been proposed in the form of "pay as you drive" insurance: conversion of insurance payments into a mileage-related charge, perhaps payable at time of refueling based on odometer readings (Bordoff and Noel 2008).

For local air emissions, a better corrective tax would be a per-mile toll whose rate depends on the emissions characteristics of the vehicle and on the extent of population exposure to those emissions, or better still a charge on the emissions themselves. This reform is less pressing, however, given that per-mile damages from local emissions are smaller than for congestion and accidents—and will become smaller still as tighter regulations work their way through the vehicle fleet.

Finally, road damage is most efficiently addressed through per-mile tolls on heavy trucks, scaled by their axle weight. Such a tax would encourage truckers to seek vehicle fleets that carry goods efficiently over more axles with much less road damage. Small et al. (1989) analyze such a system in detail. A limited version exists in the ton-mile tax in Oregon, whose rates for vehicles over 40 tons vary sharply by weight and inversely by number of axles.

In short, the ideal fiscal system for motor vehicle transport would involve charging motorists for each mile driven, where the charge is scaled according to factors affecting the congestion, accident, local pollution, and possible road damage costs imposed on others by that mile. A much reduced fuel tax component would be retained to address carbon emissions, and that component could be fully replaced by an appropriate carbon tax.

Other traditional policies

In light of where we should be headed, as just described, what are the implications for some other, more traditional, transportation policies?

Limiting road capacity. The main rationale for policies restricting road building is that the amount of motor vehicle travel adjusts to the extent of the available road network.[8] This is a direct response to what remains quantitatively as the largest motor vehicle externality: road congestion. But limiting capacity can impose enormous costs, both to individual drivers and to the efficiency of economic systems that depend on people's ability to interact. This is because such a policy largely works by keeping congestion high, rather than allowing socially desirable capacity expansions with efficient rationing of new road space as would occur with appropriate congestion pricing.

Supporting public transport. Another policy attempting to reduce motor vehicle use is the support of public transportation, especially bus and rail transit in urban areas. This policy helps to shift people from automobile trips to transit (to the modest extent that congestion is reduced), but also will attract some people to transit who would otherwise walk, cycle, or carpool.

The main drawbacks to this policy involve effectiveness and cost-effectiveness. Except for high-density urban areas during peak periods, such policies tend to be very expensive and have a quite limited impact on motor vehicle travel and a very small impact on externalities other than congestion. It has even less effect on energy consumption, because transit vehicles also use energy, particularly petroleum-based energy in the case of diesel buses, and this energy consumption can be quite high per passenger when (as is common) transit vehicles run with low occupancy. On the other hand, public transit provision may provide important travel benefits, and it is even subject to positive externalities that can warrant some level of subsidy. Specifically, by increasing ridership, promoting public transit makes it feasible to offer a denser network of service lines and more frequent service, thereby reducing other users' costs of getting to transit stops and waiting for buses and trains. Both the costs and benefits of public transit are very case-specific, however. Given its limited value for energy conservation, we suggest that policies promoting public transit be evaluated based mainly on their benefits to travelers, regardless of whether carbon emissions are taxed or not.

Policies targeting technology and behavior. The discussion above has already included a variety of financial incentives to reduce motor vehicle use and, perhaps more importantly, to change technology and driving behavior. For example, congestion pricing would encourage changes in routes or times of day of travel; accident-related mileage charges would encourage safer vehicles; local emissions charges would encourage lower-emitting vehicles and better maintenance of their emissions control systems; and heavy-vehicle charges would encourage distributing loads over more axles.

The same types of changes may be encouraged through regulations rather than financial incentives. Indeed, these are the mainstays of policies in all these areas. Regulations may include restrictions on entering certain areas, incentives to change work hours, mandates on vehicle safety features, emissions maintenance inspections, and axle-specific truck weight limits. One drawback of such regulations is that they do not raise revenues, though the flip side is that, since they do not involve a large transfer from motorists to the government, they may be more politically acceptable. Overall, they are less effective than well-targeted taxes in promoting all of the opportunities for reducing externalities, but they are often reasonably effective in the absence of pricing policies.

A good example is pollution control mandates on motor vehicle manufacturers. Such regulations, along with inspection and maintenance requirements, have dramatically reduced emissions rates over time (Figure 12.2) with substantial public health benefits. While some further gains could be achieved by the emissions-related VMT charge discussed above, it would make little sense to put previous gains at risk by rolling back emissions rate standards. Similarly, legal penalties for dangerous driving practices and weight limits on heavy vehicles would retain a useful reinforcing role, even after the introduction of VMT tolls related to accident risk and road damage.

Issues in transportation finance

The more dramatic implication of carbon taxes could be on the fiscal, rather than the environmental, side. Fuel taxes are the primary component of federal funding for highway infrastructure, and to some extent of transit infrastructure, through the Highway Trust Fund. A carbon charge on motor fuels would raise substantial revenues – far more than can be achieved by any amount of pleading on the part of infrastructure advocates or complaints by transportation analysts about the erosion of the user pay principle for transportation finance. While it is impossible to predict how a political agreement to enact a carbon tax might treat revenues, here we consider the possibilities if those revenues deriving from light-duty vehicles were allocated to expenditures on related infrastructure.

We discuss three main issues. The first is the size of the extra revenues from carbon charges on motor fuels in the context of transportation funding needs. The second is the appropriate balance between (and forms of) transportation taxes and broader taxes in financing infrastructure or public spending more generally, accounting for the dual role of transportation taxes in correcting externalities and in raising revenue. The third is to what extent revenues collected from transportation should be earmarked for that sector.[9]

The infrastructure funding gap

The problems of infrastructure finance in the United States have occupied several national commissions and untold policy commentaries (e.g., TRB 2006). Virtually all have bemoaned the failure of tax rates on gasoline and diesel fuel to keep up with perceived infrastructure needs, although they differ in their views as to whether those infrastructure needs are primarily for more highway capacity or for investments in other modes. The possibility of a large infusion of funds from a carbon tax into infrastructure finance has the potential to transform the nature of these discussions.

To gain some perspective, let us consider two of the many ways the shortfall in infrastructure funds has been quantified.

Maintaining current real expenditures. For many years, revenue sources dedicated to the Highway Trust Fund, mainly from the fuel tax on motor vehicles, have fallen short of authorized highway and transit expenditures. In 2008–2010, the gap averaged $7 billion annually (Table 12.1). As a result, the federal government has periodically needed to appropriate general funds to bolster the Fund. Furthermore, the situation will get much worse as cars become even more fuel efficient. Under anticipated trends even prior to the latest CAFE tightening, it is likely that revenues will decline 37 percent in real terms, from $35 billion in 2008 to $22 billion in 2035, thus adding another $13 billion to the current gap of $7 billion between funding and expenditures required just to maintain current expenditure levels (all in 2008 dollars).[10] The new fuel economy standards for 2017–2025 model cars will further reduce annual revenues by about 21 percent once those models constitute the bulk of the fleet, according to CBO (2012), bringing the decline in revenue to 50 percent and adding an additional $5 billion shortfall for a total of $25 billion in 2035.[11]

Maintaining current levels of service. The National Surface Transportation Infrastructure Finance Commission (2009, pp. 50–52) assessed estimates of federal highway capital investment needed to maintain current conditions and performance of highways and transit, showing average annual funding gaps of $47 billion for highways and $21 billion for transit, over the years 2008–2035, for a total of $68 billion.[12]

How much would a carbon tax help with these shortfalls? A tax of $25/ton CO_2 would raise approximately $33 billion in extra revenue from motor vehicles in 2010.[13] These revenues would be subject to roughly the same sources of decline in the tax base over time as are fuel tax revenues, due to a more efficient fleet; however, this decline would be mostly offset by a likely increase in the carbon tax rate if the latter is continually set to the damage cost of carbon emissions. Using the estimated rate of increase in the latter from the Interagency Working Group on Social Cost of Carbon,[14] revenues from the carbon tax on light-duty vehicles would decline by about 10 percent by 2035, to $30 billion.[15]

Thus, it appears that carbon tax revenues of a size we are discussing would largely fill the gap between revenues and current expenditure levels, but would not provide enough revenues to maintain current levels of service. And even this latter level

TABLE 12.1 Federal transportation spending and revenues: annual average, 2008–2010 ($ billions)

	Gasoline excise tax revenues	*Total excise tax revenues*	*Expenditures*	*Revenues less expenditures*	*Transfers to highway trust fund*
Highways	20.3	30.5	35.5	-5.0	
Transit	3.8	4.9	6.9	-2.0	
Total	24.1	35.4	42.4	-7.0	9.6

Source: FHWA (2011), Table FE-210.

of expenditure is projected by US Department of Transportation (2010) to fund only those projects with benefit-cost ratios of 2.0 or greater. Thus, a carbon tax is unlikely to offset the failure of fuel tax rates to keep up with inflation, a growing motor vehicle population, aging infrastructure, and other factors preventing our highway and transit capital stock from performing its desired purposes. Rather, such a tax would shift the focus from how to save the Highway Trust Fund from collapse to whether and how our infrastructure should be significantly upgraded to keep up with growing needs.

Balance between (and form of) transportation and broader taxes

When transportation taxes are considered in the context of an overall fiscal system of raising revenue, one needs to consider a desirable set of relative tax rates on these and other commodities. A common approach to this in public finance is to ask how the inevitable distortions to economic activity can be minimized by choosing such relative rates. When externalities are important, as in the case of the fuel tax, the question comes down to whether the best tax rates on transportation goods, such as fuel, should be set above, equal to, or below the rates appropriate for addressing externalities. To analyze this issue, we can draw on some well-known general principles of tax design for guidance.

First, final goods consumed by households (including passenger vehicles and the fuels they use) may be legitimate bases of taxation on purely fiscal grounds, but intermediate inputs such as commercial trucks and their fuels are not. Taxing intermediate inputs at rates higher than those needed to control externalities would distort the way firms do business, causing them to use too little of the taxed input, and too much of other inputs, from a cost-minimizing perspective.

Second, spending on different consumer products should be taxed under a common formula. An important component of such a formula would be its effect on labor supply: specifically, how much would an excise tax affect employment compared to broader taxes on income, payrolls, or consumption set at rates to produce equal revenues? Products that are potentially attractive targets for taxation on these grounds generally are those that have relatively inelastic tax bases.

These possibilities have been explored in the specific case of gasoline taxes for the United States, with studies finding that on balance fiscal considerations might warrant some additional taxation of gasoline in excess of levels to correct for externalities – perhaps by 25 to 50 cents per gallon. These estimates are imprecise, but the main point is valid: the case for higher fuel taxes is even stronger when they are considered within in a broader revenue-raising context, even aside from the oft-cited political advantage of creating transparency for how highway infrastructure is financed.

But gasoline taxes are not the only way to achieve these fiscal goals. Rather, fiscal considerations again seem to favor mileage-based taxes over fuel taxes. One reason is that mileage-based taxes are a more stable source of revenue – especially going forward, as rising fuel economy standards will erode the base of fuel taxes. A second reason is that mileage-based taxes, especially if they are carefully designed to efficiently manage traffic congestion, help to maximize the benefits of highway investments.

Earmarking

A frequent method of obtaining political acceptance of new sources of revenue is to promise to spend them for specific purposes, usually closely related to those sources. For transportation finance, such hypothecation, also known as "earmarking" or "ring fencing," would take the form of dedicating revenues from motor fuel taxes to related infrastructure such as highway construction, maintenance, or rehabilitation. Indeed, in the United States this principle is embedded in the Highway Trust Fund, although in recent decades the tie has been loosened by allowing some of the funds to be spent on public transit and various measures related to environmental goals. For our purposes, the principle could be applied equally well to revenues from a fuel tax, a VMT tax, or the portion of a carbon tax coming from transportation.

To the extent that these taxes are considered in part to be for environmental reasons, hypothecation might take the form of dedicating the revenues to amelioration of environmental damages. For example, revenues from a carbon tax could be dedicated to helping people adapt to the consequences of climate change.

The trouble with hypothecation is that there is no necessary relationship between the efficient level of the tax and the efficient level of spending on a particular program. This is especially true for a tax aimed at mitigating climate change, since the damage resulting from today's emissions will be mostly felt over a period of centuries. Similarly, transportation infrastructure needs are long term and may not bear a close relationship to current revenues from efficient transportation taxes. Moreover, the chance evolution of the tax rate over time, due to political considerations or unforeseen technological changes that improve fuel economy, may unduly influence the level of spending actually adopted. These concerns are not just theoretical: four decades ago the nation was consumed by debates over whether the ready availability of fuel tax revenues was producing unnecessary highway spending, abetted by the "highway lobby"; whereas today the concern is that infrastructure finance has been choked off arbitrarily because fuel tax rates have not compensated for inflation, improved fuel economy, or changing infrastructure needs.

Nevertheless, the tie between highway usage taxes (primarily but not exclusively the fuel tax) and highway spending has served a valuable political purpose in the past, allowing citizens to see a connection between their taxes and a widely appreciated road system. This connection has eroded for many reasons, leading to proposals to forge it anew, for example, by eliminating the use of fuel tax revenues for public transit. Hypothecation may turn out to be a reasonable political mechanism to achieve a sound economic goal, but the implications for spending levels need to be carefully assessed against the effectiveness of that spending.

Conclusion

It is an opportune time for a thorough re-assessment of federal transportation policy. The scourge of urban traffic congestion is set to worsen in the near term as the economy recovers, and in the longer term as population and economic growth require accommodation of resulting traffic. Pressure remains on policymakers to enact a comprehensive carbon pricing program, as extreme weather events remind us that atmospheric concentrations of greenhouse gases are rising above levels deemed safe by scientists. And highway budgets are being squeezed as far as the eye can see, as the traditional source of highway funding is eroded through failure to adjust tax rates to inflation, progressively rising fuel economy standards, and demand suppression from high oil prices.

These trends create a strong case for a system of mileage-based taxes, leaving a fuel charge only to correct for carbon emissions. Charging motor fuels for carbon emissions would not necessarily undermine the case for fuel efficiency standards, and would have little relevance for the other traditional policies, such as support for mass transit or standards governing new-vehicle emissions of local pollutants. At the same time, the growing shortfall in funding for socially desirable transportation projects, or perhaps for even more urgently needed non-transportation initiatives, underscores the high cost of squandering the potential fiscal dividend from carbon taxes.

Notes

1. Federal gasoline and diesel tax rates have been frozen in nominal terms since 1993, despite an increase of over 50 percent in the consumer price level during this time. State rates have roughly remained steady in real terms, but have not been adjusted to compensate for changes in fuel economy: see FHWA (2011), Table MF-205.
2. Between 1980 and 2010, for example, urban vehicle miles traveled increased by 132 percent, against an increase in lane-mile capacity of 76 percent (BTS 2006, Tables 1.6 and 1.36).
3. Authors' calculations using congestion delay data from TTI (2011) and a value of time equal to 60 percent of the market wage.
4. From Parry (2011), after updating for inflation and declining fatality rates.
5. This would require a value of carbon emission damage above the highest value considered by the Interagency Working Group on Social Cost of Carbon, which chose $21 per ton CO_2 as a central value and a range of $5–$65 for sensitivity analysis, all in 2007 US dollars (Greenstone et al. 2013, p. 38).
6. If manufactures were to fully comply with the new CO_2 standards solely through improvements in fuel economy of the existing fleet mix, the average fuel economy of new light-duty vehicles would rise from 29 mpg in 2012 to about 35.5 mpg in 2016 and 54.5 mpg in 2025. Actual fuel economy will likely fall significantly below these levels, however, due to some manufactures opting to pay penalties in lieu of fully meeting the standards, others shifting to larger-size vehicles (as explained in Box 12.3), and many taking an option to reduce emissions of greenhouse gases from air conditioners in lieu of some fuel efficiency improvements. There is also a systematic difference of about 20 percent between the legally rated fuel economy (as determined by specified laboratory tests) and actual on-the-road performance. Sources: EPA and NHTSA (2010), p. 25328; NHTSA and EPA (2012), pp. 3–4.

Fuel economy standards are also being phased in for heavy-duty trucks, which are expected to reduce fuel consumption rates by 7–23 percent by 2017, depending on truck category (Harrington 2011). The same sorts of considerations arise for these regulations as for those discussed above affecting light-duty vehicles. The energy paradox discussed below might be less compelling in this context, however, given already strong incentives to economize on fuel use in trucks (which are driven more intensively than cars).

7. See recent reviews in Helfand and Wolverton (2011) and Busse et al. (2013). The energy paradox may apply not only to vehicles but also to many other sectors such as home appliances and building heating and cooling technologies. See, for example, Hausman (1979). The evidence of high discount rates could be consistent with consumer rationality if consumers are credit constrained, but then the apparent myopia still constitutes a market failure.
8. This phenomenon appears especially for high-speed expressways, where some evidence (Duranton and Turner 2011) even suggests that traffic adjusts in proportion, or nearly so, to road capacity. For a representative sampling of the large empirical literature on how traffic adjusts to road capacity, see Goodwin (1996), Cervero (2003), Duranton and Turner (2011), and the reviews therein.
9. A further issue is whether it makes sense for the federal government to play such a large role in directing infrastructure investments when these projects primarily benefit the state in which they are carried out. This question, relevant to all nations with a federal structure, is covered in commentaries such as by Poole (2013) and we do not address it here.
10. National Surface Transportation Infrastructure Finance Commission (2009, pp. 101–103), especially Exhibit 4-5.
11. The combined effect of a 37 percent decline from previous policies and 21 percent further decline from the new standards for 2017–2025 is (1-0.37)*(1-0.21) = 0.50.
12. The commission's estimate comes from updating and adjusting figures from the biannual assessment of needs and performance reported in US Department of Transportation (2010).
13. Authors' (simplified) calculation, assuming the average federal and state tax on gasoline and diesel fuel (without the carbon tax) was 40 cents per gallon in 2010; the carbon tax adds 22 cents per gallon on average; and fuel consumption falls 4 percent in response to the tax. Combining this with fuel consumption of 170 billion gallons (BTS 2012, Table 4.5), implies the carbon tax increases revenue from $68 to $101 billion.
14. This damage cost is estimated by the Interagency Working Group on Social Cost of Carbon to rise by 43 percent over the 15-year period 2010–2025, under their central value (based on 3 percent discount rate); applying the same annual growth rate (2.41 percent) to 2035, damage cost per ton would therefore rise by a total of 81 percent by 2035 (since $1.0241^{25} = 1.81$).
15. That is, the combined effect of a 50 percent decline in tax base and an 81 percent increase in tax rate is (1-0.50)*(1+0.81) = 0.905.

References

Bordoff, Jason E., and Pascal J. Noel (2008), "Pay-As-You-Drive Auto Insurance: A Simple Way to Reduce Driving-Related Harms and Increase Equity," Discussion Paper 2008–09, The Hamilton Project, Brookings Institution, July. www.brookings.edu/~/media/research/files/papers/2008/7/payd%20bordoffnoel/07_payd_bordoffnoel.pdf

Brown, Stephen P. A., and Hillard G. Huntington (2013), "Estimating US Oil Security Premiums," *Energy Economics* 38: 118–127.

BTS (2006), *National Transportation Statistics 2006*, Bureau of Transportation Statistics, Washington, D.C.: U.S. Department of Transportation. www.bts.gov/publications/national_transportation_statistics/2006/index.html

BTS (2012), *National Transportation Statistics 2012*, Bureau of Transportation Statistics, Washington, D.C.: U.S. Department of Transportation. www.bts.gov/publications/national_transportation_statistics/pdf/entire.pdf

Busse, Meghan R., Christopher R. Knittel, and Florian Zettelmeyer (2013), "Are Consumers Myopic? Evidence from New and Used Car Purchases," *American Economic Review*, 103(1): 220–256.

CBO (2012), *How Would Proposed Fuel Economy Standards Affect the Highway Trust Fund?* Congressional Budget Office, Washington, D.C., May. www.cbo.gov/publication/43198

Cervero, Robert (2003), "Road Expansion, Urban Growth, and Induced Travel: A Path Analysis," *Journal of the American Planning Association*, 69(2): 145–163.

Davis, Stacy C., Susan W. Diegel, and Robert G. Boundy (2012), *Transportation Energy Data Book: Edition 31*, Oak Ridge, Tenn.: Oak Ridge National Laboratory, July.

Duranton, Gilles, and Matthew A. Turner (2011), "The Fundamental Law of Road Congestion: Evidence from US Cities," *American Economic Review*, 101: 2616–2652.

EIA (2012), *Annual Energy Outlook*, Energy Information Administration, US Department of Energy, Washington, D.C. www.eia.gov/forecasts/aeo/data.cfm#transdemsec

EIA (2013), *Petroleum and Other Liquids*, Energy Information Administration, US Department of Energy, Washington, D.C. www.eia.gov/petroleum/data.cfm#imports

EPA and NHTSA (2010), "Light-Duty Vehicle Greenhouse Gas Emission Standards and Corporate Average Fuel Economy Standards; Final Rule," Environmental Protection Agency and National Highway Traffic Safety Administration, *Federal Register* 75(88): 25324–25728, May 7.

FHWA (2011), *Highway Statistics 2010*, Federal Highway Administration, Washington, D.C. www.fhwa.dot.gov/policyinformation/statistics/2010/

Goodwin, Phil B. (1996), "Empirical Evidence on Induced Traffic: A Review and Synthesis," *Transportation*, 23: 35–54.

Greene, D. L., Patterson, P. D., Singh, M., and Li, J. (2005a), "Feebates, Rebates and Gas-Guzzler Taxes: A Study of Incentives for Increased Fuel Economy," *Energy Policy*, 33: 757–775.

Greene, D. L., Patterson, P. D., Singh, M., and Li, J. (2005b), Corrigendum to "Feebates, Rebates and Gas-Guzzler Taxes: A Study of Incentives for Increased Fuel Economy," *Energy Policy*, 33: 1901–1902.

Greenstone, Michael, Elizabeth Kopits, and Ann Wolverton (2013), "Developing a Social Cost of Carbon for US Regulatory Analysis: A Methodology and Interpretation," *Review of Environmental Economics and Policy*, 7(1): 23–46.

Harrington, Winston (2011), "Fuel Consumption Standards for Heavy Duty Vehicles," Resources for the Future Policy Commentary. www.rff.org/Publications/WPC/Pages/Fuel-Consumption-Standards-for-Heavy-Duty-Vehicles.aspx

Hausman, J. A. (1979), "Individual Discount Rates and the Purchase and Utilization of Energy-Using Durables," *Bell Journal of Economics*, 10(1): 33–54.

Helfand, Gloria, and Ann Wolverton (2011), "Evaluating the Consumer Response to Fuel Economy: A Review of the Literature," *International Review of Environmental and Resource Economics*, 5: 103–146.

National Surface Transportation Infrastructure Finance Commission (2009), *Paying Our Way: A New Framework for Transportation Finance*. http://financecommission.dot.gov

NHTSA and EPA (2012), *NHTSA and EPA Set Standards to Improve Fuel Economy and Reduce Greenhouse Gases for Passenger Cars and Light Trucks for Model Years 2017 and Beyond*, National Highway Traffic Safety Administration and U.S. Environmental Protection Agency, Washington, D.C. www.nhtsa.gov/staticfiles/rulemaking/pdf/cafe/CAFE_2017-25_Fact_Sheet.pdf

NRC (2009), *Hidden Costs of Energy: Unpriced Consequences of Energy Production and Use*, Washington, D.C.: National Academies Press.

Parry, Ian W. H. (2009), "Pricing Urban Congestion," *Annual Review of Resource Economics*, 1: 461–484.

Parry, Ian W.H. (2011), "How Much Should Highway Fuels Be Taxed?" In Gilbert E. Metcalf (ed.), *U.S. Energy Tax Policy*, Cambridge: Cambridge University Press, 269–297.
Parry, Ian W. H., David Evans, and Wallace E. Oates (2014), "Are Energy Efficiency Standards Justified?" *Journal of Environmental Economics and Management*, 67: 104–125.
Parry, Ian W.H. and Kenneth A. Small (2005), "Does Britain or the United States Have the Right Gasoline Tax?" *American Economic Review*, 95: 1276–1289.
Poole, Robert W., Jr. (2012), "Express Toll Lanes in High Gear," *Surface Transportation Innovations*, 101 (March). http://reason.org/news/show/surface-transportation-news-101
Poole, Robert W., Jr. (2013), "Funding Important Transportation Infrastructure in a Fiscally Constrained Environment," Policy Brief 102, Reason Foundation, January. http://reason.org/files/transportation_funding_budget_constraints.pdf
Small, Kenneth A. (2012), "Energy Policies for Passenger Motor Vehicles," *Transportation Research Part A: Policy and Practice*, 46(6): 874–889.
Small, Kenneth A., and Erik Verhoef (2007), *The Economics of Urban Transportation*, New York: Routledge.
Small, Kenneth A., Clifford Winston, and Carol A. Evans (1989), *Road Work: A New Highway Pricing and Investment Policy*, Washington, D.C.: Brookings Institution.
TRB (2006), *The Fuel Tax and Alternatives for Transportation Funding*, Transportation Research Board, Special Report 285, National Academies, Washington, D.C.
TTI (2011), *Urban Mobility Report 2011*, Texas Transportation Institute, Texas A&M University System, College Station, Tex. http://mobility.tamu.edu/ums
US Department of Transportation (2010), *2010 Status of the Nation's Highways, Bridges, and Transit: Conditions and Performance*, Washington, D.C. www.fta.dot.gov/about/about_FTA_5208.html

Appendix: Assessing efficient fuel taxes to address externalities

If fuel taxes are the only available fiscal instrument to address externalities, and there is no other market distortion, then the efficient level of corrective tax for gasoline is given by the following formula (Parry and Small, 2005):

[CO_2 damages per gallon]
+
[(congestion, accident, and local pollution costs imposed on others per extra mile of driving)
×
(miles per gallon)
×
(fraction of the fuel reduction due to reduced driving rather than higher fuel economy)]

Multiplying the congestion, accident, and local pollution costs per mile by miles per gallon expresses these costs in dollars per gallon, though account should be taken of how fuel economy rises with higher taxes (e.g., via manufactures incorporating fuel-saving technologies). Moreover, these mileage-related costs need to be multiplied by the fraction of the tax-induced reduction in fuel use that comes from reduced driving, as opposed to the other fraction that comes from fuel economy improvements. The smaller the first fraction, the smaller the congestion, accident, and local pollution benefits per gallon of fuel reduced.

The formula assumes that reductions in vehicle miles driven reduce local pollution, but improvements in fuel economy do not. One channel for fuel economy improvement is people shifting to smaller vehicles, but this does not obviously reduce emissions given that large and small light-duty vehicles alike must now satisfy the same emissions per mile regulations (and, at least to some degree, emissions rates are maintained throughout the vehicle life by state-level emissions inspections and maintenance programs). The other main channel for fuel economy improvement is through manufactures making technological modifications to new vehicles, such as improvements in engine efficiency, use of lighter materials, or improved aerodynamics. However, evidence for the United States suggests that any local emissions gains are offset, as manufacturers can cut back on emissions abatement equipment and still meet the same binding emissions per mile standards.

The formula above is easy enough to iterate in a spreadsheet, using the values for external costs discussed in the text. We assume a (pre-tax) gasoline price of $2.60 per gallon and fuel economy of 20 miles per gallon, both for year 2010; that each 1 percent increase in fuel price (including tax) increases fuel economy by 0.2 percent; and that half of any tax-induced fuel reduction comes from reduced driving. Finally, following Parry and Small (2005), we scale back congestion costs by 30 percent, because driving on congested roads (which is dominated by commuting) is less responsive to fuel prices than driving on non-congested roads (this further reduces congestion benefits per gallon of fuel).

13

COMPARING COUNTRIES' CLIMATE MITIGATION EFFORTS IN A POST-KYOTO WORLD

*Joseph E. Aldy and William A. Pizer**

KEY MESSAGES FOR POLICYMAKERS

- A key element of sustaining a meaningful reduction in greenhouse gas emissions among a coalition of willing countries must be comparability of mitigation effort. Countries should perceive that their policies and measures are roughly in line with one another or at least meet a minimum acceptable standard.
  - Absent comparability, it is hard to imagine participation and compliance in a global agreement.
  - Countries wanting to harmonize domestic carbon tax systems or link domestic cap-and-trade programs will demand metrics of comparability, much in the same way that metrics for evaluating trade barriers are necessary for countries wanting to expand free trade.
  - Assessments of comparability will affect decisions about whether to implement and, if necessary determine the stringency of, unilateral border measures (e.g., a border tax).
  - Policy surveillance of pledged actions and their outcomes through reviews by independent experts would inform consideration of the comparability of effort in periodic rounds of negotiations.
- A variety of measures could be used to evaluate countries' climate change mitigation effort:
  - explicit carbon prices, driven by carbon taxes or emission trading;
  - implicit carbon or energy prices, summarizing a wider range of climate-related fiscal or regulatory policies;
  - emissions rates, levels, or reductions, summarizing policy impacts and environmental outcomes; and,
  - mitigation costs, summarizing net burden on the economy.

- All metrics have their strengths and weaknesses and policymakers will likely need a suite (rather than just one) of the metrics. For example, while explicit or implicit carbon prices may provide a useful and transparent measure of incremental mitigation costs, they do not provide a robust representation of emissions reductions, the scale of mitigation-related policies, or total economic burdens.

1. Introduction

This chapter is about how international coordination over climate mitigation efforts might move forward, focusing primarily on the issue of cross-country comparability of mitigation efforts.[1] Clearly, nations are more likely to participate in, and comply with, an agreement that they consider a "fair" deal – one, in particular, where comparable countries (e.g., different industrialized countries) undertake comparable effort, and non-comparable countries (e.g., least developed countries versus advanced countries) undertake differentiated effort.[2] It is difficult to understate the importance of transparent and credible cross-country comparisons for progress in the design of international climate policy architecture.

At first glance, if the United States introduced a carbon tax, it might appear straightforward to compare it with other country efforts, at least those in the club of countries with explicit carbon prices resulting from tax or emission trading programs. But on closer inspection, the issue is far more complex.

For one thing, countries may not have a single carbon price, but instead may apply differential rates (including zero rates) to different emissions sources, and some countries may price emissions from forestry or from non-CO_2 greenhouse gases (GHGs) while others may not. For another, countries may heavily (though implicitly) impose additional carbon charges through other fiscal policies (e.g., excise taxes on motor fuels or electricity consumption) or non-fiscal measures (e.g., renewable or energy efficiency standards). To complicate matters further, these measures might be motivated at least in part by domestic problems (e.g., local air pollution). And for some countries (e.g., those with limited possibilities for shifting to low-carbon fuels) carbon pricing may provide a robust revenue source, but may do little to help the environment.

All this suggests that we need to think carefully about how to measure "the" carbon price across different countries and whether other comparability metrics might not be better, or at least provide useful additional information. Our aim here is to provide an introductory sketch of the pros and cons of alternative, comparability measures – getting into the technical details needed to develop credible measures is beyond our scope.

Besides the carbon price, mitigation effort metrics could focus on other policy parameters, such as the level of a renewable power mandate, or a sector-specific (emission) performance standard. Alternatively, the metric might focus on outcomes, such as emission levels or emission intensities, or derivatives of policy outcomes, such as implicit carbon prices, emission abatement from a business-as-usual (BAU) emission level, or abatement costs.

Benchmarks for comparing metrics may also be needed. These could take the form of historic measures of the metric, forecast future levels, or an agreed global standard. Benchmarks may also vary among nations on the basis of various factors, such as population, incomes, past or forecast future emissions. Fundamentally, these benchmarks address how the burden of mitigating climate change is shared, how much past progress is rewarded, how future growth is accommodated, and how countries with different characteristics are treated.

The 1997 Kyoto Protocol focused on an outcome metric (emission levels) and a benchmark (1990 emissions, for most countries) that in practice are quite divorced from most other measures of effort, such as emission reductions (relative to BAU emissions for that year), abatement costs, and carbon prices. The Protocol also imposed an extreme version of differentiated effort, in which industrialized countries (members of the OECD as of 1992) and most of the former Soviet bloc, took on legally binding emission targets while all other (primarily developing) countries had no specified binding or non-binding emission goals.

Following the 2009 Copenhagen Accord, and subsequent UN climate change negotiations, the question of comparability of mitigation effort arises as far more than an academic question. The second Kyoto commitment period (2013 to 2020) replicates previously negotiated emission targets for a subset of industrialized countries submitted under the Cancun Agreements, which included a varying array of mitigation policies, actions, and goals submitted by a broader set of nations.

This Copenhagen model represents a new framework in which nation-states propose or pledge actions, policies, and goals unilaterally (the one constraint being that developed countries agreed to submit pledges in the form of economy-wide emission limits). In this context, there is considerable scope for stakeholders and other countries to critique other party's pledges as inadequate and/or to attempt to inspire greater action. Absent some notion of comparability and a credible system of transparency and review, this new model could break down.

Beyond the UN-sponsored climate negotiations, the comparability of actions can affect domestic climate policy design in several important ways. Some countries pursuing more aggressive mitigation may take actions, including imposition of border measures on those viewed as making insufficient action (warranted or not) to protect domestic industries. This demands some objective measure of whether other countries are doing enough.

The evolution of bottom-up coordination also requires an assessment of comparability. Take two examples. First, countries that may consider harmonizing domestic carbon taxes would seek assurance that such efforts are, in practice, comparable in light of other policy instruments, including other energy taxes or policies, that could undermine any explicit carbon tax. Second, countries seeking to link their domestic cap-and-trade programs need to make judgments about whether proposed partners would unfairly benefit from the linkage. An analog to each of these examples is consideration of lower tariffs and non-tariff trade barriers in the context of bilateral free trade negotiations.

To illuminate the considerations of the comparability of effort in designing international climate change policy, we first discuss the pros and cons of various alternatives. Next, we apply these insights to the prospect of coordinating domestic carbon tax policies among a set of nations. Finally, we discuss how specific national circumstances can impact the benchmarking of metrics and further policy implications.

2. Metrics for comparing effort

This section describes the advantages and disadvantages of prices, emissions, costs, and their assorted variants, as metrics for comparing effort, leaving aside (for now), how to account for differences in country income and other factors.

Some useful analytical work, discussed in Box 13.1, could serve as a foundation for developing climate mitigation comparability metrics – some of it applied specifically to the consequences of various climate agreements and proposals. The importance of this kind of analysis can only grow as we evolve from the Kyoto model of negotiated targets and essentially pre-established metrics to the Copenhagen-Cancun model of pledge-and-review.[3]

BOX 13.1 EXAMPLES OF ANALYTICAL WORK ON COMPARABILITY OF EFFORT METRICS

The Stanford Energy Modeling Forum organized an exercise with a dozen or so modeling teams to evaluate the Kyoto Protocol (Weyant and Hill, 1999). Through this exercise, the modeling teams produced estimates for a variety of metrics, including carbon prices ($ per ton of CO_2) and abatement costs (expressed as a percent of GDP) assuming cost-effective mitigation policies among all Annex I countries. Emission levels were explicit in the Kyoto Protocol commitments, and the modeling teams' forecasts of BAU emissions would permit an estimate of emission abatement. Estimated emission intensities (CO_2 per unit of GDP) could be inferred from these analyses.

Recent evaluations of the emission commitments under the Copenhagen Accord have focused on a similar set of metrics. For example, employing a global energy-economic model, McKibbin et al. (2011) compare emissions, emission reductions, carbon prices ($ per ton CO_2), and costs (percent changes in GDP and percent changes in consumption) among countries, again assuming cost-effective mitigation policies. Dellink et al. (2010) estimate emission reductions – both relative to baseline and a specified base year, carbon prices ($ per ton CO_2), fiscal revenues (assuming a carbon tax or a cap-and-trade program with an auction), production among energy-intensive industries, and costs (changes in GDP and changes in household income). Houser (2010) also compares emission reductions under Copenhagen Accord emission targets, goals, and policies among participating countries.

The OECD (2013) has recently published estimates of effective energy tax rates – expressed per units of energy and per ton of CO_2 – for all member countries. The report finds significant variation across countries and within countries in effective carbon tax rates in the developed world. This work provides an illustration of the data and methods necessary to construct national average effective carbon tax rates.

2.1 Prices

Under a carbon tax or cap-and-trade program, the explicit carbon price is often viewed as the benchmark of stringency.[4] Ultimately, however, the delivered price of fossil energy reflects a combination of global and local resource costs, other tax and subsidy policies, and any explicit carbon price. This leads us to consider not just explicit carbon prices but energy prices more broadly.

2.1.1 Carbon prices

The observed carbon price is a natural benchmark for effort, as it measures the incremental cost of CO_2 mitigation. When carbon prices differ across sectors and fuels, they need to be weighted by emissions, abatement, sectoral output, or some other means to obtain one aggregate measure of price.

In contemplating comparability, there is a natural question of whether to expect the same carbon prices for every country and how one might explain or accommodate differences. For example, should countries with relatively high fossil energy prices (e.g., due to resource constraints or other policies) be asked to seek even higher prices to reflect carbon content, or should countries with low prices be first asked to raise theirs? How should we view a policy that raises gasoline prices in the United States, but not enough to reach EU levels? This question tends to lead us to more general questions of energy price comparability which we discuss below.

A second issue is converting prices into a common currency. Market exchange rates are the most relevant for competitiveness concerns and traded goods. However, purchasing power parity exchange rates allow a comparison of domestic costs in terms of domestic goods.

Advantages

A carbon price represents the marginal cost for emitting a ton of CO_2 among sources covered by a country's market-based climate change programs – all mitigation opportunities that are less expensive than the carbon price should be undertaken by households and firms. In this way a national carbon price measures the degree to which a country is undertaking less expensive or more expensive mitigation efforts. Comparing carbon prices across countries provides an indicator of how hard each country is trying to reduce emissions, at the margin. If countries face similar opportunities to reduce emissions, this would also be an indicator of a country's total effort to reduce emissions.

Importantly, for countries with emissions trading programs, carbon prices also indicate what will happen if countries link their systems. Large price discrepancies indicate that there are considerable gains to trade, but also indicate that one country will be financing emission reductions in the other country: firms in the pre-linking high carbon price country will buy allowances, and hence finance emission reductions, from firms in the pre-linking low carbon price country.

Disadvantages

There are a number of reasons why carbon prices may not reflect mitigation effort.

A carbon price may be too narrow a measure of a country's efforts to mitigate GHG emissions. It may only cover a subset of emissions (e.g., only large emitters, such as in the EU Emission Trading Scheme). There is also the risk that a country may undermine the effectiveness of the carbon price by adjusting other taxes downward (or increasing subsidies) for firms covered by the carbon price – typically referred to as "fiscal cushioning."

In principle, broader tax and subsidy provisions affecting energy and transportation sectors can be combined, along with explicit carbon prices, into a measure of "effective" CO_2 price, which can be used to measure fiscal cushioning. Nonetheless, there are still some thorny analytical issues – particularly the choice of weights (mentioned earlier) to aggregative over fiscal provisions, and the possibility that these other measures (e.g., fuel taxes) are addressing other (local) problems (like pollution) – see Chapter 12.

Moreover, carbon prices (whether effective or not) fail to account for other, non-price policies that reduce GHG emissions. Efficiency standards or regulations supporting renewable energy can have significant emission consequences and represent significant effort, but are not reflected in carbon prices (in fact, they imply lower prices when also covered by cap-and-trade systems).

Finally, the effort represented by a carbon price – in terms of resources expended – depends on both the price and the amount of emissions reduced. For a country with relatively few opportunities to reduce emissions, a high carbon price may largely amount to a within-country wealth transfer without affecting behavior much or changing emissions.

2.1.2 Energy prices

Energy prices – including both fossil fuels and electricity – are what matter for both the supply and demand for energy along with investment in energy technologies by businesses and households. Higher overall energy prices will drive more investment in energy efficiency, and higher relative prices for more carbon-intensive energy sources (such as coal and petroleum products) will spur investment in low- and zero-carbon technologies (such as natural gas, wind, solar, and nuclear). If one is concerned about fiscal cushioning, or more generally the culmination of multiple policies that affect energy prices, it may make sense to contemplate energy prices themselves as a metric for comparability. Energy prices may be compared directly or in terms of a carbon price equivalent.

While there is a tendency to focus energy price comparisons across countries on the current policy and tax differences, emissions comparisons are often based on changes vis-à-vis an historic base year. Similarly, one might contemplate a price comparison that included both policy-related price changes as well as absolute price levels.

Advantages

Most energy prices are transparent and measurable with high frequency. Energy prices may permit a net assessment of policies, and thus can mitigate concerns that a country engages in fiscal cushioning by simultaneously imposing a carbon tax and source-specific tax relief. They can also capture the effect of some non-price regulations, such as utility sector energy efficiency and renewable power mandates, that get built into the cost of developing and producing energy. An energy price metric would highlight the effort made by those countries that reduce or eliminate their fossil fuel consumption subsidies, which may not be evident in some other metrics (e.g., see discussion of costs below).

Disadvantages

Not all energy price differences across countries represent policy choices, which can make it difficult to reconcile different fossil energy prices with a particular carbon price and isolate mitigation effort. Different resource endowments, labor costs, and transportation constraints can lead to significant regional disparities in coal and natural gas prices, and price changes over time can reflect fundamental supply and demand shifts unrelated to policy changes. Focusing on prices would not capture some prominent regulatory policies either, such as those intended to reduce energy consumption (including appliance efficiency and fuel economy standards) or regulate CO_2 emission rates. The costs of these policies are largely reflected in product (e.g., vehicle and appliance) prices rather than energy prices.

The energy price metric is also less relevant for some countries where the electricity sector is largely non-fossil fueled and/or where a large fraction of their GHG emissions occur beyond the energy sector (e.g., Brazil and Indonesia due to land use change or New Zealand due to agriculture). In these cases, additional metrics to compare action will be necessary.

2.2 Emissions

Recent climate negotiations (including the 2009 Copenhagen and 2010 Cancun talks), have spurred consideration of alternative emission-based commitments, including emission intensity goals (e.g., CO_2/GDP), estimated emission abatement from BAU forecast emission levels, and emission mitigation policies. All three of these new metrics represent the outcome of policy actions (and potentially other drivers of emissions – see below). They are typically translatable: for example, an intensity target can be converted into an emission target or reductions from BAU based on economic and emission forecasts. A key question for intensity or more general indexed-based targets is whether and how such targets are updated as new information arises about the index. Such updating may be helpful in terms of accommodating growth if emissions are highly correlated with the index (see Newell and Pizer, 2008).

2.2.1 Emission levels versus historic base year

The centerpiece of the Kyoto Protocol and the focus of most environmental stakeholders is emission levels. The Kyoto Protocol framed comparability in terms of percent reductions in emissions from a historic base year (1990 for most countries) for most industrialized countries. The question of baselines and benchmarks among countries is, in part, a question of rewarding leaders or supporting followers – for countries that have already (intentionally or inadvertently) undertaken significant efforts that reduce emissions, should their baseline include those efforts or not? This is not an easy question.

Advantages

Measuring a country's emissions – at least energy-related CO_2 emissions – is relatively straightforward, related mainly to fossil fuel combustion. Industrialized countries report their annual total GHG emissions to the UNFCCC typically within two years.

The benchmark of a historic base year facilitates evaluation given the relative ease of measurement and verification. Establishing a base year for some period of time that predates the negotiations also removes the prospect that a country could game the system by increasing its emissions to create a higher benchmark level for evaluation.

Disadvantages

Emission trends may vary from country to country for a number of reasons beyond policy actions to mitigate GHG emissions. For example, significant heterogeneity in economic and population growth among countries suggests that the effective stringency of a common percentage reduction from a common base year could differ dramatically among countries.[5]

Measuring annual emissions may be more problematic for developing countries, considering some of the largest developing countries have reported no more than one emission inventory per decade to the UNFCCC since the 1992 Earth Summit.[6] This may also be challenging for countries that have land use change as a significant contributor to their GHG emissions, given the technical challenges and typically lower precision in estimating these types of emissions than for energy-related CO_2 emissions – though metrics related to these activities would be equally problematic to characterize using any other metric.

In addition to the challenge in assessing effort based on a simple aggregate of emissions, such an approach by itself does not promote learning about policy effectiveness. A variety of factors beyond the control of government policies could typically impact national emissions – with a pure cap-and-trade policy being the exception. Therefore, an assessment of effort using such a measure as the sole metric would do little to identify sources of emission growth and the impact of current policies. Such information about sources of emissions growth and current policy impact could inform the next round of negotiations and other countries' future climate change programs.

Emissions are also a more natural metric when the policy of choice is an emission trading program. As domestic policymakers consider traditional forms of regulation or carbon taxes, emissions become a less obvious metric.

2.2.2 *Emission abatement – emission levels versus future emission forecast*

Recently, interest among some large developing countries has turned to emission goals specified as percent reductions – effectively, comparing actual emissions with an estimated BAU forecast.[7] For example, Brazil, Indonesia, Korea, and South Africa have established emission targets relative to their forecast BAU levels under the Copenhagen Accord.

Advantages

Estimated emission abatement could represent a much more precise measure of mitigation *effort* than emission levels, since it takes into account what emissions would have been in the absence of the GHG mitigation program. Compared to emission levels, it does not automatically penalize countries that have growing population or economies.

Disadvantages

Given the uncertainty in forecasting BAU emissions, especially for developing countries that are expecting faster but more volatile economic growth, it is easy to imagine how such efforts might be gamed. Countries would certainly have the incentive to assume faster economic growth to ensure a higher baseline emission level.

The use of such forecasts also raises philosophical questions about the basis for entitlement to a particular emission level or limit: is the basis for an emission target tied to population or a level of economic attainment? Is it tied to some effort to deviate from a notional BAU? What defines BAU? Or should all countries seek to quickly turn and reduce emissions? Regardless of the preference for base years or future forecasts (discussed below), future forecasts are harder in practice to establish. They cannot be measured and reasonable analysts may make different but equally plausible assumptions to produce significantly different emission forecasts, particularly as one peers further into the future.

2.2.3 Emission rates

Emission rates (e.g., tons of CO_2 per unit of economic output, in aggregate or by sector) have long been heralded as a way to divorce discussions of reducing emissions from concerns that absolute emission limits constrain growth in a practical way (Pizer 2005). In the run-up to the 2009 Copenhagen talks, China and India each proposed emission goals structured as percentage improvements in emissions to GDP. The George W. Bush administration proposed such an intensity-based emission goal for the United States in 2002 and the Government of Argentina proposed in 1999 an emission commitment specified as a function of economic output and requested consideration for joining Annex B of the Kyoto Protocol with this emission target.

Advantages

Energy-related CO_2 emissions and GDP are each fairly straightforward to measure. A variety of related data collected in numerous contexts (UN collection of fossil fuel data, IMF collection of economic data, private collection of financial and economic data, etc.) can be used to verify the quality of emissions and GDP data reported by countries.

It is also possible to further devolve such measures to sectors or activities. For example, the power sector could target emissions per kilowatt-hour and transport could target emission per vehicle mile.

Such an approach can ensure that a country does not appear as a climate leader simply because of economic decline, or another does not appear as a climate laggard simply because of faster-than-expected economic growth (though intensity may more than compensate for unexpectedly higher or lower growth as noted below).

Applied to sectors, emission rates could be a reasonable measure of non-market policy activity, such as fuel economy standards, energy efficiency standards, or technology and performance regulation more generally. The effect of such policies would be hard to measure with prices or emissions alone.

Disadvantages

Emissions will continue to grow unless the reduction in emission intensity exceeds the economic growth rate. Since many countries (once industrialized) experience a decline in emission intensity as their economies grow – reflecting a natural tendency towards lower energy intensity and higher efficiency – it is also possible that a declining emission rate target could be set such that it requires little or no effort for compliance.

Some analysis has shown that emission intensity targets become more stringent if a country grows slower than expected and less stringent if it grows faster than expected.[8] From a policy design perspective, it would be more appealing to design a policy that requires more emission mitigation effort for those countries that grow faster, and hence are wealthier, than they expected to be, instead of the opposite (Aldy, 2004). This could lead to an "indexed" rather than intensity-based approach (Newell and Pizer, 2008), where deviations from the expected economic growth rate translate into a deviation from the expected emission target based on a fractional adjustment, as an alternative to the strictly proportional intensity-based approach.

It may turn out that tracking intensity, emissions, and economic activity could be used jointly for comparability in a way that avoids penalizing economic growth without creating the perverse outcome that the target is harder in the face of slower growth.

2.3 Costs

An alternative to measure emissions abatement or carbon prices is to combine them into an estimate of abatement costs. That is, the amount of resources being diverted towards emission mitigation by a country. Of course, just as emissions are easily observed but reductions are not, prices are easily observed but costs are often not. But even more than estimating reductions, cost estimation requires additional economic assumptions and detailed frameworks for evaluating economic changes in specific sectors and national economies.[9]

One question this immediately raises is whether to focus on simply the costs of abating emissions or the net cost to the economy of a climate policy program. For example, some have suggested that a tax swap – imposing a tax on carbon while reducing tax rates on labor and/or capital – would lower the net effects of a climate policy program, and could even make the economy grow faster than it would have otherwise depending on the distortions on the pre-existing factor taxes and the magnitude of the carbon tax. Should such economic benefits reduce how we view a country's contribution to climate change mitigation?

Advantages

Ultimately, it is costs that most closely align with effort. For that reason, it is an intrinsically appealing metric. Expressed as a share of national income, or per capita, it could be scaled to be comparable across countries of vastly different sizes.

The concern about the costs of combating climate change represents one of, if not the most, significant impediments to serious action by countries around the world. A metric to compare effort based on costs could promote confidence that the international effort is fair by ensuring that comparable countries bear comparable costs from their actions. It could also highlight the potential advantages of some policies (with lower costs) over others.

Disadvantages

The design of such a metric risks rewarding inefficient policies. A country with high costs, but little emission reductions, as a result of poorly designed command-and-control regulations would be awarded "more effort" than one that implements a cost-effective emission mitigation program.

The cost metric has trouble rewarding the impacts of negative-cost policies in reducing GHG emissions. For example, a reduction in fossil fuel consumption subsidies should result in a net negative cost for a country, but could meaningfully reduce CO_2 emissions. Related to this, focusing on overall costs ignores the distribution of costs and benefits among members of society. This may represent a high political cost, or even welfare cost, even as a country's income as whole may rise.

As noted above, measuring costs is problematic, typically requiring a model-based approach, many layers of assumptions, an accounting of overlapping fiscal provisions and non-climate regulations, and an estimate of baseline emissions.

3. US implementation

If the United States implemented a domestic carbon tax, it would raise a variety of questions about the design of international climate agreements beyond assessments of the comparability of effort. In particular, four questions merit consideration:

- Should mitigation commitments be structured as agreements over carbon price?
- How would a domestic carbon tax square with quantitative emission goals, such as those enumerated for developed countries in the Copenhagen Accord and Cancun Agreements?
- Would a domestic carbon tax enable the United States to explore partnering with a small number of large emitters on a price-based emission mitigation agreement?
- How would a U.S. domestic carbon tax impact non-mitigation elements of the international climate framework, such as financing for technology transfer, adaptation, and reduced deforestation?

We take each of these issues in turn.

3.1 Structuring international agreements on price

In the early years of the international climate change negotiations, some countries advocated for "harmonized policies and measures." An obvious example of this could be a common carbon tax among nations participating in an international climate agreement. Such an agreement would have to deal with significant and thorny questions: On the one hand, there would be significant challenges in determining the appropriate, comparable carbon tax value in the context of pre-existing energy taxes and energy pricing regimes. For example, should a $20 per ton CO_2 tax on gasoline be treated the same in Europe (with high gasoline taxes), in the United States (with relatively low gasoline taxes), and in most energy-exporting countries (with subsidized prices for gasoline)? On the other hand, assuming a decision to proceed based on market exchange rates, how would an international agreement deal with currency devaluation? For example, in the Mexico peso crisis, Mexico's currency devalued by 1/3; Korea's currency devalued by 1/2 during the Asian financial crisis. Would such events in a carbon tax regime require a dramatic increase in their carbon tax?

By 1997, negotiators had steered away from such an approach and had opted for agreements based on emission goals. In effect, the Kyoto Protocol in 1997 (and the recent extension of the second period emission commitments) delivered an agreement on emission goals but provided parties to that agreement the discretion to design their own domestic policies necessary to realize those goals. But the clear metric for comparison was emissions.

Efforts to broaden participation to emerging economies and other developing countries have resulted in greater variation in the forms of emission commitments. Under the Copenhagen Accord and Cancun Agreements, nations reached agreement on commitments structured as economy-wide emission goals, economy-wide emission intensity goals, emission reductions relative to a forecast business as usual, energy efficiency programs, transportation projects, forestry conservation investments, and wind and solar power capacity goals, among other forms. While the Kyoto Protocol second commitment period covers nations responsible for less than 15 percent of annual global GHG emissions, this varied commitment regime under the Copenhagen Accord and Cancun Agreements covers more than 80 percent of annual global emissions.

In addition to the above issues, this evolution away from designing emission commitments in terms of quantitative emission targets suggests that it may be even more difficult to structure future international agreements on price-based mitigation commitments. First, policy lock-in and the interests of those stakeholders who prefer the status quo may make it infeasible to transfer over to an exclusive price-based commitment regime. Second, some countries have now pledged mitigation commitments in terms of forestry or efficiency policies that may be easier to implement, given their limited institutional capacity, than a tax-based policy. Thus, they may not be inclined to sign on to an agreement oriented around a carbon tax and attempts to do so may decrease participation.

If agreement on price-based mitigation commitments could be reached, then it would facilitate a comparison of effort across nations. To the extent prices were ultimately harmonized, it would also promote cost-effective emission abatement. Such an agreement could also address concerns about disparate competitiveness impacts of those participating in the agreement, since trade partners covered by the agreement would presumably face the same carbon price. Despite these benefits, the challenges of securing an agreement on a carbon tax regime among 190+ governments – or even a small subset – may be a tall challenge. In lieu of agreement on carbon tax levels, it may be possible to advance agreement on a carbon tax floor that would permit variation in carbon tax rates above that floor.

3.2 Domestic carbon tax policy within an international agreement on quantitative emission targets

Since the UNFCCC, Kyoto Protocol, and Copenhagen Accord reflect deference to sovereign policy decisions regarding the implementation of emission goals and commitments, a national government could choose a carbon tax as its means (or one of its means) for implementing an international agreement premised on quantitative emission targets. While there may be an apparent disconnect between a quantitative national objective and a domestic price-based policy employed to attain that objective, the experience over the past two decades illustrates the typically limited connection between form and implementation of commitments.

For example, the UNFCCC established a non-binding quantitative emission goal for developed countries to return their GHG emissions to 1990 levels by 2000. Several European nations, including Denmark, Finland, Norway, and Sweden, implemented carbon tax regimes in the 1990s as their primary policy efforts to comply with such aims.

Even in the case of domestic cap-and-trade programs implemented in response to the Kyoto Protocol, there is an incomplete relationship to the national quantitative emission commitments under the UNFCCC process. The EU Emission Trading Scheme is the largest, but not sole, policy implemented to ensure EU member states' compliance with emission targets under the Kyoto Protocol. The EU ETS covers about half of EU GHG emissions. Thus other non-quantity policies, such as tax incentives for renewables, efficiency standards for vehicles, and carbon taxes, contribute to EU member states' compliance under Kyoto. If an EU member fails to meet its emission commitment through the suite of policies implemented within its borders, then it could either (a) purchase emission offsets or assigned amount units from other countries; or (b) acknowledge non-compliance. As Canada demonstrated in December 2011, a Kyoto party may simply acknowledge non-compliance and then withdraw from the agreement.

In thinking about a carbon tax operating under a quantitative emission agreement, it is useful to recall that a carbon tax would not cover every source of GHG emissions. Even if the United States implemented a carbon tax, for example, there would likely exist uncovered (perhaps fugitive sources, such as landfill or coal-mine methane) along with policies that deliver emission mitigation for those uncovered sources. There would also be policies that affect sources covered by the tax (e.g., fuel economy standards, state renewable mandates, agricultural conservation subsidies, etc.). Whether it's the 2020 emission goal under the Copenhagen Accord or a post-2020 emission goal that could be negotiated under the Durban Platform, the United States would demonstrate attainment of these goals based on its aggregate emissions and it may need to consider modifications of, or supplementary efforts to, a carbon tax to ensure compliance.

3.3 Carbon price agreement club

In the event that the United States implements a carbon tax, then coordinating action among a small group of large (perhaps neighboring) developed and developing nations may be a natural next step. In this case, countries could agree on a carbon tax regime that provides some common characteristics and some minimums (e.g., tax floors). This would be akin to negotiations over import tariffs in the global trade context and, within the European Union, efforts both to coordinate value-added tax policy and to ensure that tax preferences in the corporate tax code do not provide undue trade advantages. Such a loose, possibly evolving coordination could parallel bottom-up linking of domestic cap-and-trade programs. Absent a strong international agreement, such efforts could emerge as the de facto international climate policy. Over time, such a club could be expanded to new members, increasing the environmental benefits and cost-effectiveness.

3.4 Domestic carbon tax and non-mitigation commitments in international agreements

The increasing breadth of topics addressed in recent international climate negotiations – mitigation, financing, adaptation, technology transfer, reduced deforestation, capacity building, among other issues – makes clear that even if a carbon tax were to be part of the design of future international agreements, then non-tax issues would need to be addressed as well. In addition, a carbon tax approach may have different implications for these other issues, in particular finance and reduced deforestation, than a Kyoto-style approach with emission trading. Domestic cap-and-trade programs that permit emission offsets from developing countries have an explicit mechanism for enabling private sector financial flows from developed to developing countries, and could serve as a meaningful factor in delivering on the Copenhagen objective of providing at least $100 billion in international climate finance by 2020. These flows have been contentious, however, as developing countries express concern about whether developed countries are going to claim "double credit" for both the financial flow and the offsetting emission reductions as counting towards developed countries' commitments.

For a carbon tax regime to yield similar financial transfers, there would either need to be (a) an offsets program implemented through tax credits, which would implicitly channel carbon tax revenues to developing countries to pay for offsets; or (b) an explicit channeling of a significant fraction of carbon tax revenues to developing countries. On the plus side, such efforts would not raise the question of double credit – they would not impact developed country emissions. On the other hand, developed country finance ministries may be reluctant to grant credits for activities in foreign countries overseen by international organizations. The transparent nature and magnitudes of the transfers, especially for this second option, may also undermine political support in many developed countries. For this second option, however, this effect might be partially or wholly offset by the improved ability of the government – as opposed to private actors participating in offset markets – to channel funds to the poorest or most politically appealing recipients (if they have the capacity to receive it).

For example, funding for reduced deforestation in developing countries has been a particularly appealing subset of envisioned financial flows and, sometimes, offsets. Reduced deforestation often involves global biodiversity benefits and occurs in regions and sectors that raise less of a competitiveness specter than efforts, say, to improve industrial energy efficiency. The question remains, however, whether developed countries have the political means to direct (and recipients the capacity to receive) a much more significant volume of public funds to slow and stop deforestation in developing countries.

4. National circumstances and policy implications

In reviewing these various metrics, and even imagining a scenario where a large group of countries pursue carbon taxes alone or in concert, a single enduring benchmark for any of the proposed metrics is unlikely. Even the Kyoto Protocol specified country-specific deviations from 1990 emissions. The history of the UNFCCC is one that includes a sharp and often insurmountable distinction between rich and poor nations. While that history was reversed to large degree in Durban (Aldy and Stavins, 2012), it is difficult to imagine undifferentiated benchmarks. As a matter of parsimony, however, metrics that come as close as possible to providing a universal benchmark, or one with small deviations, are desirable.

In this way, it is not a matter of choosing emissions, or reductions, or carbon price, or emission intensity as a metric and then identifying a global standard for comparison. Rather, it is a matter of thinking about how each metric might be adjusted to reflect national circumstances. The larger the adjustments, in some sense, the more problematic the metric.

Yet, underlying the choice of metrics or the adjustments are fundamental questions about what sorts of concerns are valid. For example, countries like the United States, Canada, and Australia have all faced annual population growth on the order of 1 percent per year. Europe and Japan have not. China, India, and other emerging economies are not as wealthy on a per-capita basis as the United States; how should their comparability metrics be recalibrated? Some countries are endowed with plentiful fossil resources; others are not.

And, as we have mentioned, one must decide whether "fairness" means rewarding countries that have been early leaders or supporting countries that have yet to engage. This is not entirely a philosophical or ethical question: international negotiations require nations to agree to (or at least accept, ex post) outcomes. Leaning too far in one direction or another will likely thwart a winning coalition.

4.1 Linking

An important area for policy consideration is whether jurisdictions will link trading systems or explicitly coordinate carbon taxes or other policies. Here, it may be more obvious that carbon prices are the key metric. Absent similar carbon prices, linking two trading systems will lead to significant flows of allowances in one direction and payments in the other. To the extent this kind of exchange is palatable – for example, between a rich and poor country – that may make sense. But for many situations, it would be a sign of imbalance.

Similarly, the explicit nature of the carbon price in a carbon tax regime suggests that efforts to harmonize carbon prices will need to explicitly tackle why those prices might deviate. Perhaps, some countries are expected to lead. Perhaps there is a recognition of an unequal starting point, in terms of underlying fossil energy prices or other policies. Those would have to be contemplated.

4.2 Border measures

Border measures arise in the context of domestic policy debates when jurisdictions are worried about emission leakage and preserving international competitiveness. As this typically arises in relation to the status quo, the question is tied quite closely to *new* carbon pricing or other regulation. Here, the benchmark is a bit more obvious – namely, changes against levels in the recent past. Among carbon prices, fossil energy prices, and/or electricity – it is electricity and fossil energy prices that matter the most. Electricity prices might be the most amenable to allowing comparisons across policies; electricity is what matters for many manufacturing activities and – for regulations in the power sector – summarily measures the impact on end users. Here, it is clear that market exchange rates, as opposed to purchasing power parity, are relevant for comparing prices in different currency.

To the extent that border measures are viewed not as a vehicle to address emission leakage but as a penalty to laggards in order to encourage action, the metric and benchmark becomes less obvious for justifying action. This could reflect the general array of metrics and adjusted benchmarks discussed throughout.

4.3 Policy surveillance

National governments are more likely to participate in an international agreement to combat climate change if they believe they will be making a fair contribution to the global effort alongside others' efforts. Many nations do not have the resources or capacity to evaluate other nations' commitments and performance, and they may be suspicious of self-assurances by nations themselves. Thus, an independent cadre of experts could provide a legitimate assessment of the effort pledged – and outcomes achieved – by nations to mitigate GHG emissions. To be effective, ensuring comparability requires a professional, regular, and independent assessment of countries' policies, actions, and emissions that can inform the periodic rounds of international climate negotiations.

Several models for such policy surveillance exist (see summary and review in Aldy, 2013). The International Monetary Fund undertakes so-called Article IV consultations of member governments' economic, fiscal, and monetary policies. Under the World Trade Organization, the Trade Review Policy Board evaluates the trade policies of WTO members, with greater frequency for the largest trading nations. While these treaty organizations created professional bureaucracies to undertake such policy surveillance, the G-20 tasked international organizations (the World Bank, OECD, International Energy Agency, and OPEC) to identify fossil fuel subsidies and evaluate the performance of G-20 nations in reducing their fossil fuel subsidies pursuant to the 2009 Pittsburgh G-20 Leaders' Agreement. Under the UNFCCC, ad hoc groups of experts evaluate the emission inventories submitted by industrialized nations and their national communications.

5. Conclusions

Metrics and benchmarks to compare climate change actions across countries are increasingly relevant as we transition to unilateral pledges of domestic action and policy within international negotiations. The negotiations no longer provide a revealed preference for particular choices; instead countries will state what they intend to do, other countries and various stakeholders must make decisions about adequacy, and then everyone will react accordingly. This reaction may be in the venue of international negotiations; it may also relate to decisions to cooperate, harmonize, or link domestic systems; and it may arise in situations where countries unilaterally (or mini-laterally or multi-laterally) decide to act against laggards.

When we contemplate metrics for comparability, a number of relatively deep differences emerge. First, some metrics are relatively easy to observe (emissions, prices) but are one or more steps removed from key concepts we care about – effort and underlying incentives. Second, there are in reality a number of important dimensions – such as effort, incentives, and (of course) emissions/environmental outcomes. Third, there are a number of possible benchmarks that may or may not make adjustments for resource endowments, historic behavior, or future growth. These benchmarks can be further differentiated in an ad hoc or formulaic matter.

The notion of a particular benchmark choice – whether global, historic, or based on future forecasts – has important consequences for how past progress – intentional or inadvertent – is rewarded and, conversely, whether and how the lack of progress is penalized. This is more than an ethical judgment; it is a practical constraint to finding a winning coalition of actors.

Developing metrics and benchmarks for assessing comparability of effort, compiling data and related information in light of these metrics, and reporting the results of the assessments will require a serious, professional, transparent, and legitimate mechanism for policy surveillance. The identification of benchmarks inherently reflects value judgments, and thus could involve extensive negotiations among countries. In the meantime, an array of metrics, such as those presented in section 2 above, could be developed and data collected by existing international organizations to facilitate comparisons in the near term. Feedback on the feasibility, integrity, and precision of various metrics could be solicited to enable further refinement of metrics and to inform the deliberations over metrics and benchmarks going forward.

Notes

* Aldy is affiliated with Harvard University, Resources for the Future, and the National Bureau of Economic Research. Joseph_Aldy@hks.harvard.edu; 617-496-7213; Harvard Kennedy School, 79 JFK Street, Mailbox 57, Cambridge, MA 02138. Pizer is affiliated with Duke University, Resources for the Future, the National Bureau of Economic Research, and the Center for Global Development. billy.pizer@duke.edu; 919-613-9286; 190 Rubenstein Hall, 302 Towerview Drive, Box 90311, Duke University, Durham, NC 27708. We thank Ian Parry, Adele Morris, Rob Williams, Warwick McKibbin, and participants of the November 2012 AEI/Brookings/IMF/RFF "The Economics of Carbon Taxes Conference" for productive comments and suggestions.

1. Although comparability of effort could be important in other international climate policy contexts (e.g., adaptation, climate finance, geo-engineering, research and development), our focus here is limited to mitigation.
2. This principle is often referred to as "common but differentiated responsibility and respective capabilities" (see Article 3.1 of the UN Framework Convention on Climate Change).
3. In addition to these quantitative and modeling-based evaluations of effort, Fischer and Morgenstern (2010) discuss comparability metrics and review the wide array of GHG mitigation policies identified by Australia, the European Union, Japan, the United Kingdom, and the United States in recent national communications submitted to the UNFCCC.
4. For example, lower-than-expected carbon prices in the Regional Greenhouse Gas Initiative (the northeast U.S. power sector cap-and-trade program) were interpreted as a weaker-than-anticipated program, with regulators now poised to tighten the system (RGGI 2013).
5. If one simply reviewed industrialized countries' emissions relative to their Kyoto base year, then one would reach the inappropriate conclusion that Russia and Ukraine are the world's leaders in combating climate change as their 2010 emission levels were 34 and 59 percent below 1990 levels (UNFCCC. 2012). Their emissions reductions, however, are a product of dramatic economic restructuring following the end of the Cold War rather than an extensive climate mitigation program. See also discussion of the "IPAT" identity (Chertow 2001).
6. For example, in 2004, China submitted an emission inventory for 1994, and in November 2012, an emission inventory for 2005.
7. It is possible to compute emissions abatement by aggregating up individual projects; however, even for individual projects, it is necessary to establish a baseline. Individual projects also raise the questions of project boundaries. Does a project to reduce electricity consumption, for example, count reduction in emissions by a power plant? If so, which power plant? What if electricity consumption goes up elsewhere? What if it was already declining? These kinds of questions have led many analysts to express concern about the environmental effectiveness of the Clean Development Mechanism, an extensive program to provide project-level credits for emission abatement under the Kyoto Protocol.
8. This finding holds for those economies whose incremental growth in economic activity is less carbon-intensive than the average of the economy. For example, this is generally true for those countries as they experience structural change as services take a larger share and manufacturing takes a smaller share of economic output.
9. If estimates of both carbon price and emission reductions are available, one simple approach is to estimate costs as ½(price)×(emission reductions). This assumes the average cost of emission reductions is one-half the observed price, which represents the cost of the most expensive reductions, and ignores other issues discussed below.

References

Aldy, Joseph E. 2004. "Saving the Planet Cost-Effectively: The Role of Economic Analysis in Climate Change Mitigation Policy," In *Painting the White House Green: Rationalizing Environmental Policy Inside the Executive Office of the President,* ed. R. Lutter and J. F. Shogren. Washington, DC: Resources for the Future, pp. 89–118.

Aldy, Joseph E. 2013. "Designing a Bretton Woods Institution to Address Climate Change," In *Handbook on Energy and Climate Change,* ed. R. Fouquet. Cheltenham, UK: Edward Elgar, pp. 352–376.

Aldy, Joseph E. and Robert N. Stavins. 2012. "Climate Negotiators Create an Opportunity for Scholars." *Science,* 337, pp. 1043–44.

Chertow, Marian. 2001. "The IPAT Equation and Its Variants." *Journal of Industrial Ecology,* 4(4), pp 13–29.

Dellink, Rob, Gregory Briner and Christa Clapp. 2010. "Costs, Revenues, and Effectiveness of the Copenhagen Accord Emission Pledges for 2020." OECD Environment Working Papers No. 22. http://dx.doi.org/10.1787/5km975plmzg6-en.

Fischer, Carolyn and Richard D. Morgenstern. 2010. "Metrics for Evaluating Policy Commitments in a Fragmented World: The Challenges of Equity and Integrity," In *Post-Kyoto International Climate Policy: Implementing Architectures for Agreement,* ed. J. E. Aldy and R. N. Stavins. Cambridge: Cambridge University Press, pp. 300–342.

Houser, Trevor. 2010. "Copenhagen, the Accord, and the Way Forward." Peterson Institute for International Economics. www.piie.com/publications/pb/pb10–05.pdf.

McKibbin, Warwick J., Adele C. Morris and Peter J. Wilcoxen. 2011. "Comparing Climate Commitments: A Model-Based Analysis of the Copenhagen Accord." *Climate Change Economics (CCE),* 2(02), pp. 79–103.

Newell, Richard G. and William A. Pizer. 2008. "Indexed Regulation." *Journal of Environmental Economics and Management,* 56(3), pp. 221–33.

OECD. 2013. "Taxing Energy Use: A Graphical Analysis." Paris: Organisation for Economic Co-operation and Development.

Pizer, William A. 2005. "The Case for Intensity Targets." *Climate Policy,* 5(4), pp. 455–62.

Regional Greenhouse Gas Initiative, Inc. (RGGI). 2013. "RGGI States Propose Lowering Regional CO_2 Emissions Cap 45%, Implementing a More Flexible Cost-Control Mechanism." www.rggi.org/docs/PressReleases/PR130207_ModelRule.pdf.

United Nations Framework Convention on Climate Change. 2012. "Greenhouse Gas Inventory Data – Detailed Data by Party." http://unfccc.int/di/DetailedByParty.do.

Weyant, John P. and Jennifer Hill. 1999. "The Costs of the Kyoto Protocol: A Multi-Model Evaluation, Introduction and Overview." *The Energy Journal,* Special Issue on the Costs of the Kyoto Protocol: A Multi-Model Evaluation, pp. vii–xliv.

CONCLUSION

Adele Morris, Ian Parry, and Roberton C. Williams III

Although the future extent and effects of global climate change remain uncertain, the expected damages are not zero, and risks of serious environmental and macro-economic consequences rise with rising atmospheric greenhouse gas concentrations. Despite the uncertainties, reducing emissions now makes sense, particularly by using flexible policy approaches that can adjust as further scientific evidence accumulates. At the same time, historically high debt-to-GDP levels in the United States (and other countries), along with rising health care and social security spending, and calls to reform personal and in particular capital income taxes, suggest that new revenue sources (in addition to spending cuts) are an inevitable part of the fiscal future. Carbon taxes help to address both of these challenges.

It is well established that carbon taxes (or emissions trading systems) are the most effective instruments for promoting the full range of mitigation options, including longer-term mobilization of clean technology investments, so long as they target the right base (the carbon content of fossil fuels and perhaps also other greenhouse gases). They also achieve a given emissions reduction at lowest overall cost to the economy, in the sense of providing the same marginal incentive for emissions reductions across different sources. In contrast, a fragmented system of regulations is far less cost-effective, because it pushes too hard on some sources while exempting others. Indeed, a carbon tax could provide opportunities for eventually scaling back a range of climate-related, federal- and state-level regulations, as well as suspending upcoming initiatives at the U.S. Environmental Protection Agency (e.g., limits on power plant emissions) under the auspices of the Clean Air Act – efforts that would become unnecessary in the presence of a sufficient carbon price. Aligning the carbon price appropriately also provides an automatic balance between environmental benefits and economic costs.

Carbon taxes also raise substantial new revenues, and if these are put to good use – lowering the burden of other taxes, reducing the budget deficit, funding high-value public spending – the overall costs of a (reasonably scaled) carbon tax are modest. Moreover, carbon taxes that build off long-established fuel excise taxes are among the easiest of tax policies (or climate policies) to administer. Indeed, finance ministries around the world have a potentially important role to play in implementing and administering carbon taxes, ensuring the appropriate balance between energy taxes and broader taxes in the fiscal system, and reforming fiscal incentives (e.g., tax preferences for fuel suppliers) that create unwarranted distortions in energy markets.

In short, a carbon tax could be a central component of a broad policy package that replaces less efficient regulations, lowers other taxes, eliminates tax loopholes, and reduces the federal budget deficit.

Comprehensive carbon pricing in the United States would also promote international dialogue over similar pricing initiatives among other large emitting countries (20 countries account for about 80 percent of global carbon emissions). Negotiations around carbon pricing could add a useful economic dimension to climate talks, by building on other multilateral economic fora on issues such as international tax, finance, and trade policies.

A carbon tax can work in the United States. This volume shows how, by laying out sound design principles, opportunities for broader fiscal and regulatory reforms, and feasible solutions to specific implementation challenges.

INDEX

Page numbers in italic designate figures and tables.